I0055636

PROGRESS
IN
IRON SMELTING

A CRITICAL REVIEW

JAMES R WARREN

BLOXWICH
2024

Published in the United Kingdom in 2024 by Midland Tutorial Productions

First Edition 1 November 2024

Text copyright © James R Warren, 2024
Photographs and diagrams copyright © Various Contributors 2024
Diagrams © James R Warren 2024
Cover Design: James Warren

The right of James Randolph Warren to be identified as the author of this work has been asserted in accordance with the Copyright, Designs and Patents Act 1988

All rights reserved.

No part of this publication may be reproduced, stored or transmitted in any form or by any means (including photocopying or storing it in any medium by electronic means and whether or not transiently or incidentally to some other use of this publication) without the written permission of the copyright owner, except in accordance with the provisions of the Copyright, Designs and Patents Act 1988.

This book may not be lent, resold, hired out or otherwise disposed of by way of trade in any form of binding or cover other than that in which it is published, without the prior consent of the Publishers.

The ┼ Cross Potent device is Registered Trademark UK00003785735

File Prefix Code: FESM

ISBN 978 1 915750 10 5

Midland Tutorial Productions Publishers
31 Victoria Avenue
Bloxwich
Walsall
WS3 3HS
United Kingdom

MIDLAND TUTORIAL

PROGRESS IN IRON SMELTING

A CRITICAL REVIEW

James R Warren

MIDLAND TUTORIAL PRODUCTIONS
BLOXWICH

Other Books By James R Warren

Boscawen-Ûn
Beyond Tourist Britain
Gleanings as I Pass
Exordium
Meditations
Gamma Solution
Moddeshall Hydropower
Unreasonable Mathematics
Mathematical Explorations
Researches: Volume One
Researches: Volume Two
Researches: Volume Three
Researches: Volume Four
Pi and Phi
Four Famous Numbers

WITH GREAT GRATITUDE TO THE

Late
Mr Don Braggins

OF CLARE COLLEGE AND OF METALS RESEARCH LIMITED

AND IN MEMORY OF DON
AND
ALL THE PIONEERS OF MICROCOMPUTING AND PHOTONIC TECHNOLOGIES

**A Macrocrystalline Iron Bar Gifted to the Author by
Mr Don Braggins of Metals Research Limited
Cambridge (England) 1964**

TABLE OF CONTENTS

Page

PUBLICATION INFORMATION   2
TITLE PAGE   3
EPIGRAPH   5
TABLE OF CONTENTS   7

PREFACE
Page 11

CHAPTER ONE
HISTORICAL SUMMARY
Page 13

CHAPTER TWO
FUNDAMENTAL SCIENCE
Page 35

CHAPTER THREE
STOICHIOMETRIC PRINCIPLES
Page 59

CHAPTER FOUR
THERMOCHEMICAL PRINCIPLES
Page 81

CHAPTER FIVE
THE PRINCIPLES OF PROCESS ENERGY BALANCE
Page 95

CHAPTER SIX
AN IMPRESSION OF ACTIVATION ENERGY
Page 111

CHAPTER SEVEN
THE PRINCIPLES OF ELECTROLYTIC SCIENCE
Page 119

CHAPTER EIGHT
THE IRON SMELTING BLAST FURNACE
Page 141

CHAPTER NINE
GAS DENSITY AND HEATING VALUES
Page 163

CHAPTER TEN
THE THERMAL POWER OF A BLAST FURNACE
Page 179

CHAPTER ELEVEN
THE HYDROGEN DIRECT REDUCTION FURNACE
Page 195

CHAPTER TWELVE
THE DIRECT ELECTROLYSIS FURNACE
Page 241

CHAPTER THIRTEEN
WAYS AHEAD
Page 267

EPILOG
Page 299

APPENDIX A
BASIC ELECTRICS
Page 301

APPENDIX B
ALTERNATING CURRENT AND DIRECT CURRENT
Page 309

APPENDIX C
RECTIFICATION AND POWER LOSS AT AN ALUMINUM
SMELTER
Page 323

APPENDIX D
THE IRON CONTENT OF MODERN CONTAINER SHIPS
Page 329

BIBLIOGRAPHY AND REFERENCES
Page 341

INDEX
Page 367

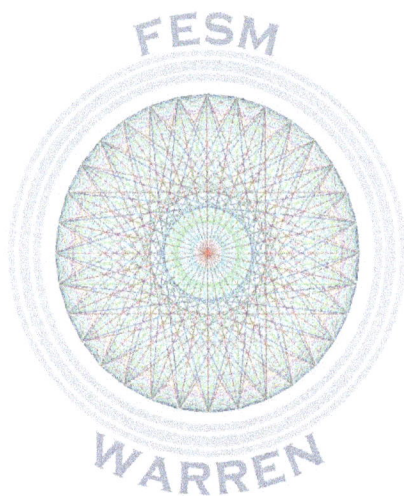

PREFACE

I appreciate that my readers are not doctrinaires, subversives or faddists. They are all people of superior intelligence and sometimes altruistic. We are ordinary patriotic men and women who strive to conserve what we have, both spiritual and material, and delight to see our communities and nations prosper in wealth and safety.

Accordingly my readers do not see a vast steel complex as a "satanic mill" but, properly controlled, an asset to culture and civilisation, a complement to the landscape, and a guarantee of defence and advance.

I also appreciate that at least as far as Britain is concerned, the overwhelming majority of my readers are not scientifically trained, though many of them are Humanities graduates, among them a number of active policy makers.

Therefore I have tried to express difficult technical concepts (in so far as I understand them myself) in words of few syllables, trying to avoid higher mathematics. It can be taken for granted that I have leaned heavily upon Wikipedia (https://www.wikipedia.org/) and I ask readers kindly to search that splendid website if they need clarification of terms or statistics.

The text is accessible to sixth-formers, undergraduates and apprentices. There is a fair amount of repetition, due to the need to develop concepts briefly introduced earlier in the text.

Elsewhere, I have eschewed the passive voice.

This text contains very little original thought on my part. Almost all of the material is the fruit of the research, experience and hard rational labor of other men and women.

When key concepts or substances are introduced, and for a while thereafter, I capitalise the initials of their names. This is for educational reasons, and is reflected in index entries.

At school and college the masters and mistresses thought I would probably spend my adult life as a journalist, and accordingly they provided plenty of poetry, which I did not understand, and even more précis, which I did. But when I could I chose to return to my first and most loyal love, Science, and that is what I cleaved to throughout university. So I evaded a life of stuffy courtrooms and windy stadia to try my hand and mind at something I hoped less tedious.

As for many men my youth was full of darkness, frustration, ignorance, bullying and skitterings along the cold sharp edge of madness. In the spring of 1964 I was twelve, and on reading Mother's copy of Lowry and Cavell I was fascinated by the chemistry of metals, their cunning if archaic methods of extraction, and their important if unlikely applications. I attempted to acquire a collection of metals and approached Metals Research Limited for a price list.

The Late Mr Don Braggins invited me to the premises of his partnership then above Percival's Garage at 91 King Street, Cambridge, all now long gone. Don gave me a splendid and very valuable collection of metal offcuts including germanium, silicon and tantalum that he and his colleagues had used in pioneering experiments in zone refining and the earliest investigations of microelectronic fabrication chemistry and photonics, now as you know the bases of global megaindustries. The assemblage has pride of place in my cabinet to this day.

Bloxwich
Good Friday
2024

CHAPTER ONE

PART I
HISTORICAL SUMMARY

On the 16 February 1923 my Late Mother was 279 days old and Howard Carter broke a hole through a dusty doorway in the desert.

Like all other British families, including Carter's, my Mother's family grieved their brothers lost, as they would for the remainder of the century.

But Carter's antic was to prove one of the few public crises of the age that had a happy outcome, comparable to Hilary and Tensing on Everest, or, later, the Moon Landings.

George Herbert, fifth Earl of Carnarvon, stood behind Carter. We think of Howard Carter as an Egyptologist or perhaps an archaeologist, but this is not how contemporaries necessarily saw him. Carter was a technical artist, and a very excellent one, arguably the last European artist to enjoy an individual aristocratic patron, in distinction to the State.

"Can you see anything?", asked Carnarvon.

"Yes, wonderful things", replied Carter.

Through the hole, in the darkness and the dust, gold glinted everywhere. Gracile golden cheetahs exquisitely wrought in wood; gilded boxes and bedsteads. Golden wheels of dis-assembled chariots. Whether the latter had been disassembled by disturbed robbers, or by the Bronze Age pious in assistance of the angels of a certain afterlife where chariots were assumed useful was not immediately clear to Carter and Carnarvon. And it is unclear to you and I, and ever shall remain so, unless an Agency later enlightens us.

A far portal occluded by a perfect plaster wall was flanked by two armed but elegant ebony Nubians, silent and constant guardians for 3245 years.

Carnarvon soon died in lurid circumstances, but for the next few years Carter meticulously photographed and cataloged the contents of the tomb.

The tomb was the long-lost and hard-sought resting place of eighteen-year-old Tutankhamun (1341-1323BC), Pharoah of Egypt, and son of the heretic pharaoh Akhenaten.

In 1925 the boy king's iron dagger was identified. We earnestly hope the lad had no cause to use it in life. In tunnels and chambers of long-forgotten wonders, this was a wonder beyond conjecture or anticipation.

Because we are speaking of the Bronze Age.

Scholars have now identified several rare artifacts of iron, preserved by desiccated desert airs, and worked to shape in the fires of the Age when the eponymous copper alloy was almost the only metal of tools and weapons.

Iron reduces and melts at high temperatures. It would be some two hundred years later, at the dawn of the Iron Age, that humans would discover the technology to smelt iron. Like many crucial inventions, it would seem that this skill emerged suddenly in disparate and dispersed cultures.

So whence came King Tut's iron dagger and similar iron trinkets?

Non-destructive chemical analyses have demonstrated that the dagger's blade is about 88.4% Iron (Fe), 11% Nickel (Ni), and 0.6% Cobalt (Co). All three are metallic elements, are chemically related, and their alliance is suggestively similar to the composition of iron meteorites. Iron falls out of the sky. Iron falls out of the sky, and may persist as bright metal for centuries in the dry heat of the deep Sahara, though it soon disappears into the soil in the damp European climate.

Figure 1.1[1.1]
Tutankhamun's Dagger
To the Boy King, Iron was more precious than Gold, to him a commonplace

The Bloomery

To be technical, an iron ore seldom or never reduces directly to iron metal. The ore makes a stage-wise transition through several distinct chemical reaction pathways before it arrives at the end-product. Usually, a reduction is effected by the partial combustion gas called Carbon Monoxide (CO) which takes an oxygen atom to become the waste gas Carbon Dioxide (CO_2) which passes up the chimney. In modern Blast Furnace complexes the waste gas first heats the Hot Blast in Cowper Stoves (large ceramic gas burners and heat exchangers). In a modern furnace an unstable intermediate iron oxide forms at a high temperature. This intermediate is Wüstite (FeO) and is usually that which degrades to the ultimate metallic iron. But in 1840, Russo-Swiss chemist Germain Hess showed that such intermediate pathway cascades were in sum thermodynamically-equivalent to a direct reaction of iron ore and carbon to form metal and carbon dioxide. This is a corollary of The Law of the Conservation of Energy, which we shall shortly discuss.

The bloomery is a rough beehive-like oven quickly built to chest height and made of crude baked clay or stone. It is not a permanent structure and was often dis-assembled when cool. A batch of iron ore (hematite, magnetite or limonite) was mixed with a roughly equal amount of charcoal (wood roasted anoxically beneath a loose earth cover to remove volatiles).

In addition to this carbon as chemical-reagent, some carbon burnt as fuel to raise the working temperature.

Charcoal presents as a lightweight, jet-black substance, nearly pure elemental Carbon (in its "graphitic" allotropy), preserving the structure of tree limbs. Charcoal is too fragile for large modern furnaces, even where ecologically acceptable.

A bloomery is built with a hole at the bottom, a tuyere, through which a bellows can manually operate to provide a Cold Blast which oxygenates the reductive fire within.

Bloomeries are Occidental. They may have originated as early as 3000BC, within the European or Near Eastern Neolithic (New Stone Age, the age of settled agriculture).

At the completion of the smelt the semi-molten "sponge" mass of ashy, vesicular, impure iron was extracted, and whilst hot violently beaten by hand upon a stone anvil. This sufficed to beat out impurities, especially, excess carbon, to give Wrought Iron which subsequently could be re-heated in a forge and beaten into the forms of tools and weapons.

The whole process is small-scale and expensive, but iron was very precious to the Ancients, semi-sacred, and the extent to which slave labor was employed in the cruder phases of production is not clear.

Wrought Iron is a robust, tensile, weldable, high-class material resistant both to fracture and corrosion.

The Blast Furnace

The Blast Furnace (BF) is a large permanent structure, historically lined with refractory ceramic, structurally supported exteriorly by stone ashlar, and in modern times by cooled steel.

The crucial distinction from the bloomery is that the BF is engineered to output liquid iron which is cast in sandy moulds as Pig Iron (crude Cast Iron). Liquid slag is tapped separately.

Blast furnaces were and are continuously charged from the top with charcoal or coke. It is of the continuous operation of Blast Furnaces that burning fuel and melting materials slowly gravitate to the tap zone. Once snuffed or "blown-out" BFs are slow and expensive to re-start, so they operate continuously.

Blast furnaces were also equipped with one or more tuyeres through which water-driven bellows forced a Cold Blast to inflame the carbon fuel, raise the furnace temperature and thus assist iron ore reductions. Modern BFs use a Hot Blast to accelerate throughput and purification.

Coke is an analog of charcoal in so for as it is an anoxically roasted form of coal, preferably special-quality Coking Coal. Similarly, coke is more-or-less pure carbon. Coking Coal was, like charcoal, roasted in open heaps at the start of the English Industrial Revolution, but special Retorts and Coking Ovens were soon developed to manufacture it, as Town Gas, phenol and hydrocarbon tars became important by-products.

The Chinese moved directly to Blast Furnace iron smelting, missing out the Bloomery stage. Their earliest cast iron dates from the fifth century before Christ.

The Blast Furnace was introduced to England in 1491AD.

The Pig Iron product still needed beating and rolling in order to forge Wrought Iron. In the late Eighteenth-Century the beating was replaced by "puddling": A lengthy process of "boiling" the liquid pig in a specially-lined shallow vat. The output was taken from the puddling furnace as a large bloom and repeatedly hammered or rolled to a desired size of wrought iron rail or plate. Such beatings, rollings and foldings could easily pass through fifty cycles to forge Best Iron Bar.

Photo: James R Warren

Figure 1.2
Charlcotte Cold-Blast Charcoal-Fired Blast Furnace
am 24 March 2003
OS Grid Reference: SO 63845 86098
Lat 52°28′18″ N: Long 2°32′02″ W

       Figure 1.2 shows the charcoal-fired blast furnace at Charlcotte in Shropshire, England. It is built of local Permo-Triassic Ferruginous Sandstone which has weathered in the four centuries or so since it was built. Little is known of Charlcotte but it may have been abandoned in 1802 when charcoal was becoming very expensive and trees were protected by the Government for use by The Royal Navy. In any case, iron smelting by coke had already been well-established in Shropshire for nearly a hundred years.

       The Charlcotte blast bellows was operated by water-power from a nearby stream and the hot pig may have been pre-processed with a trip hammer, conceivably driven from the same axle as the bellows.

The Coke Blast Furnace

Photo: James R Warren

Figure 1.3
Coalbrookdale Cold-Blast Coke Blast Furnace
pm 16 November 2007
OS Grid Reference: SJ 66751 04862
Lat 52°38′26″ N: Long 2°29′35″ W

During the course of the 17th Century many experiments were made into the feasibility of smelting iron with coal. None were successful as the pig product was polluted with Sulfur and Phosphorous from the coal and could not further be worked.

In 1709, Abraham Darby I (14 April 1677 – 5 May 1717), a Quaker brass-founder of Bristol, took over an existing charcoal furnace at Coalbrookdale beside the River Severn in Shropshire. (The year 1777 painted on the lintel in the picture possibly adverts to Abraham Darby III, grandson of the 1709 ironmaster).

Here Abraham Darby I became the first profitably to smelt useable iron employing coke as the fuel and reducing agent. Charcoal iron smelting ceased in England, except in heavily-forested areas such as The Forest of Dean and The Sussex Weald, where it lingered for another hundred years.

As a coke furnace, Darby brought the furnace back into blast on 10 January 1709 and sold 81 tons (probably English Long Tons) of iron goods in the first coke year.

Darby's coke process was the essential enabling trigger of the Industrial Revolution and in due course it would permit the production of precision machinery, reliable ordnance and indeed further useful research.

As the Industrial Revolution gathered pace coke-based iron smelters proliferated.

In his seminal "Economic History of the British Iron and Steel Industry", Alan Birch[1,2] advises:-

> *"...In 1788 its 85 blast furnaces produced during the year some 68,000 tons; by 1815 the production of about 200 furnaces was between 344,000 and 373,000 tons. In 1850 the industry's output was roughly 2,500,000 tons from some 450 of a total of over 600 furnaces. This, indeed, was a rapid rate of progress, but the second half of the century eclipsed this in a way which is readily discernible from the statistics of output. In 1865 the output of the British industry was not far off double the 1850 figure— 4,825,254 tons—whilst the number of furnaces available and in blast had increased by some 300 and 200 respectively. This is striking evidence of the increased capacity of the blast furnaces resulting from technical innovations and the greater skill and knowledge of the ironmasters."*

When Britain lost the American colonies it also lost a major source of cheap iron. Native coke-smelted iron was instrumental in enabling Britian to maintain its national sovereignty in the face of the much larger and more advanced French Empire, and to drag the British and Irish peoples out of the sloughs of poverty in which they suffered.

Moira Furnace was built by The Earl of Moira in 1804 at the height of The Napoleonic Wars when Continental Europe was occupied by the French and it seemed that Britain would require even the smallest asset or utmost exertion to maintain its independence.

Moira was a steam-powered cold-blast installation that utilised abundant local coal, ironstone and limestone brought along the Ashby Canal at (0.9, 0.5) in the photograph, but hidden by its canal embankment. On a local exposed coalfield, part of the broken Leicestershire Coalfield, Moira was an emergency provision, ill-founded and inexpertly managed. Brought to blast in 1806, it was abandoned in 1811, six years after Nelson had brought the French Navy under control at the Battle of Trafalgar. Archaeological digs have established that the furnace, abandoned with its final melt *in situ*, suffered from high-sulfur coal in addition to its other disadvantages.

It appears that the premises was inhabited by at least two families.

Photo: James R Warren

Figure 1.4
Moira Cold-Blast Coke Blast Furnace
pm 16 November 2007
OS Grid Reference: SK 31430 15145
Lat 52°43′59″ N: Long 1°32′10″ W

So far we have neglected the important intermediate Cast Iron, which stands between the furnace-fresh Pig Iron and

Wrought Iron. Cast Iron is essentially solidified Pig Iron, not beaten or rolled to give it purity and fabric, but allowed to cool to an apparently amorphous but microcrystalline mass.

Cast Iron is strong in compression and was useful for structural columns: Wrought Iron is strong in tension and useful for horizontal structural girders and wires, for example railway rails and elements that were used in bridge construction or shipbuilding. Eighteenth and nineteenth century structures, including ships like the SS Great Britain, seem to survive indefinitely, and some serve today.

For example, Abraham Darby III, grandson of the 1709 pioneer, sought the advice of Thomas Farnolls Pritchard to bridge the then busy River Severn at at place now known as Ironbridge but then called Hodgebower. In 1773 Pritchard in turn consulted "Iron Mad" Wilkinson of nearby Broseley and the three asked Parliament to let them throw a single high span over the river gorge. Parliament usually insisted upon very high spans if the straight or river was navigable in case the Royal Navy needed to navigate the waterway with tall ships. Permission received, work started and on 2 July 1779 the 100 feet 6 inches (30.63 meters) span was completed.

This very elegant structure, the World's first *compressive* iron bridge is notably a culminating example of the *joiner's* art, rigged as it is with mortice and tenon, albeit in cast iron. There is no single bolt or rivet in the entire crossing, and of course, since this is a Cast Iron structure, there is no weld.

Photo: James R Warren

Figure 1.5
The Iron Bridge in Shropshire, England
am 4 November 2009
OS Grid Reference: SJ 67241 03396
Lat 52°37′39″ N: Long 2°29′08″ W

The Steel Era

Ordinary Mild Steel is essentially a purified iron with an apparently amorphous but microcrystalline structure. Mild Steel is a low-carbon steel which contains 0.05% to 0.25% carbon by weight. Better class Japanese cold-rolled steels tend to be 0.0015% carbon. On the whole, however, mill-fresh steel is a low-class material, readily subject to corrosion and even biological attack in marine contexts, as exemplified by the sorry remains of the *RMS Titanic*. On the other hand, steel is cheap and tough whilst simultaneously malleable, ductile and machinable. Desirable properties can be enhanced by alloying with related trace elements, by reworking, or corrosion can be circumvented by coating. Alloying purified iron with 0.015 to 0.10% carbon, 16 to 21% chromium, 6 to 26% nickel, and or 0 to 7% molybdenum yields various grades of Stainless Steel which is soft but highly corrosion resistant, cheaper than platinum, and robust enough for many chemically-demanding applications.

Steel is the central strategic resource, essential for ship construction, railway applications and major civil works such as rail and road bridges. A nation without native steel production is not able to sustain its sovereignty in wartime, and for that reason governments and indeed steel companies are very jealous of any intelligence which bears upon steel production, willingly trumpeting only general statistics, if they portray their organisation in a favorable light.

Steel production was a very expensive artisanal process before the invention of the Bessemer Converter in 1856AD, and for low-grade iron ores mass-production of the alloy had to await the development of refractory and chemically-suitable furnace linings.

Mass steel production today is mediated by the Basic Oxygen Furnace, essentially a sophistication of the Converter.

After Iron has been smelted, it can be allowed to cool to ingot form for sale, but this engages horrific expense when the steel is re-heated by the customer for inevitable re-working. Accordingly, all ironworks are today neighbored by steelworks which pre-process the liquid or semi-solidified iron to form semi-fabricated merchandise such as railway rails, slabs, wire, or coiled sheet steel for car or white goods makers, or indeed the manufacture of armaments.

<u>The Abbey Works</u>

Figure 1.6 is an aerial photograph of Britain's Abbey Works, an Integrated Steel Works at Port Talbot in South Wales. Correspondingly, Figure 1.7 presents the much smaller iron works with its two remaining (2024AD) Blast Furnaces. The third picture, Figure 1.8, gives an impression of the footprint of the total complex from a range of about 6.2 kilometers. I was not permitted to approach further, and in any event an unbridged river was in my way!

Figure 1.7 is a superb picture that I have failed to do justice to. It was taken using excellent equipment from a range of about 2.2 kilometers. A powerful telephoto lens was used as the Devon coast of England, which looks so near across the sea, is actually 52.8 kilometers distant!

My own distant picture, Figure 1.8, was taken at 11:50 on 23 April 2012 using a Casio Z1200, from around 6.2 kilometers to the South. The Z1200 has unfortunately been discontinued by the maker, but for my money was the best compact of its type.

Photo: Chris Shaw[1.3] (adapted by the author)
Figure 1.6
Aerial View of Part of the Abbey Works

Photo: Lewis Clarke[1.4] (adapted by the author)
**Figure 1.7
The Ironworks in the Northern Part of the
Abbey Works complex**

Photo: James R Warren

**Figure 1.8
The Abbey Works from Kenfig Beach**

The Abbey Works ironworks was first erected in 1902AD upon the site of the old Morfa Colliery. The word "morfa" is Welsh for "marsh", especially lagoonal coastal duneland of the kind known in Scotland as "links". The Abbey referred to is the Medieval and now ruinous Cistercian house of The Blessed Virgin Mary, founded in 1147AD. The abbey's monks were adept at iron smelting and refining. In 1536AD, the English and Welsh monasteries were confiscated from the Church of Rome by King Henry VIII, King of England but a Welshman, and Margam was given to the Mansel family, who then passed it to the Talbot family by way of dowry. Secular iron working continued on the former Abbey premises.

The modern steelworks is about three and a half kilometers in length from North to South, and the remains of Margam Abbey are around two kilometers from its southernmost parts.

As recently as fifty years ago The Abbey Works was the largest steelworks in Europe, but today, with an estimated annual production of some five million tonnes or maybe little more than three million, it qualifies as a particularly small establishment. It employs around 4000, down from 18000 in the Sixties of the last century. There are two surviving blast furnaces. Another two are in use by a different company at Scunthorpe in England, and a fifth, large, blast furnace survives at Redcar in Northern England, but is blown out and its attaching steelworks is mothballed. Taken together, these five furnaces are the only furnaces in the British Isles which are capable of smelting "virgin" iron from ore, of the more than nine hundred that existed in the late nineteenth-century.

At Port Talbot, steel making, as opposed to iron making, was commenced by the Margam Iron and Steel Works, which started production in 1927AD.

The plant was nationally owned between 1947 and 1951. The Steel Company of Wales took over the site and itself commenced iron smelting in 1951AD. The UK-wide steel industry was re-nationalised as part of the British Steel Corporation in 1967AD. British Steel was again de-nationalised as it was sold to the Anglo-Dutch Corus Group plc. This sale took place in 1999AD but by 2007AD Corus was disbanded, and its assets sold to the Indian firm Tata Group which now trades in Europe as Tata Steel UK and Tata Steel Netherlands.

As I write in April 2024 the future of Port Talbot and its Abbey Works continue to be in doubt. Tata is proposing to scrap its two remaining blast furnaces and replace them with Electric Arc

Furnaces that can melt scrap steel but are not suited to the reduction of iron ore *ab initio*.

World Iron Production Today

As I have remarked elsewhere it is very difficult to obtain meaningful data about international iron production, at least for the last sixty years or so.

We can probably be provisionally confident that wrought iron production is not quite extinct. I think it very likely to survive as a small-scale artisanal activity in parts of India. The last UK wrought iron maker was Thomas Walmsley and Sons of Bolton in Lancashire who ceased production in 1975AD.

Some claim that China produces more steel than the rest of the World put together.

Whatever the particulars, a few recent trends in iron smelting output can be decerned:-

(A)   Inferior ore Sources in Europe and elsewhere have been abandoned.

(B)   Geographical Divisions between Ore Sources and Iron Processing factories have intensified across the World.

(C)   Intermediate Powers such as Britain and Venezuela have had serious Difficulties maintaining Iron and Steel Facilities. In both of those countries political problems have damaged investor confidence in the Continuing Viability of installations, and impeded Exports.

(D)   Australia, South Africa and Argentina are further intermediate powers where native iron smelting is problematical, though Australia is a major exporter of high-quality ore to China.
At least in the short term, South Africa has successfully addressed iron industry problems with politically-driven Industrial and Infrastructure Demands and with Protectionism.

(E)   Energy Supply is a Critical Criterion.
        Traditional Hot Blast requires a Supply of Coal
        that is not only Reliable in Quantity, but of the
        Appropriate Type. Metallurgical Coal (Coking
        coal) is low-ash, low-moisture and high carbon
        content. Most coking coal is found in Russia,
        Australia and the USA.

(F)   Paradoxically, War has led to special problems
        in the massive Steel Industries of both the
        Ukraine and the Russian Federation. War has
        also badly affected Libyan production.

(G)   Nations with both large Population and large
        Area, especially in East Asia, have risen from
        little or nothing to Global Iron Smelting
        prominence, though Japan is in decline.

According to the World Steel Association global steel
output was 850 million tonnes in 2000AD, rising to 1435 million
tonnes in 2010AD and 1882 million tonnes in 2020AD.

This implies respective annual growth rates of about
4.4%, 3.3% and 1.6%, continuing a secular decline of the steel industry
since around 1950AD.

The UK was the 31st most prolific producer, sending
either 5 or 3.859 million tonnes of steel to market in 2022AD, slightly
less than Pakistan and slightly more than Australia.

PART II
SELECTED IRON SMELTING MINERALS DESCRIBED

Hydrogen (H_2)

Hydrogen is the most primitive chemical element whose atom comprises a single proton orbited by a single electron (the isotope Deuterium has a nuclear proton and a neutron paired in its nucleus and one orbital electron; Tritium has a proton and two neutrons with one orbital electron: Both are rare, and both behave chemically as ordinary hydrogen).

At ambient temperatures (STP) and at the hottest furnace temperatures hydrogen presents as a very limpid and penetrative ("fugacious" would be a good word for it, but unfortunately that would be easily confused with a technical term used in thermodynamics) colorless gas. It is highly-reactive and potentially explosive with O_2 and any halogen, so special and expensive measures must be taken for its safe use.

Hydrogen is not directly used by Hot Blast Furnaces (BFs) or by MOE furnaces, but is a central intermediate in the H_2DRF.

Oxygen (O_2)

Oxygen is an essential component of all commercial iron ores and is an intermediate product in the H_2DRF and a copious end product in the MOE.

Like hydrogen, oxygen is a gas at STP and furnace temperatures and as such is subject to the Universal Gas Law, and also like hydrogen oxygen is diatomic.

Essential in modern steel production, at the iron smelting stage oxygen can be used to enrich the hot blast and is the effective component of the BF hot blast (highly heated air circa 900°K).

Water (H_2O)

Water is a polar covalent compound of two hydrogen atoms and one oxygen in the molecule as shown by its well-known chemical formula.

Water is stable, non-flammable, and as a polar liquid at STP capable of dissolving a wide variety of other chemicals, especially acids and readily forms alkalis with Alkali or Alkali Earth metals.

At furnace temperatures water absorbs heat of vaporisation to become steam, a gas which exhibits complex thermal properties. Steam raised by furnace jacket cooling or by H_2DRF processes is potentially useful in energy recovery and recycling, though corrosive contaminants currently frustrate turbine applications. Combined with atmospheric or artificial oxygen, steam rapidly corrodes steel industrial equipment including furnace parts.

Water is the intermediary electrolyte in the H_2DRF, but is a dielectric (insulator) unless slightly ionised by the addition of small amounts of acid or some other polar solute.

Carbon (C)

Carbon is, with iron ore, the key input to the BF, in the form of coke (roasted coal). In the BF, carbon burns in hot air to provide most of the chemical energy that facilitates ore reduction, and itself chemically participates in the actual reduction of ores to metal.

At STP carbon presents in its graphitic allotrope as a brittle black somewhat slippery or darkly lustrous non-metal, and unusually for a non-metal is a good conductor of electricity. It does not melt at furnace temperatures but decomposes directly into product gases.

Coke is a light, floatable, gray, course but robust residue of anoxic coal roasting, pitted more than an asteroid, so to say, due to its violent expulsion of coal volatiles like benzene and gases during the roasting process.

Carbon is also the key thermogenic component of a vast range of so-called organic chemicals including gasoline, kerosene and benzene, as well as truly organic fuels such as wood and fat.

Graphite carbon is the most promising long-term solution to the problem of commercial H_2DRF and MOE electrodes.

Iron (Fe)

Iron is the prime objective of iron smelting, a process that yields a wide range of other products, some of them saleable. Iron is a robust hard gray transition metallic element that can give a variety of natural and manipulated crystal fabrics.

Iron is the second most abundant metal in the Earth's Crust at about 5.63% (Aluminum is 8.23%), but has a distinct tendency to corrode in the environment. Notwithstanding, the classic products cast iron, and especially fibrous wrought iron, are resistant to salt and fresh water corrosion. Modern mild steel is an inferior material, requiring additives or coatings to resist the weather.

During the Industrial Revolution, a feature of Cast Iron was its high resistance to compressive stress forces, making it ideal for bridge piers, or structural columns in low-rise buildings. On the other hand, Wrought Iron, though no longer made, was ideal for making long-aspect wires, rails or struts subject to high tensile forces, especially in maritime applications.

Converted to steel by the reduction of carbon and phosphorous content, and by the addition of small amounts of related metals, iron becomes ductile, tough and refractory, the mainstay of industrial production of myriad types of durable products and structures, including machine tools, vehicles, ships, bridges, and tunnel supports. Only steel weapons are capable of enduring enemy action.

Calcium (Ca)

Calcium is a soft, sectile, silvery, Alkaline Earth metal readily corrodible and effervescent in cold water to evolve calcium hydroxide that enters solution, and hydrogen which rises as gas.

Calcium is an important constituent of furnace flux and the resulting slag. Besides reducing the viscosity of the melt an important role of furnace fluxes is to reduce the melting temperature of the mix so as to conserve energy and assist management. Fluxes are not essential but are desirable and limestone (calcium carbonate) in particular is sufficiently cheap and plentiful almost always to be used.

Silicon (Si)

Silicon is a light, gray non-metal with metalloid properties that render it useful as an electrical semi-conductor. It has a very high melting point and sublimates rather than melts at ambient pressures. In the smelting process it participates as sand and other impurities that engage in slag formation. There are several iron silicates but none are of themselves economic ores of iron. Sand is Silicon Dioxide (SiO_2).

Silicon is the second most abundant element in the Earth's Crust (28.2%) after Oxygen (46.1%).

Hematite (Fe_2O_3)

Hematite is the primary iron ore of the modern world, and good-quality hematite is 70% iron. However, quartz is a common pollutant of delivered hematite (British-language: Haematite) and is removed at smelt-time as silicious slag.

Hematite has a number of habits (forms) geologically. In the Copeland district of Britain it sometimes occurred as a red, heavy botryoidal (grape-like) "kidney ore" which may have precipitated on free surfaces in hydrothermal veins. Commercial hematite today is dug open-cast from extensive flat strata of massive rock.

Red Ochre is hydrated hematite ($Fe_2O_3.nH_2O$). Other conformations of ochre involving iron oxy-hydroxides are sometimes found.

Magnetite (Fe_3O_4)

Magnetite is a secondary good-quality iron ore and is stoichiometrically 72.36% iron.

Magnetite usually occurs as a dark gray, massive, striated rock, heavy and magnetically attractive to iron filings or, in the geological survey field, to a penknife blade. (I am talking about my day: The modern geologist uses a XRF gun of course!).

Economic magnetite occurs as thin beds or laminae in extensive sandstone formations, apparently paleo-depositional relicts of the wave-sorted magnetite-quartz beach sands observed by geographers. Like hematite, magnetite also offers macrocrystalline hydrothermal habits.

Limonite ($FeO(OH).nH_2O$)

Limonite, "bog iron ore", is formed in cold humid climates and is a rusty yellow or rust-colored apparently amorphous (microcrystalline) heavy rock classically obtained from marshes in Sweden and similar climes.

Formed from the hydrous oxidation of magnetite and hematite, it is often seen as an orange rusty precipitate in mine waters and tailings, where it may arise from the breakdown to oxides of vein sulfides, etcetera. It is not, of itself, ecologically toxic.

The best limonite from the iron production viewpoint is that which contains only one molecule (n) of the associated "water of crystallisation" and stoichiometrically this is 52.25% iron. Limonite deposits are sometimes mined for nickel and gold, rather than iron.

Limonite occurs in extensive massive beds in the Jurassic of England, Luxemburg and elsewhere in Western Europe but is highly phosphatic and its exploitation had to await the development of the Gilchrist-Thomas process in the Late Nineteenth-Century.

Siderite ($FeCO$)

Siderite is the classic "spathic blackband ironstone" abundant as black seat earth mudstone in the collieries of the Industrial Revolution, where it was often associated with coals and calcareous material that assisted reduction. However, siderite has to be pre-roasted to oxides as it cannot of itself be richly used without snuffing a blast furnace.

Without blackband many early industrial efforts would not economically have been viable.

Stoichiometric siderite is 48.2% iron.

Wüstite (FeO)

I include Wüstite only because it is a chemical intermediate in some phases and spacial zones of Blast Furnace operation. It is only stable at temperatures around 800-900°K and is reduced by carbon and carbon monoxide to liquid iron and carbon dioxide.

Limestone ($CaCO_3$)

Limestone is a calcareous residue of prehistoric marine organisms, and it forms massively-thick, landscape forming strata in several temperate countries.

It is used as a flux in iron reduction to condition the melt and economise on coal or other fuel.

Limestone presents as massive white microcrystalline rock that readily fractures into strong twenty-centimeter shards tough enough for furnace use, with the exception of the crumbly ultrafine variety called chalk which is disfavored.

Limestone is cheap and abundant and assists purification of the melt by removing impurities as slag.

Flourite (CaF_2)

Flourite is a good class *steel* flux mineral which typically occurs in non-ferrous flats and veins as a gangue material.

It frequently presents as beautiful yellow or purple free-formed macrocrystals, slow to weather in surface tip heaps.

In Britain, two thousand years of intensive lead mining bequeathed vast hills of "rubbish" excavated by long-dead miners. Much of this waste was flourite and was taken for flux by the steel-makers of the Twentieth Century, especially during the two World Wars when importation was frequently dangerous or impossible.

Larnite (Ca_2SiO_4)

Larnite is Calcium Olivine and occurs naturally in its type locality at Scawt Hill, near Larne in County Antrim. Thus it occurs in the Irish part of the British Tertiary Igneous Province exposed elsewhere in the Giant's Causeway, and the Isle of Staffa in Scotland. At Scawt Hill, TIP basaltic lavas overlie chalk and at the contact thermal metamorphism has created the calcium olivine from elements of both strata.

I am quoting the mineral here as the nominal representative of BF slag in case the formula is useful in stoichiometric or thermochemical illustrations.

Like many chemicals, Larnite has both natural and synthetic varieties.

In nature Larnite presents as a white marmoreal mass of amorphous appearance: It is actually crystalline of a monoclinic prismatic type. Ex furnace Larnite is a dirty, vesicular, coke-like but incombustible nodular substance. Silicious slag is used as cement (or rather in the production of Portland Cement), road sealant and a treatment for acid mine effluvia. It is also an agricultural fertiliser.

CHAPTER TWO
FUNDAMENTAL SCIENCE

The Conservation of Energy

In Antiquity, and again throughout the Age of Enlightenment and the Nineteenth Century scientists believed that matter could be neither created nor destroyed except perhaps by God in Person.

Before 1843, James Prescott Joule demonstrated experimentally that mechanical friction and heat are equivalent and we now know that 4.1868 joules of heat energy is required to raise the temperature of one gram of water by 1°K.

During the opening years of the last century this was qualified as it was realised that matter could be turned to energy within the atomic nucleus.

Notwithstanding, chemists are able to do the most exacting work on the assumption that Mass and Energy are both conserved during reactions involving whole atoms.

Mass and Energy[2.1,2.2]

Mass is the entity that exerts force. Energy is the agency that gets work done. These congeries of words are hardly helpful, I know.

Heat is a form of energy. So is electricity.

Most of us know instinctively what sorts of things Force, Area, Pressure and Volume are. The concept of Mass is perhaps a little more arcane. But we know that Mass is distinct from Weight, if only because a spaceship keeps its existence whilst being weightless, whilst the same vehicle parked on an airfield has enough Weight not to be blown away by a stray squall.

Speaking of instinct, we all know that Area is measured in square meters and that a meter is a Length. Therefore it is not preposterous to say that Area has the dimensions of length squared, or L^2 for short.

Or analogously that Volume may be quoted in Cubic Meters, otherwise known as m^3, or the dimensionality L^3. That L^3 is mensurationally non-committal, and the L concerned could be measured in meters, ells or Angstroms.

Physical Dimensional Analysis

We can exploit mathematics to detect new facts about Physics and Chemistry. Physics is the science of Energy and it transformations. Chemistry is the science of Matter and its transformations. Mass is a property of all matter.

There are seven Physical Dimensions currently recognised by international law. By the way, modern scientific practice strongly approves of the *Systéme International* (SI) convention of referring masses to the Kilogram; lengths to the Meter; time to the Second; and other basic physical quantities to homologous units. But you can convert old American and British units to their SI equivalents for consistency and convenience.

These are the seven Fundamental SI-recognised Physical Dimensions:-

(a)   Mass, M
Amount of Matter expressed as Kilograms

(b)   Molarity, mol
Amount of Elementary or Composite Chemical Substance expressed in terms of grams representing the molecular, unit cell, or elemental Weight composition

(c)   Length, L
Extension (of matter) in Meters.

(d)   Time, T
Duration (of action) in Seconds.

(e)   Temperature, θ
Thermodynamic potential measured in Degrees Kelvin, K°. The Freezing Point of Water at Standard Pressure is 0°C or 273°K (approximately). Accordingly the Boiling Point of Water is about 373°K.

(f)   Electric Current, I
Measured in Amperes: The flow of Electric Charge through a surface. Or at a more practical level, the Amount of Charge passing in a given Time.

(g)   Luminous Intensity, J
       Measured in Candelas

None of these seven fundamental units may be expressed in terms of each other. But *derived* entities and exertions may be expressed as *products of exponentiated dimensions*, and by implication combinations of measures.

For example, Acceleration, a, is not a fundamental physical entity, but is the dividend of length by the square of time, in other words the rate of change of Velocity, v, which is length divided by time.

To put it in mathematical terms:-

$$v = \frac{Length}{Time} = LT^{-1}$$
Equation 2.1

and:-

$$a = \frac{Length}{Time^2} = LT^{-2}$$
Equation 2.2

Calculus practitioners often write:-

$$v = \frac{ds}{dt}$$
Equation 2.3

or:-

$$a = \frac{d^2s}{dt^2}$$
Equation 2.4

which is mathematically much the same.

So what is Force? Force is defined to be the product of mass and acceleration. Where there is no acceleration there is no force. A star or a planet like The Earth constantly exerts a Gravitational Acceleration by virtue of its mass (and that of the attracted object). It is the mass of the object and the gravitational acceleration that produces the adhesive force that allows people, cars and indeed parked spaceships to rest on the planet's surface without drifting into outer space.

Newton's Law of Gravitational Attraction defines *and quantifies* the strength of this force:-

$$F = G\frac{Mm}{d^2} = MLT^{-2}$$
Equation 2.5

where G is (Newton's) Universal Gravitational Constant which has the value 6.67430×10^{-11} Nm²kg⁻², where the Newton, N, is the SI metric of force.

Hence G has the dimensions $MLT^{-2} \times L^2 . M^{-2} = M^{-1}L^3T^{-2}$.

M is the Mass of the Larger Attractant; m the Mass of the Smaller Attractant; and d the Distance separating the Centers of Gravity of the two bodies.

Since:-

$$\frac{Mm}{d^2} = M^2L^{-2}$$
Equation 2.6

It may readily be verified that $M^{-1}L^3T^{-2} \times M^2L^{-2} = MLT^{-2}$.

By way of example, the Radius of the Earth is approximately 6.3781×10^9 meters; the Mass of the Earth is 5.9722×10^{24} kilograms and the Mass of our parked spaceship is 10000 tonnes or 10^7 kilos.

If the Intercentric Distance d can be approximated to the Earth's Radius then we may write:-

$$F = G\frac{Mm}{d^2} = 6.6743 \times 10^{-11} \times \frac{5.9722 \times 10^{24} \times 10^7}{(6.3781 \times 10^9)^2}$$
$$= 9.798451 \times 10^7$$
Equation 2.5

So the force that binds the parked spaceship to the Earth is around 98 million Newtons.

Now we already know that force is the product of mass times acceleration:-

$$F = ma$$
Equation 2.7

so we may rearrange Equation 2.7 to give:-

$$a = \frac{F}{m} = \frac{9.798451 \times 10^7}{10^7} = 9.798451 \; ms^{-2}$$
Equation 2.8

This figure for a is known as the (Terrestrial) Acceleration due to Gravity, g, and is *approximately* the same for any object standing on or near the Earth's Surface. My calculated value for g is a smidgin out as the modern agreed average for g is 9.80665 m/s^2.

Energy

The smelting of metal requires Energy. Work is done and work requires energy.

So what is Energy?

In the simplest terms Energy is the product of Force and Distance. If you have ever shoved a 25 kilo pallet of carburetors two meters across a concrete floor you will instinctively know what I mean.

So in those terms:-

$$E = Fd = MLT^{-2} \times L = ML^2T^{-2}$$
Equation 2.9

Where E is Energy and d is Distance. Energy is measured in Joules: mass times distance times acceleration. It seems, then, that an object at rest has no energy but we shall see that such thoughts are generally not realistic.

Mechanical Power is Energy divided by Time t:-

$$P = \frac{E}{t} = ML^2T^{-3}$$
Equation 2.10

Power is measured in Watts. Joules and Watts are both tiny: It is more convenient to discuss lots of 1000 Joules (KJ) and 1000 Watts (KW).

For example, if I drive my 97KW thirty-five year old sports sedan for an hour (3600 seconds) I consume 3.49×10^8 Joules-worth of gasoline energy. This optimistic figure neglects road friction, air drag, and the sheer age and inefficiency of both the car and its driver. All these considerations, intelligently adapted and applied, affect iron smelting.

You can readily ascertain that Energy is measured in SI terms as kilogram.meters²/seconds². This combination of dimensions is general to *all* expressions of energy *anywhere* in the universe and is sound as a matter of Celestial Mechanics, but we may (and shall) find it convenient to introduce the supplementary dimensions of Molarity, mol; Temperature, θ, and Electric Current, I, during the course of our discussions of iron smelting.

For example, electrical Power is frequently assessed as the product of Electromotive "Force", V volts, multiplied by Current, I. Again the unit is the Watt or the Kilowatt. It is convenient to introduce Current, I, as a physical dimension.

The analysis of Voltage, V, offers:-

$$V = \frac{E_{Pot}}{Charge} = \frac{J}{C} = \frac{Kg.m^2.s^{-2}}{A.s} = Kg.m^2.s^{-3}.A^{-1} = ML^2T^{-3}I^{-1}$$

Equation 2.11

Therefore:-

$$P = VI = Kg.m^2.s^{-3}.A^{-1} \times A = ML^2T^{-3}I^{-1} \times I = ML^2T^{-3}$$

Equation 2.12

and so the dimensionality of mechanical power, the same concept in all discussion, stands recovered.

In context, E_{Pot} is Potential Energy and C is Charge. Sometimes Charge is noted as Q, and often treated as the electrical dimension, though not in the *Systéme International.*

Kinetic Energy

Kinetic Energy is contained in a body by virtue of its motion relative to something else. For example, a car at rest on the Earth's Surface has no kinetic energy by convention, but acquires kinetic energy as it begins to move.

Kinetic Energy is specified by:-

$$E_k = \frac{1}{2}mv^2 = M \times (LT^{-1})^2 = ML^2T^{-2}$$

Equation 2.13

where E_k is Kinetic Energy and v is Relative Velocity.

Note that the mathematical scale constant 1/2, that arises from calculus, is not relevant to elementary dimensional checking, but the scale constant *is* relevant to actual quantifications.

Of course, a car at rest on the planet's surface is whirling round at many meters per second due to the Rotation of the Earth, but we can ignore this as irrelevant to arguments about speed of travel.

For example, is it more dangerous to drive at thirty miles per hour than at ten? If so why, and if not why not, as they used to say in the Navy.

First note for consistency's sake that ten miles an hour is 4.4704 m/s and that thirty miles an hour is accordingly 13.4112 m/s. This is not necessary to the integrity of the argument: We could use *arbitrary and relative* comparators of speed such as 1 league per diem and 3 leagues per diem.

The mass of the car, at 1100 kilos, is assumed to be constant.

Let E_{10} be the SI energy of the vehicle at ten miles per hour and E_{30} be the energy at thirty.

Then:-

$$E_{10} = \frac{1}{2} \times 1100 \times 4.4704^2 = 10991.46 \, Joules$$

Equation 2.14a

and:-

$$E_{30} = \frac{1}{2} \times 1100 \times 13.4112^2 = 98923.16 \, Joules$$

Equation 2.14b

Therefore, should our car hit an immovable object at 30 mph then nine times as much energy as at ten mph must almost instantaneously be surrendered. That extra energy is expended to crush ourselves as well as our car and possibly demolish whatever else we strike.

To confirm our Proof of Concept we may ignore the unchanging system constants, the integral co-efficient ½, and the mass, and use mere proportionally:-

$$\frac{E_{30}}{E_{10}} = \frac{\frac{1}{2} \times 1100 \times 3^2}{\frac{1}{2} \times 1100 \times 1^2} = \frac{3^2}{1^2} = 9$$

Equation 2.15

Potential Energy

Potential Energy, E_p, is energy that inheres in an object by virtue of its position in a force field. Very often the force field in question is the gravitational force field of The Earth, but this is not necessarily the relevant force field.

Terrestrial Gravitational Potential Energy may be given as:-

$$E_P = mgh = M.LT^{-2}.L = ML^2T^{-2}$$

Equation 2.16

where Ep is Potential Energy, m is Mass, g is the Acceleration Due to Gravity and h is Height, or to be more helpful, a difference in height. So a better expression is:-

$$E_P = mg.\Delta h = M.LT^{-2}.L = ML^2T^{-2}$$

Equation 2.17

where the dot period represents multiplication as an alternative to ×, and Δ means "Difference in ..."

For example, a lorry leaves a 50kg sack of potatoes on the pavement below a cathead on a warehouse. Your job is to lift the sack of potatoes to the second floor at an elevation of 20 feet or 6.095999805 meters. Of course, this is technically a lift between the geographic altitude of the pavement and that of the second floor, but we will for convenience declare the pavement altitude to be datum zero. Also we will ignore frictional energy losses and other vagaries of warehouse work.

We will also assume that the 6.095999805 meters ought to read 6.096 meters, because the British foot is itself defined today in terms of the *Systéme International* as 0.3048 meters exactly.

Accordingly:-

$$E_P = mgh = 50 \times 9.80665 \times 6.086 = 2984.163595 \text{ Joules}$$

Equation 2.18

So the work in raising the sack of potatoes is about 2984 Joules before losses.

As a second example in a slightly more complex circumstance we shall examine a small hydroelectric aluminum smelter. The surface of the smelter's reservoir is at an elevation (h_2) of 250 meters whilst the turbines are at 20 meters (h_1). The System Efficiency (η, a dimensionless fraction) is 0.95. Also the Mass Flow Rate, r_m, is equal to ρD, where ρ is the Density of Water at 9°C (999.85 kg/m^3) and D is the Discharge of Water in Cumecs which is 38.27 (m^3/s).

What is the Shaft Power in Megawatts?

$$P = \eta r_m g(h_2 - h_1) = \eta \rho D g(h_2 - h_1) = \eta \rho D g . \Delta h$$
Equation 2.19

Substituting data:-

$$P = 0.95 \times 999.85 \times 38.27 \times 9.80665 \times (250 - 20)$$
$$= 82826931.863085$$
Equation 2.20

Therefore the Power available at the smelter is about 82.8 MW.

Allow that K_{sys} is a constant for any given hydropower system, such as an unmodified installation like our aluminum smelter. We could write:-

$$K_{sys} = \eta \rho g(h_2 - h_1)$$
Equation 2.21

where η, ρ, g and (h_2-h_1) are all sub-constants. We can check that K_{sys} has the dimensions $M^0L^0T^0 \times ML^{-3} \times LT^{-2} \times L = ML^{-1}T^{-2}$

Energy has the dimensions $ML^2 T^{-2}$. Whatever it is that may readily vary, or be made to vary, it has the dimensions of E/K_{sys} = $ML^2T^{-2}/ML^{-1}T^{-2} = L^3$, that is to say cubic meters. So energy is proportional to the volume of water passed through the turbines.

Now energy is Power times Time, so that in an hour:-

$$E = W = Pt = 82826931.863085 \times 3600$$
$$= 2.98176954707106 \times 10^{11} \, Joules$$
Equation 2.22

10^9 Joules is a GigaJoule, so the shaft work done in an hour is about 298 GJ.

Also Pressure is Force divided by Area which thus has the dimensions $MLT^{-2}/L^2 = ML^{-1}T^{-2}$:-

$$p = \frac{F}{A} = \frac{ma}{A}$$
Equation 2.23

where p is Pressure (of water at the turbine nozzles).

You can check that K_{sys} has the dimensions of Pressure: In fact K_{sys} *is* Pressure, because it never varies at the nozzle as long as the reservoir and turbines stay where they are and the density of water is about the same (it varies a little with temperature). K_{sys} may also be vitiated if the penstocks roughen hydraulically or if moving parts wear.

pV is $ML^{-1}T^{-2} \times L^3 = ML^2T^{-2}$ is energy, where V is the Volume (of water passed through the system). Therefore:-

$$E = W = pV$$
Equation 2.24

So it is also the case that $K_{sys} \times D \times t = ML^{-1}T^{-2} \times L^3T^{-1} \times T$ = ML^2T^{-2}, so that:-

$$E = W = K_{sys}Dt$$
Equation 2.25

Combining Equations 2.24 and 2.25 we may write:-

$$pV = K_{sys}Dt$$
Equation 2.26

The Equation 2.20 figure of about 82826931.863085 Watts is, to be even more conservative, around 82 Megawatts (MW). Something like 27609 three-bar electric fires shining-forth all at once. An old-fashioned Rolls-Royce Merlin piston engine as fitted to the Spitfire and four-fold to the Avro Lancaster bomber had a rated power of one megawatt. A modern Rolls-Royce Trent 600 gas turbine for use on land is 58MW, somewhat shy of our hydropower system. Modern Rolls-Royce turbofans for aviation generate around 52MW each. The large US aircraft carrier warship USS Nimitz uses two nuclear reactors to yield 190MW with shaft power being mediated by steam turbines. A contemporary road-legal Rolls-Royce Ghost Prism sedan car (made by BMW) has a maximum power of 0.42 MW. My old Volvo 460GLE sports sedan has 0.097MW, as previously noted.

The power of the typical Blast Furnace of 2024AD is surprisingly small but will be systematically addressed later. The preferred metric of Blast Furnace performance is tonnes of iron output per annum.

The Universal Gas Law

The Universal Gas Law is relevant to iron smelting because all of the current actual or proposed methods of reducing iron ore to metal either use or generate large volumes of gas.

A gas is a phase of matter subject to Kinetic Theory, meaning that the individual particles (molecules) of gas are treated like individual randomly-jostling tiny projectiles that have no mutual attractions but occasionally impact one another transferring kinetic energy whose bulk directional sum is zero. Temperature is proportional to the particle kinetic energy. Therefore, in this discussion it is now convenient to introduce to the three mechanical dimensions MLT, a fourth dimension of Temperature θ to give the Thermal System MLTθ. We may also have cause to dimensionalise the Mole in distinction to Mass M.

The probability of an inter-molecular strike also increases with gas density, as also with pressure: So does the energy within a given parcel of gas.

Accordingly the result of the labors of several was the Universal Ideal Gas Law first articulated by Benoît Paul Émile Clapeyron in 1834:-

$$pV = nRT$$
Equation 2.27

where p is Pressure; V is Gas Volume; n is the Amount of Matter (in Kilograms or Moles); R is the Universal Gas Constant; and T is the Gas Temperature in Degrees Kelvin K°.

The Universal Gas Constant R has a value of 8.31446261815324 JK^{-1}mol^{-1} so that it has the dimensions ML^{2}T^{-2}×θ^{-1}×mol^{-1} = ML^{2}T^{-2}θ^{-1}mol^{-1}. The Mol is essentially a standardised form of chemical mass: The Molecular Weight of a substance expressed as grams. It is sensible to treat it as other than M.

The Universal Gas Constant R is itself a product of Boltzmann's Constant k$_B$ and Avogadro's Constant, N$_A$. Boltzmann's Constant is 1.380649×10^{-23} J/K°. Boltzmann's Constant gives the average thermal (kinetic) energy of a gas molecule at a stated Temperature.

Avogadro's Constant is of especial interest as the count of the Number of Molecules in a Mole of Substance. In 1812 Amadeo Avogadro demonstrated that "equal volumes of all gases, at the same temperature and pressure, have the same number of molecules." The technology of the time did not permit Avogadro to actually determine this number, though he knew it must be incredibly big.

Avogadro's Constant, otherwise called Avogadro's Number, is now known to be 6.02214076×10^{23} mol^{-1}, or in layman's terms there are about 602 sextillion O_2 molecules in a parcel of one mole of diatomic oxygen. This would weigh about 32 grams.

The theta's on the RHS of Equation 2.27 cancel out so you can check that both the LHS and RHS of Equation 2.27 are mechanical energy and that the equation balances as all equations must.

The Atom

Avogadro showed us that whatever the details, an atom of a chemical element, or even a molecule compounded of several elements' atoms, must be very little by the standards of our humanly intelligible world.

Atoms are made of Baryons and Leptons. Baryons congregate in the atom's nucleus. Baryons can either be Protons or Neutrons. They are about the same size: Protons "weigh" 1.67262192369×10^{-27} kilograms and Neutrons 1.67492749804×10^{-27} kilograms. As you can see the Neutron is a smidgeon heavier than the Proton.

There are very roughly equal numbers of Protons and Neutrons in a stable, natural nucleus, though some "top heavy" nuclei have an excess of neutrons and tend to break down either naturally or with artificial encouragement.

The Number of Protons in a nucleus determine the Chemical Element. The Number of Neutrons determine the Isotope of the Element.

In orbit about the nucleus are Leptons, specifically Electrons. Electrons at rest weigh $9.1093837015 \times 10^{-31}$ kg. If you have a pocket calculator, or maybe MathCad® handy you may readily confirm that a Proton is about 1836 times more massive than an electron.

However, each electron has an Electric Charge, specifically minus $1.602176634 \times 10^{-19}$ Coulombs (-e or e⁻) whilst this is balanced by each of the Protons in the Nucleus which have a charge of $+1.602176634 \times 10^{-19}$ C. Since there are as many Electrons as Protons in the whole atom the aggregate intraatomic charges cancel and the undisturbed atom is electrically neutral.

Whilst it is clear that an individual atom consists of almost nothing, it is not always helpful to try and illustrate one. Baryons are not really like billiard balls, and electrons are not flies, though in several scientific models they buzz around the nucleus at ridiculous speeds describing a "volume of probability" which type of illustrates their position in space.

Early conceptions of the atom saw it as a more or less uniform Euclidean sphere. As the vision of electrons and nuclei began to resolve at the end of the Nineteenth-Century people started to picture the "Bohr Model" as a planet-like nucleus about which electrons transited like moons through orbits, which apart from being at different heights, began to ossify into different "quantum levels".

By the Fifties of the Twentieth-Century American chemists and others began to see the atom as aggregations of cigar-shaped zones of reactive probability, similar in shape, if you will, to the toy dogs that children's entertainers wrangle using cylindrical toy balloons.

The truth is that the atom and especially its force fields are inaccessible to the vision of man, though important features of the atom and its properties may be inferred.

You are entitled to some provisional volumetric estimate of an atom's size. The simplest atom is that of the gas Hydrogen. In particular, the isotope that has a sole proton orbited by a single electron, designated ^1_1H. H is the Chemical Symbol of the Element that holds a single proton; the Subscript designates the Atomic Number, i.e. the proton count determinative of the Chemical Element; the superscript 1 is the Total Baryon Count of the nucleus that defines the relevant Isotope.

The Sparsity of Matter

According to the so-called Classical Model, the "Bohr Radius" of the Hydrogen Atom, a_0, is given by:-

$$a_0 = \frac{4\pi\epsilon_0\hbar^2}{e^2.m_e}$$

Equation 2.28

a_0 is the Bohr Radius of the Hydrogen Atom; ϵ_0 is the Permittivity of Free Space; $\hbar$ is the Reduced Planck Constant; e is the Charge on the Electron and m_e is the Rest Mass of the Electron.

The value of a_0 is $5.29177210903\times10^{-11}$ meters.

A notional Euclidean sphere representing the atom would be $(4/3)\pi a_0^3$ or $6.20714667913693\times10^{-31}$ cubic meters (V_{bohr}).

So the "Bohr Density", ρ_{bohr}, of the Hydrogen atom is around:-

$$\rho_{bohr} = \frac{m_P + m_e}{V_{bohr}}$$
$$= \frac{1.67262192369 \times 10^{-27} \times 9.1093837015 \times 10^{-31}}{6.20714667913693 \times 10^{31}}$$
$$= 2696.13873905239 \ kg/meters^3$$

Equation 2.29

My Late Father did not believe in ghosts. Strangely, when I was a very small boy, our conversation once turned to whether or not an object could pass through walls, and Father conjectured that this was theoretically possible because the atom was almost empty space.

But according to Bohr the atom has a density about the same as ordinary rock. What a conundrum! But we now know that many tiny particles such as neutrinos (electrically-neutral electrons) can pass unhindered through solid matter thousands of meters thick because they do not interact with the charged particles in atoms. Even charged particles like ions can make limited progress through matter.

The Boulby Polyhalite Mine has driven to more than 1100 meters depth through the Permian evaporitic marls of Northern Yorkshire, England. The Palmer Laboratory has been instituted at 1100 meters under the overburden of the mine. This large research facility is able to detect negatively-charged muons (cosmic ray

$1.883531627 \times 10^{-28}$ kg leptons) from outer space. There is also realistic anticipation that anti-neutrinos (emanating from the Hartlepool Nuclear Power Station some 40 rock kilometers distant) will prove detectible at the Palmer Laboratory.

Perhaps the Bohr Model is oversimplified?

Nuclear and Chemical Reactions

Nuclear reactions only concern the baryons in their nucleus. Chemical reactions only concern the orbiting electrons. Chemical reactions are about bonding atoms by stealing or sharing their orbital electrons. All reactions transact energy.

Many thousands of different nuclear and chemical reactions are possible.

We have now discriminated Nuclear and Chemical reactions, but Albert Einstein showed that Matter and Energy are interchangeable. Many modern workers even view matter as merely superdense energy. The relevant equation could hardly be simpler:-

$$E = mc^2$$
Equation 2.30

where E is Energy; m is mass; and c is the Celerity of Light (in ordinary terms the Speed of Light: the Celerity of a wave differs from the Group Velocity of its constituent particles [if any]). But in ordinary language the Celerity of Light is a velocity and accordingly has the dimensions LT^{-1}, so that it is measured in meters per second.

The value of c is 299792458 m/s.

For example, we established for our small aluminum smelter works that it consumed 82MW of shaft power. Therefore, how many Joules of energy is consumed in a 24-hour day of smelter operation? And how much matter must be annihilated to yield this energy?

$$E = Pt = 820000000 \times 24 \times 60 \times 60 = 7.0848 \times 10^{12} \ Joules$$
Equation 2.31

By transposition of Equation 2.30:-

$$m = \frac{E}{c^2} = \frac{7.0848 \times 10^{12}}{299792458^2} = 0.000078829031171$$
Equation 2.32

So 0.078829031171 grams of matter is destroyed to power our aluminum smelter for 24 hours. Call it 79 milligrams. About the mass of 800 of the laboratory Fruit Fly *Drosophila melanogaster*.

Please note that this argument has got nothing to do with nuclear power or the (ir)relevant fact that our smelter is water-powered.

Perhaps the matter annihilation was the result of nuclear fusion taking place in The Sun, whose heat evaporated water from the surface of the sea, and blew it as clouds of rain to the mountains, where the rain fell to plenish our reservoir?

Nuclear Reactions

Contemporary (2024AD) nuclear power stations are worked by nuclear fission which typically involves the break-up of ^{235}U (Uranium 235) atoms. There are several possible avenues through which this may be achieved but the following pathway is usual:-

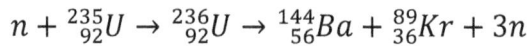

$$n + {}^{235}_{92}U \rightarrow {}^{236}_{92}U \rightarrow {}^{144}_{56}Ba + {}^{89}_{36}Kr + 3n$$
Nuclear Pathway 2.1

The relevancy of this pathway is that it is basic science and that it is mathematically an equation whose structure resolves to present an energy yield. This structure and outcome is identical in principle to the *purely chemical* equations that control iron furnace processes.

I may have omitted several hard-to-detect reaction products from the RHS outcomes (for example, anti-neutrinos) and this leads inevitably to underestimations of energy yield.

Uranium is a dense, shiny gray metal, very slightly radioactive in nature. Barium (Ba) is a so-called Alkaline Earth metal, malleable and lustrous but rapidly corroding in air and rapidly effervescent in water as it chemically reacts to form barium hydroxide and hydrogen gas. Krypton is a so-called Nobel or Inert Gas, a relative of Neon and Xenon. Barium is very closely related to Calcium and Strontium metals and is sometimes accepted by the human body in substitution of Calcium. All of the metals and gases mentioned are chemical elements defined by the proton count in their atomic nuclei.

You have decerned that the Pathway 2.1 reaction generates three neutrons for each neutron consumed. These daughter neutrons can be manipulated similarly to break further ^{235}U atoms. It is this effect that produces the sustainable nuclear chain reaction, and thus the exponential growth of energy release.

The (Unified) Atomic Mass Unit or Dalton

It is inconvenient to process nuclear equations in kilograms because the objects involved are unimaginably tiny, having masses of the order $\sim 10^{-27}$ kilograms.

Therefore, we arbitrarily declare that the isotope Carbon 12 (^{12}C) has a mass of exactly 12 Da (Daltons) and "weigh" other atomic particles against it.

Of course, modern computers being what they are arithmetic is not nearly as daunting as it was, but it can still be argued that amu (atomic mass units = Daltons) should be cataloged and the kilogram-level implications economically consigned to post-calculative reporting.

So:-

$$m_u = \frac{1}{12} m(^{12}_6C) = 1 \; Dalton$$

Equation 2.33

This implies that the mass of a Dalton (amu) is very roughly the same as that of a proton or neutron.

The actual mass of one Dalton is $1.660539066600000 \times 10^{-27}$ kilograms.

The Electron Volt (eV)

The Electron Volt (eV) is the energy imparted to one electron accelerated from rest in vacuo by an EMF (Electromotive "force") of one Volt.

The Charge on the Electron (Q_e) and the Mass of the Rest Electron (m_e) are respectively $-1.602176634 \times 10^{-19}$ Coulombs and $9.1093837015 \times 10^{-31}$ kg.

The Electron Volt is a measure of Energy.

The agreed value of the Electron Volt is $1.602176634 \times 10^{-19}$ Joules.

Discount the sign of the charge and the electron mass.

Energy is the product of charge and voltage potential difference.

Therefore:-

$$E = QV = 1.602176634 \times 10^{-19} \times 1 = 1.602176634 \times 10^{-19}$$
Equation 2.34

where E is Energy (i.e. the Electron Volt); Q is Charge; and V is the Potential Difference in Volts. The Electron Volt is numerically equal to the Charge on the Electron.

Given that the dimensions of any Electric Charge are T^1I^1 and the dimensions of Voltage are $ML^2T^{-3}I^{-1}$ you may readily confirm that the Electron Volt is dimensionally Energy ML^2T^{-2}.

Can we measure the Electron Volt in terms of the Dalton?

Well according to Einstein's Mass-Energy Equivalence Equation as transposed at Equation 2.32:-

$$m = \frac{E}{c^2} = \frac{1.602176634 \times 10^{-19}}{299792458^2}$$
$$= 1.7826619216279 \times 10^{-36} \, kilograms$$
Equation 2.35

This is exactly the same result as Wikipedia contributors calculate in their article "Electronvolt".

The Dalton is $1.660539066600000 \times 10^{-27}$ kg and accordingly there are $9.31494102417143 \times 10^{8}$ electron volts in a Dalton.

The Calculation of Nuclear Yield

We can estimate the energy yield of a single nucleus fission *provided that we take most or all of the daughter particles into account* by using simple tallies of the parental and daughter weights in Daltons.

This is shown in Tables 2.3 and 2.4 which present the Entry and Exit pathway tallies.

Before we review those it may be convenient by way of preparation to list the key physical constants implicated:-

Entity	Symbol	Value	Units
Avogadros Number	N_A	6.02214E+23	mol-1
The Celerity of Light	c	299792458	Meters per Second
The Charge on the Electron	Q_e	−1.602176634E-19	Coulombs
The Mass of Electron at Rest	M_e	9.1093837015E−31	Kilograms
Electron Volt	eV	1.602176634E-19	Joules
Molar Mass of ^{235}U	$^{235}U_{mol}$	235.0439299	Grams
Density of ^{235}U	ρ_{235}	19050	Kg per Cubic Meter

Table 2.1
Nuclear Physical Constants

and the relevant particle weights in Daltons:-

Entity	Daltons (amu)	Kilograms (calculated)	Kilograms (published)	PSD(kg_c,kg_p)
Definition	1	1.660539066600000E-27		
Electron	0.000548579909065	9.109383701543080E-31	9.109383701500000E-31	4.7291E-10
Proton	1.007276466621000	1.672621923690980E-27	1.672621923690000E-27	5.8658E-11
Neutron	1.008664915950000	1.674927498043780E-27	1.674927498040000E-27	2.2572E-10
U^{235}	235.043929918000000	3.902996279960320E-25		
Ba^{144}	143.922952853000000	2.389896857928360E-25		
Kr^{89}	88.917630581000000	1.476511992892570E-25		

Table 2.2
The Dalton and Selected Particle Weights

ENTRY				
Particle	Atomic Number	Isotope	Number of Neutrons	Isotopic Mass (Da)
Neutron	0	1	1	1.00866491595
U	92	235	143	235.04392991800
Total (Da)				236.05259483395

Table 2.3
The ENTRY (LHS) Parental Particles of the
2.1 Fission Pathway

EXIT				
Particle	Atomic Number	Isotope	Number of Neutrons	Isotopic Mass (Da)
Ba	56	144	88	143.92295285300
Kr	36	89	53	88.91763058100
3 Neutrons	0	1	1	3.02599474785
Total (Da)				235.86657818185

Table 2.4
The EXIT (RHS) Daughter Particles of the
2.1 Fission Pathway

You have perceived straight away that taken together the daughters are about 0.2 Daltons lighter than their parents. This "mass defect" is liberated as energy.

More accurately to summarise:-

0.18601665210	Mass Defect (Da)	
3.08887917850E-28	Mass Defect (Kg)	
2.77614615817E-11	Energy Defect (J)	3.24×10^{-11} wiki
173.27341438253	Energy Defect (MeV)	202.5MeV wiki
		+8.8MeV as anti-neutrinos
16.71834293484	Molar Energy Defect (TJ/mol)	19.54 wiki
71.12858834802	Molar Energy Defect (TJ/Kg)	83.14 wiki

Table 2.5
Mass and Energy Defects for the ^{235}U Fission of
2.1 Fission Pathway

As you can see, my results are roughly 15% shy of the Wikipedia findings, doubtless because I have failed to take the daughter anti-neutrinos into account, as well as other unknowns.

To compute the Mass Defect in kilograms we multiply m_D, the Mass Defect in Daltons, by the Actual Mass of One Dalton, m_u, ($1.6605390666 \times 10^{-27}$ kg):-

$$m_{kg} = m_D m_u$$
Equation 2.36

To obtain a human-scale mass of parent nuclei consumed we need to multiply by Avogadro's Number N_A ($6.02214076 \times 10^{23}$) which yields the *molar weight in grams*:-

$$m_{mol} = N_A \frac{m_{kg}}{10^3}$$
Equation 2.37

A mole of ^{235}U (whether gas, liquid or solid) is equal to its Atomic Weight *in grams* (^{235}U$_{mol}$) which is 235.0439299 grams. The Energy Defect in Joules, E_J, is given by:-

$$E_J = \frac{m_{kg}}{c^2}$$
Equation 2.38

where c is the Celerity of Light.

Fission Power and Smelting

As of 2024AD nuclear fission is used to generate electricity and electricity is used to smelt aluminum. The smelting of iron using electricity is experimental.

In order, so to say, tie-in nuclear power with smelting we can look again at our exemplary aluminum smelter, the one consuming about 82MW.

We computed at Equation 2.31 that the daily energy consumption of the aluminum smelter was about 7.085×10^{12} Joules:-

$$E = Pt = 820000000 \times 24 \times 60 \times 60 = 7.0848 \times 10^{12} \ Joules$$
Equation 2.31

We already know that this quantity of mass is 78.8 milligrams of *anything*:-

$$m = \frac{E}{c^2} = \frac{7.0848 \times 10^{12}}{299792458^2} = 0.000078829031171 \ kg$$
Equation 2.39

so that:-

$$l = \sqrt[3]{V_{235U}} = \frac{m_{kg235U}}{\rho_{235U}}$$
Equation 2.40

where l is the Side of the Representative Metal Cube; V_{235U} is the Volume of Metal; m_{kg235U} is the Total Mass of Metal Consumed by Fission; and ρ_{235U} is the Density of ^{235}U in kg/m^3.

Noting that the Atomic Weight of ^{235}U is 235.043929918 grams, the *Kilogram* mole, $^{235}U_{mol}$, of ^{235}U is 0.235043929918 kg; and that the Mass Density, ρ_{U235}, of ^{235}U is 19050 kg/m^2, we may move forward to compute:-

$$Mole\ Of\ 235\ Uranium = \frac{m}{U_{mol}^{235}} = 0.000335379991301$$
Equation 2.41

and:-

$$l = \sqrt[3]{V} = \sqrt[3]{\frac{m}{\rho_{U235}}} = 0.001605451054231$$
Equation 2.42

Therefore, the ^{235}U that would theoretically be consumed by a nuclear reactor to fuel our aluminum smelter for one day is represented by a cube of ^{235}U metal of side length 1.605 millimeters.

Please note that these figures are indictive only: The transmission of electricity from any source is prone to losses, losses that magnify with any transformation of the current.

Bear in mind that a block of pure ^{235}U would be extremely expensive and thoroughly dangerous. A 20% pure mass of ^{235}U would only be considered in the manufacture of detonators for hydrogen (fusion) bombs, or conceivably as submarine propellant ("enriched uranium"). Commercial nuclear power station fuel is around 5% ^{235}U. Natural uranium is 0.7% ^{235}U, most of the rest being inert ^{238}U

We will consider the partial performance of reaction pathways in a subsequent chapter.

Energy versus "Energy"

Mystics often speak of subjective sensations as "energy" or "energies", especially if they are localised or are judged supernaturally-significant, as in the phrase "since entering this room I have detected a malign energy lurking in that corner". They may even propose a thermodynamic condition in that corner such as it being distinctly chilly. This is all very well as metaphor, as long as their auditors accept it as metaphor, in default of known metrics of assessment.

Much more reprehensible is the cant of politicians who too often prattle about "green energy" or renewable energy or something without knowing what they are talking about, many of them knowing that they do not know or care, but hoping to fool you, and divert your money to their own purposes.

I have written this book to help you against them.

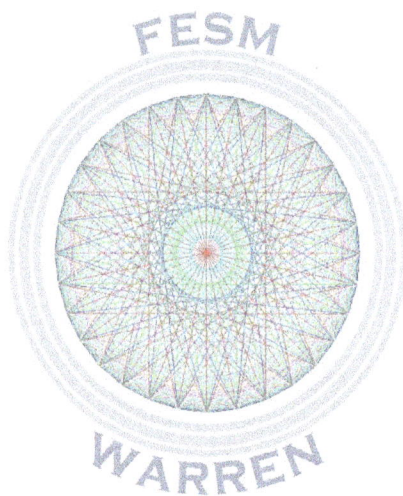

CHAPTER THREE
STOICHIOMETRIC PRINCIPLES

Stoichiometry is the science of weights (more properly masses) of reagents (reactants) that must be engaged to make products, without undue waste or dearth. Stoichiometry depends upon knowledge of reagent unit (formula) weights and valences.

The Volumetric Visualisation of Mass

Allow that m_x is the Given Mass of Substance x, and that M_x is the Molar Mass of Substance x. Meanwhile, the Mass Density of Substance x in g/cc is denoted ρ_x.

Then:-

$$Moles\ of\ x = \frac{m_x}{M_x}$$
Equation 3.1

and:-

$$V_x = \frac{m_x}{\rho_x}$$
Equation 3.2

where V_x is the Volume of the Given Substance x.

To visualise this mass of substance as a cube of material of Side Length l (letter el) take the cube root of the volume as below:-

$$l = \sqrt[3]{V_x}$$
Equation 3.3

For example, a tonne of iron has a Density ρ_{Fe} of 7.874 g/cm³ and contains $10^3 \times 10^3 = 10^6$ grams of matter. The Formula Weight of Iron is 55.845

How many moles is this?

$$Moles_{Fe} = \frac{10^6}{55.845} = 17906.70606142$$

What is the Volume of Iron we have in cm³?

$$V_{Fe} = \frac{10^6}{7.874} = 127000.254000508 \ cm^3$$

Accordingly, the Cube Side Length is:-

$$l = \sqrt[3]{127000.254000508} = 127000.254000508^{\frac{1}{3}}$$
$$= 50.2652904633508 \ cms$$

So think of a tonne of Iron as a cube about half a meter wide.

The Concept of Atomic Weight

Strictly speaking, Atomic Weight is actually a Mass with the dimension M, but it is perhaps more useful to think of it as the Formula (Unit) Mass in Daltons *expressed in terms of grams per mole*.

This Unit Mass applies to the fundamental repeated assembly of atoms in the case of a crystalline solid (for example, the repeating substructure in Larnite Ca_2SiO_4 comprises seven atoms, that is two of calcium, one of silicon and four oxygen atoms). In the case of liquids true Molecules of independent atom clusters are the Unit of reference: For example the eight carbon and eighteen hydrogen molecule of Octane, C_8H_{18}. Some elemental gases are monoatomic, but stable Oxygen O_2 and stable Hydrogen H_2 are diatomic. The gaseous compound Acetylene C_2H_2 is clearly tetra-atomic.

Given, then, that Atomic "Weight" is reconned in g/mol we can check that it is a dimensionless number in these terms:-

$$\frac{Atomic}{Weight} = \frac{grams}{moles} = \frac{M}{M} = M^0 \equiv M^0 L^0 T^0$$
Equation 3.4

My excellent doctoral supervisor, the late Professor DIH Barr, insisted that such mathematical objects were "non-dimensional numbers" whereas I have always known them as "dimensionless numbers". I suppose I just try to avoid hyphens where I can. They confuse most foreigners, and they usually confuse me. But as I remark elsewhere a great Mercian poet once asked "What is in a

Name...". It is for certain that numbers of any species are more reliable than words.

By reference to a table of Atomic Weights we can see that Calcium has an AW of 40.078, Silicon an Atomic Weight of 28.0855, and diatomic Oxygen (O_2) 31.9988

To obtain the Formula Weight of Larnite Ca_2SiO_4 we must use simple arithmetic to add these seven atom values together:-

$$FW(Larnite) = 2 \times 40.078 + 1 \times 28.0855 + 2 \times 31.9988$$
$$= 172.2391$$
Equation 3.5

So the Formula Weight of Larnite is 172.2391 Daltons. If we want that in grams we have to call it grams, bearing in mind that there are Avogadro's Number of formula units in a mole, that is to say 172.2391 grams of Larnite..

The reason I multiplied the oxygen weight by 2 rather than 4 in Equation 3.5 is because the AW for oxygen is already listed for the *diatomic* form.

In general:-

$$FW = \sum_n \lambda_i . AW_i$$
Equation 3.6

where FW is the Formula Weight; n is the Number of Different Elements or Dimers in the Formula (implicit); λ_i is the Stoichiometric Coefficient for that Element; AW_i is the relevant Atomic Weight; and i is the relevant Elemental Subscript.

As the sum of products of dimensionless numbers, FW is itself dimensionless. Therefore, it may be used directly to scale Mass, Energy or anything else.

The decisive advantage of Equation 3.6 is that we can use it compute the required mass of each input substance, given a desired amount of output. (Assuming 100% efficiency of the chemical reaction).

For example, if we want a kilogram of Larnite we will need to produce 1000/172.2341 = 5.80588263640486 mols (more than five-figure accuracy is not realistic, but I include the trailing digits to facilitate your checking of the working principle).

Accordingly, we need by scaling:-

$$CW_{Ca} = \lambda_{Ca} \times AW_{Ca} \times mols_{Ca2SiO4}$$
Equation 3.7

Correspondingly, we need to scale the input chemical elements using:-

$$CW_j = \lambda_j . AW_j . MM_k$$
Equation 3.8

where CW_j is the Contributory Mass (in grams) for each of the $j = 3$ elements involved, and MM_k is the Molar Mass (in mols) for the Product denoted k (i.e. Larnite, Ca_2SiO_4). I respectfully invite you to check that the sum $CW_1 + CW_2 + CW_3$ is indeed 1000 grams. For your intermediate reference we require to react 465.376328603668 grams of Calcium, 163.061116784749 grams of Silicon, and 371.562554611584 grams of Oxygen.

Table 3.1 presents selected stoichiometric and thermochemical data for elements and compounds relevant to iron smelting:-

t °K	Substance Name	Formula	Formula Weight (Da)	ΔH_{form} KJ/mol	Heat Capacity C_p (J/mol**°K)	Latent Heat Of Fusion L_f (KJ/Kg)	Latent Heat Of Fusion L_{fmol} (KJ/mol)	Latent Heat Of Fusion $L_{fmolcalc}$ (KJ/mol)	PSD (L_f,$L_{fmolcalc}$)
298.15	Hydrogen	H_2	2.01588	0	14.304	58.8	0.05868	0.059267	-1.0001227
298.15	Oxygen	O_2	31.9988	0	32.97616924	13.9	0.444	0.444783	-0.17642342
298.15	Water	H_2O	18.0153	-241.826	33.589249	334	6	6.01711	-0.28517
298.15	Nitrogen	N_2	28.0134	0	29.125	25.7	0.7204	0.719944	0.06324542
298.15	Flourine	F_2	37.9968063	0	31	13.2	0.2552	0.250779	1.73239744
298.15	Carbon	C	12.0107	0	8.517		117		
298.15	Iron	Fe	55.845	0	25.1	247	13.81	13.79372	0.1179218
298.15	Calcium	Ca	40.078	0	25.929		8.54		
298.15	Silicon	Si	28.0855	0	20	1790	50.55	50.27305	0.54788328
298.15	Aluminum	Al	26.9815386	0	0.897	321	10.79	8.661074	19.7305478
2500	Hydrogen	H_2	2.01588	0	35.84051015	58	0.11736	0.116921	0.37402863
2000	Oxygen	O_2	31.9988	0	37.74686625	13.9	0.444	0.444783	-0.17642342
2000	Water	H_2O	18.0153	-286	50.67546275	334	6	6.01711	-0.28517
298.15	Hematite	Fe_2O_3	159.688	-825.5	103.7752759				
1100	Hematite	Fe_2O_3	159.688	-825.5	115.108889				
1100	Magnetite	Fe_3O_4	231.533	-1120.89	200.8320001	596			
1100	Wüstite	FeO	71.844	-249.532	68.1992	334.1			
298.15	Carbon Monoxide	CO	28.0101	-110.527	29.1				
298.15	Carbon Dioxide	CO_2	44.0095	-393.522	37.135	184	9.02	8.097748	10.2245233
298.15	Water	H_2O	18.0153	-241.826	33.589249	334	6	6.01711	-0.28517
298.15	Methane	CH_4	16.0425	-74.6	35.69	510	8.19	8.181675	0.10164835
298.15	Octane	C_8H_{18}	114.2285	-250	255.68	181	20.74	20.67536	0.31167551
298.15	Benzene	C_6H_6	78.1118	49.04	134.48	126	9.92	9.842087	0.78541532
1273	Calcite	$CaCO_3$	100.087	-1207	85.2	57.35			
1273	Quicklime	CaO	56.077	-635					
1273	Wollastonite	$CaSiO_3$	116.162	-1630					
1273	Quartz	SiO_2	60.0843	-911					
1000	Larnite (slag)	Ca_2SiO_4	172.2391	-1630	140.9775295				

Table 3.1
Selected Stoichiometric and Thermochemical Data

The Concept of Valency

Valency is the (integer) number of attachments that one atom can hold out to another. It is sometimes termed "Combining Power".

Valency is a chemical parameter dependent upon the outer shells (levels) of orbital electrons only.

Many diatomic gaseous elements are univalent meaning each atom offers one "hand" only to grasp another atom in partnership. For example, H_2, and all the Group 17 Halogens such as Cl_2 and F_2. At the opposite end of the Periodic Table in Group 1, all the Alkali Metals like K (potassium) and Na (Sodium) are univalent.

Therefore, a chemical like Common Salt is feasible, because one atom of Na can combine with one of Cl to form the crystal mineral NaCl. Salt and salt-like compounds are stable crystalline solids at STP. Alkali Metals are strongly electropositive and Halogens strongly electronegative so they form very strong polar electrostatic bonds. But most are soluble in water to some extent, and all are fusible at high temperatures and pressures, enabling electrolytic dissolution of the bonds to recover pure metals and pure halogens (or if relevant, pure oxygen).

Divalent elements include the Group 2 Alkali Earth Metals such as Calcium (Ca) and Strontium (Sr) and also Group 16 non-metals like O_2 (oxygen) and S (sulfur).

Nitrogen (N) and Phosphorus (P), both Group 15 elements are trivalent.

Carbon (C), Silicon (Si), Lead (Pb) and Tin (Sn) are in Group 14 and all are quadrivalent.

In Groups 3 to 12, the Transition Metals exhibit a more complex picture including multiple valency capabilities. From our viewpoint, only Iron (Fe) is interesting and its valances are either +two or +three.

The Inert or Noble Gases like Argon (Ar) and Neon (Ne) are, with Helium (He), in Group 18 and are classified as having zero valences because they do not naturally combine with themselves or any other atom.

Atomic Number	Element Symbol	Unit Mass (Da)	Valency	Contexts
1	H_2	1.008	1	reduction
8	O_2	31.998	2	ore, flux
7	N_2	28.014	3	blast
9	F_2	37.996	1	flux
6	C	12.011	4	fuel, flux, slag
26	Fe	55.845	2,3	ore, metal
20	Ca	40.078	2	flux, slag
14	Si	28.085	4	slag

Table 3.2
Selected Valences

Chemical Reaction Pathway Notation

Key to the traditional Hot Blast method of iron smelting for steel production is the oxidation (combustion) of carbon in the form of coke. This reaction is one of those capable of step-wise or cascaded series of distinct reactions. For simplicity we will consider a single equivalent pathway:-

$$C + O_2 \rightarrow CO_2$$
Pathway 3.1

The symbol $\rightarrow$ tends to be favored by chemists because unlike a traditional Recordian equals sign = it implies that the Carbon Dioxide gas CO_2 is the *outcome* of combining carbon and oxygen. Sometimes the fastidious regard even this symbol as inappropriate, considering $\rightleftarrows$ to be better, because it implies that the product can break down to recover separate carbon and oxygen under appropriate thermodynamic conditions.

For our purposes this symbolic care is not always helpful and we will normally use the = sign, the equality operator, for mathematical and physical equations, and $\rightarrow$ for a chemical reaction pathway.

<u>Stoichiometric Balance</u>

In general:-

$$\sum_{products} \lambda_j . FW_j - \sum_{reagents} \lambda_i . FW_i = 0$$
Equation 3.9

where λ_i is the Stoichiometric Coefficient of input element or compound i; FW_i is the Formula Weight of element or compound i. With regard to product outputs: λ_j is the Stoichiometric Coefficient of output element or compound j; FW_j is the Formula Weight of element or compound j. The RHS is zero because in stoichiometry as opposed to thermochemistry the two sides of the pathway must balance as a true equation.

Hence:-

$$LHS = RHS$$
Equation 3.10

The number of reagent species and separately the number of products can in theory be large finite (but integer) values. However, in practice the great majority of feasible reactions can be expressed in terms of one or two reactants and one or two products.

Accordingly, we will confine ourselves to very simple stoichiometric equations of the following pattern:-

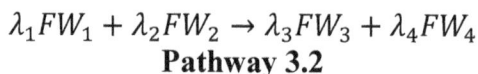

$$\lambda_1 FW_1 + \lambda_2 FW_2 \rightarrow \lambda_3 FW_3 + \lambda_4 FW_4$$
Pathway 3.2

where λ is a Stoichiometric Coefficient (i.e. the integer number of mols of Component n) and FW is the Formula Weight of the attaching Component. The Subscript i = 1,2 is the serial of the first and second reagents and j = 3,4 the serial of the two product species.

Item 4, the second product, may well be absent.

Such simple four-component pathways are justifiable because, according the chemical law formulated by Germain Hess in 1840AD the energy of a series of chemical reactions based upon the products of a previous reaction may be algebraically summed to give the same heat as a direct reaction. This predates the formal statement of The Law of the Conservation of Energy, but is a corollary of that famous maxim.

For example, we propose a balanced pathway equation for the reduction of the iron ore Hematite (Haematite in British English) with the chemical Formula Fe_2O_3 using Carbon C, presumably in the form of coke though charcoal was used in days gone by.

We consider that under desirable conditions of concentration, pressure, temperature, and porosity (particle surface areas), the products metallic iron Fe and carbon dioxide gas CO_2 will issue.

Any correct pathway will have a (possibly) different integer number of the component elements iron, carbon and oxygen on the LHS and the *same elemental count* on the RHS.

In this example, the simplest pathway formulation is:-

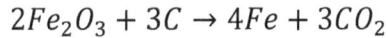

$$2Fe_2O_3 + 3C \rightarrow 4Fe + 3CO_2$$
Pathway 3.3

In layman's terms: "Take two parts of hematite and three of coke to make four parts of iron and three of chokedamp".

The λ values *do not* additively balance LHS-RHS: To check balance of atoms we need to multiply λ by the number of relevant atoms in the formula. Hence:-

LHS		RHS	
$2 \times 2 = 4$ of Fe		$4 \times 1 = 4$ of Fe	
$2 \times 3 = 6$ of O		$3 \times 2 = 6$ of O	
$3 \times 1 = 3$ of C		$3 \times 1 = 3$ of C	

Check that there are 13 atoms of all kinds on both the LHS and the RHS.

Secondly, we need to quantify and balance the Formula Weights FW using the known Atomic Weights of the Elements AW_x.

Given that AW_{Fe} is 55.845, AW_O is 15.99905 and AW_C is 12.011 we have:-

LHS $Fe_2O_3 = 2 \times 55.845 + 3 \times 15.99905 = 159.68715$
$C = 1 \times 12.011 = 12.011$

RHS $Fe = 55.845$
$CO_2 = 1 \times 12.011 + 2 \times 15.99905 = 44.0091$

Therefore in summary:-

$$FW_{Fe2O3} = 159.68715 \qquad \textbf{Equation 3.11a}$$
$$FW_C = 12.011 \qquad \textbf{Equation 3.11b}$$

$$FW_{Fe} = 55.845 \qquad \textbf{Equation 3.11c}$$
$$FW_{CO2} = 44.0091 \qquad \textbf{Equation 3.11d}$$

By using λ values and the just-computed FW values into Pathway 3.2:-

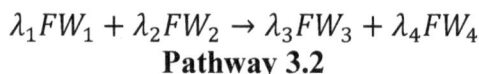

$$\lambda_1 FW_1 + \lambda_2 FW_2 \rightarrow \lambda_3 FW_3 + \lambda_4 FW_4$$
$$\textbf{Pathway 3.2}$$

By numerical substitution we arrive at:-

$$2 \times 159.68715 + 3 \times 12.011 = 4 \times 55.845 + 3 \times 44.0091$$
$$\textbf{Pathway Equation 3.12}$$

From which:-

$$355.4073 = 355.4073$$
$$\textbf{Pathway Equation 3.13}$$

Accordingly, 355.4073 grams of reagents will yield 355.4073 grams of products, thus confirming the Law of the Conservation of Mass.

Requirements and Results:- The Required Product is Fe

In the following discussion of stoichiometric balance I shall employ the following notation for short:-

P = Amount of Product
R = Requirement (mols or Kg)
Q = Result (mols or Kg)

The Number of Moles of Iron Fe in Pathway 3.3 is λ_{j1} = 4 and we have seen that the FW_{Fe} is 55.845. To establish the Amount of Product (i.e. iron) in mols we use:-

$$P_{molsFe} = \lambda_{j1} . FW_{Fe} = 4 \times 55.845 = 223.38$$
Equation 3.14

whilst the Fraction by Weight of Iron Produced, $P_{fraction}$, is:-

$$P_{fraction} = \frac{\lambda_{j1} . FW_{Fe}}{RHS} = \frac{4 \times 55.845}{355.4073} = 0.628518322499285$$
Equation 3.15

Having this information we can compute that the Number of Mols of Hematite that we need to make one kilogram of Iron using this Pathway 3.3 route, $P_{molsForKg}$ as:-

$$P_{\lambda molsForKg} = \frac{1}{P_{fraction}} = 1.59104351329573$$
Equation 3.16

As a progress check we can employ:-

$$LHS_{Kg} = \frac{LHS \times P_{\lambda molsForKg}}{1000} = RHS_{Kg} = \frac{RHS \times P_{\lambda molsForKg}}{1000}$$
$$= 0.565468479242949$$
Equation 3.17

We may now move forward to establish the Requirement for Hematite Fe_2O_3 in mols using:-

$$R_{Fe2O3mols} = \frac{\lambda_{i1} \times FW_{Fe2O3}}{LHS} = 0.898614913087041$$
Equation 3.18

also, and similarly, the Requirement for Carbon R_{Cmols} is:-

$$R_{Cmols} = \frac{\lambda_{i2} \times FW_C}{LHS} = 0.101385086912959$$
Equation 3.19

Note that this is carbon used for *chemical reduction only*. It does *not* include the fuel used for raising the mixture to working temperatures.

If we wish to know the Reagent Requirements in kilograms we use:-

$$R_{Fe2O3Kg} = P_{\lambda molsForKg} \frac{\lambda_{i1}.FW_{Fe2O3}}{LHS} = 0.898614913087041$$
Equation 3.20

$$R_{CKg} = P_{\lambda molsForKg} \frac{\lambda_{i2}.AW_C}{LHS} = 0.101385086912959$$
Equation 3.21

As a further progress check I propose:-

$$Q_{FeKg} = R_{Fe2O3Kg} + R_{CKg} = 1\ kilogram$$
Equation 3.22

and as a fourth check:-

$$\frac{\lambda_{i1}FW_{Fe2O3} + \lambda_{i2}AW_C}{LHS} - \frac{\lambda_{j1}AW_{Fe} + \lambda_{i2}FW_{CO2}}{RHS} = 0$$
Equation 3.23

Table 3.3 presents these several inputs and outcomes for a number of current and proposed chemical reactions relevant to iron smelting.

STOICHIOMETRIC BALANCE

t °K	x = Product Identity (formula)	REAGENT 1 FW (g)	v1	REAGENT 2 FW (g)	v2	PRODUCT 1 FW (g)	v3	PRODUCT 2 FW (g)	v4	LHS (g)	RHS (g)	x	P_{fmol}	$P_{Fraction}$	$P_{i,moles.rxng}$	$R_{FWt1mols}$	R_{FWtkg}	$R_{FWt2mols}$	R_{FWt2kg}	Q_{avg}
1100	$3Fe_2O_3+CO=2Fe_3O_4+CO_2$	159.688	3	28.0101	1	231.533	2	44.0095	1	507.0741	507.0755	Fe	463.066	0.913209	1.095039	0.944761	0.055239	1.034551	0.060489	1.09504
1100	$Fe_3O_4+CO=3FeO+CO_2$	231.533	1	28.0101	1	71.844	3	44.0095	1	259.5431	259.5415	Fe	215.532	0.830434	1.20419	0.892079	0.107921	1.074233	0.129957	1.20419
1100	$FeO+CO=Fe+CO_2$	71.844	1	28.0101	1	55.845	1	44.0095	1	99.8541	99.8545	Fe	55.845	0.559264	1.788065	0.71949	0.28051	1.286495	0.501571	1.78807
550	$Fe_3O_4+CO=3FeO+CO_2$	231.533	1	28.0101	1	71.844	3	44.0095	1	259.5431	259.5415	Fe	215.532	0.830434	1.20419	0.892079	0.107921	1.074233	0.129957	1.20419
	Mean																			
1100	$1.5Fe_2O_3+0.333CO=Fe_3O_4+0.333CO_2$	159.688	1.5	28.0101	0.5	231.533	1	44.0095	0.5	253.5371	253.5378	Fe	231.533	0.913209	1.095039	0.944761	0.055239	1.034551	0.060489	1.09504
1100	$0.333Fe_3O_4+0.333CO=FeO+0.333CO_2$	231.533	0.33	28.0101	0.33	71.844	1	44.0095	0.33	86.51437	86.51383	Fe	71.844	0.830434	1.20419	0.892079	0.107921	1.074233	0.129957	1.20419
1100	$FeO+CO=Fe+CO_2$	71.844	1	28.0101	1	55.845	1	44.0095	1	99.8541	99.8545	Fe	55.845	0.559264	1.788065	0.71949	0.28051	1.286495	0.501571	1.78807
550	$0.333Fe_3O_4+0.333CO=FeO+0.333CO_2$	231.533	0.33	28.0101	0.33	71.844	1	44.0095	0.33	86.51437	86.51383	Fe	71.844	0.830434	1.20419	0.892079	0.107921	1.074233	0.129957	1.20419
	Mean																			
1000	$Fe_2O_3+3CO=2Fe+3CO_2$	159.688	1	28.0101	3	55.845	2	44.0095	3	243.7183	243.7185	Fe	111.69	0.458275	2.182098	0.655215	0.344785	1.429744	0.752354	2.1821
	Mean																			
	$6C+3O_2=6CO$	12.0107	6	31.9988	3	28.0101	6	0	0	168.0606	168.0606	CO	168.0606	1	1	0.428799	0.571201	0.428799	0.571201	1
	$6CO+3O_2=6CO_2$	28.0101	6	31.9988	3	44.0095	6	0	0	264.057	264.057	CO2	264.057	1	1	0.636456	0.363544	0.636456	0.363544	1
	$6C+6O_2=6CO_2$	12.0107	6	31.9988	6	44.0095	6	0	0	264.057	264.057	CO2	264.057	1	1	0.272912	0.727088	0.272912	0.727088	1
	$C+O_2=CO_2$	12.0107	1	31.9988	1	44.0095	1	0	0	44.0095	44.0095	CO2	44.0095	1	1	0.272912	0.727088	0.272912	0.727088	1
	Mean																			
	$2Fe_2O_3+3C=4Fe+3CO_2$	159.688	2	12.0107	3	55.845	4	44.0095	3	355.4081	355.4085	Fe	223.38	0.628516	1.591049	0.898618	0.101382	1.429745	0.161304	1.59105
	$0.5Fe_2O_3+0.75C=Fe+0.75CO_2$	159.688	0.5	12.0107	0.75	55.845	1	44.0095	0.75	88.85203	88.85213	Fe	55.845	0.628516	1.591049	0.898618	0.101382	1.429745	0.161304	1.59105
	$3Fe_2O_3+9H_2=6Fe+9H_2O$	159.688	3	2.016	9	55.845	6	18.0153	9	497.208	497.2077	Fe	335.07	0.673903	1.483892	0.963508	0.036492	1.429742	0.05415	1.48389
	$Fe_2O_3+3H_2=2Fe+3H_2O$	159.688	1	2.016	3	55.845	2	18.0153	3	165.736	165.7359	Fe	111.69	0.673903	1.483892	0.963508	0.036492	1.429742	0.05415	1.48389
	$3Fe_2O_3+H_2=2Fe_3O_4+H_2O$	159.688	3	2.016	1	231.533	2	18.0153	1	481.08	481.0813	Fe	463.066	0.962552	1.038904	0.995809	0.004191	1.034551	0.004354	1.0389
	$Fe_3O_4+H_2=3FeO+H_2O$	231.533	1	2.016	1	71.844	3	18.0153	1	233.549	233.5473	Fe	215.532	0.922862	1.083585	0.991368	0.008632	1.074232	0.009354	1.08359
	$FeO+H_2=Fe+H_2O$	71.844	1	2.016	1	55.845	1	18.0153	1	73.86	73.8603	Fe	55.845	0.75609	1.322595	0.972705	0.027295	1.286495	0.0361	1.32259
	$2H_2+O_2=2H_2O$	2.016	2	31.9988	1	18.0153	2			36.0308	36.0306	H2O	36.0306	1	1	0.111904	0.888096	0.111904	0.888096	1
	$C+O_2=CO_2$	12.0107	1	31.9988	1	44.0095	1		0	44.0095	44.0095	CO2	44.0095	1	1	0.272912	0.727088	0.272912	0.727088	1

Table 3.3
Stoichiometric Balance Data for
Iron Smelting

Comminution and Concentration

You know that even the most primitive open fire needs skilled preparation and control if it is effectively to suit your needs.

This applies in spades to any furnace.

In particular, the payload of a furnace must allow space for reductive gases and other reagents to contact and process the ore mineral, despite the weight of rapidly-decomposing overburdens.

The fragmentation of the solid feedstocks is known as Comminution and its corollary is Concentration, which is the amount of a relevant substance like Iron in a given mass or volume of the active bed. The chemical concentration of iron in, say, hematite is a different issue. Chemists prefer to discuss Particle Surface Area as a factor in reactions, but this is physically a function of comminution.

Comminution is mathematically assessed as Porosity which can be measured in several ways. Porosity is the proportion of free space in a fluid-solid mixture like the hot charge of a furnace. It is perfectly reasonable to assess Porosity as a mechanical Mass Density in terms of the fraction or percentage of solid material per cubic meter of charge volume:-

$$\phi_M = \frac{Solid\ Mass}{Unit\ Volume} = ML^{-3}$$
Equation 3.24

where ϕ_M is Densimetric Porosity.

Most workers, however, favor the assessment of Porosity in terms of some dimensionless number, a dividend of volumes, ϕ_V:-

$$\phi_V = \frac{V_S}{V_U} = \frac{L^3}{L^3} = L^0$$
Equation 3.25

where ϕ_V is the Volumetric Porosity of the fluid-solid mixture; V_S is the Volume of Solid; and V_U is some suitable, local, Unit Volume.

For example, in a part of the active charge of interest the Fractional Volume of Iron Ore, f_{ore}, is 0.23 m^3, whilst the Fractional Volumes of gas, flux, and slag are respectively f_{gas}=0.60, f_{flux}=0.07 and f_{slag}=0.10.

Accordingly the *Porosity*, ϕ_V, is:-

$$\phi_V = \frac{f_{gas}}{f_{gas} + f_{ore} + f_{flux} + f_{slag}} = \frac{0.6}{1} = 0.6$$

Equation 3.26

But The Proportion of Iron Ore in the Mix, Φ_{ore}, is:-

$$\Phi_{ore} = \frac{f_{ore}}{f_{gas} + f_{ore} + f_{flux} + f_{slag}} = \frac{0.236}{1} = 0.23$$

Equation 3.27

It is vital to discriminate ϕ_V and Φ_{ore}.
The value of Porosity determines important parts of:-

(a)   The **Rate** of Metallic Iron Dropdown and hence Production

(b)   The Ultimate **Yield** of Metallic Iron in terms of metal per tonne of charge

Concentration z complements porosity.
Concentration z expresses the mass or volumetric fraction of pay-metal in the smelted ore.
We will also state the *different and chemical* mass proportion in terms of the Formula Weight ($2 \times FW_{Fe}$) of elemental Iron in a pure hematite ore (FW_{Fe2O3}). Hence:-

$$z_{Fe} = \frac{\lambda_{Fe} \times FW_{Fe}}{FW_{Fe2O3}} = \frac{2 \times FW_{Fe}}{FW_{Fe2O3}} = \frac{2 \times 55.845}{159.688}$$
$$= 0.699426381443816$$

Equation 3.28

So the mass percent of iron in hematite is nearly 70%.
z_{Fe} is the Concentration of elemental Iron in Hematite. FW_x is in mols but z_{Fe} is dimensionless. Further, we may find it useful to have to hand a summative fractional figure for the total potentially-recoverable iron in a unit of furnace charge, z_{Charge}. This can be achieved by multiplying Φ_{ore} and z_{Fe}. Hence:-

$$z_{Charge} = \Phi_{ore} \times z_{Fe} = 0.23 \times 0.699426381443816$$
$$= 0.160868067732078$$
Equation 3.29

So the amount of Iron in the furnace charge *at the relevant location in the furnace* is 16%.

The Equilibrium Constant $K_{r(T)}$

$K_{r(T)}$ is one of the summative parameters that defines a particular chemical reaction at a particular temperature (in Degrees Kelvin). It is defined as:-

$$K_{r(T)} = \frac{k_j}{k_i} = \frac{\prod[FW_j]^{\lambda_j}}{\prod[FW_i]^{\lambda_i}} = \frac{\prod z_j{}^{\lambda_j}}{\prod z_i{}^{\lambda_i}}$$
Equation 3.30

where $K_{r(T)}$ is the Reaction Equilibrium Constant at the given temperature $T°K$; z_i and z_j are respectively the Reagent and Product Concentrations for the ith Reagent and the jth Product; and λ_i and λ_j are the respective Stoichiometric Coefficients of the ith Reagent and the jth Product.

Chemical literature often uses square brackets to denote a concentration thus [...]

k_j and k_i are Rates of Reaction, proportional to the mathematical products: When k_j and k_i are equal $K_{r(T)}$ is obviously unity and the chemical reaction system is in thermodynamic equilibrium.

Note that this very simplest Equilibrium Equation 3.30 is only appropriate for reactants and products in a common phase, typically all gases.

Because we are confining ourselves to two-reagent and one or two product pathways of the pattern:-

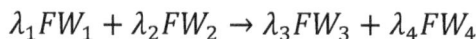

$$\lambda_1 FW_1 + \lambda_2 FW_2 \rightarrow \lambda_3 FW_3 + \lambda_4 FW_4$$
Pathway 3.2

It is adequate to employ the formulation:-

$$K_{r(T)} = \frac{[FW_3]^{\lambda_3} \times [FW_4]^{\lambda_4}}{[FW_1]^{\lambda_1} \times [FW_2]^{\lambda_2}} = \frac{z_3{}^{\lambda_3} \cdot z_4{}^{\lambda_4}}{z_1{}^{\lambda_1} \cdot z_2{}^{\lambda_2}}$$
Equation 3.31

The Furnace Melt Stack as a Hydraulic Problem

When we use mathematical models such as CD-MELT[3.1] it is, with due discretion, possible to simulate the partly-solid, partly-liquid, quasi-buoyant and convective active bed in terms of a chemically-decomposing semi-fluidised hot mass subject in part to Newton's Second Law and in part to the Navier-Stokes Equations[3.2,3.3].

In practice, both physical laws are subject to computerised numerical solution.

Key to this is the Phase Volume Fraction, α, (i.e. the solid part of the load including ore, etcetera); and β the Mass Loss Fraction (the liquid part, molten iron, molten slag, gases, etcetera).

Simple algebra declares:-

$$\alpha + \beta = 1$$
Equation 3.32

because α and β are exclusive fractions.
Also in terms of Volumetric Porosity[3.2]:-

$$\phi_V = 1 - \alpha \equiv \beta$$
Equation 3.33

Hydraulic calculations, including computer models, typically depend upon semi-experimental Dimensionless Numbers that relate measurable physical quantities. These hydraulic numbers include the Reynolds Number Re(); the Prandtl Number Pr(); and the Nusselt Number Nu().

The Reynolds Number

The Reynolds Number controls the transition between laminar and turbulent flow involving mechanical parameters only:-

$$Re \equiv \frac{Inertia}{Viscosity} = \frac{uL}{\nu} = \frac{\rho u L}{\mu}$$
Equation 3.34

where Re is the dimensionless Reynolds Number; u is the Flow Velocity $[MT^{-1}]$; L is the "Characteristic" Length (in the context of experiments involving buoyed-up spheres this is usually sphere diameter or sphere radius); ν is Kinematic Viscosity of the Fluid $[M^2T^{-1}]$; and μ is the Dynamic Viscosity of the Fluid $[ML^{-1}T^{-1}]$.

The Prandtl Number

The Prandtl Number is the ratio of Momentum Diffusivity to Thermal Diffusivity:-

$$Pr = \frac{\nu}{\alpha} = \frac{\mu/\rho}{k/c_p} = \frac{c_p.\mu}{k}$$
Equation 3.35

where α, the Thermal Diffusivity is given by:-

$$\alpha = \frac{k}{\rho.c_p} = [L^2T^{-1}]$$
Equation 3.36

k is the Thermal Conductivity measured in Watts per (meter.°K) which has the dimensions $[MLT^{-3}\theta^{-1}]$.
c_p is the Specific Heat of the molten mass which is measured in J per (Kg.°K) which has the dimensions $[L^2T^{-2}\theta^{-1}]$.

Nusselt Number

The Nusselt Number is the ratio of the Conductive Heat Transfer (CDHT) and the Convective Heat Transfer (CVHT); divided by the Conductive Heat Transfer (CDHT).
This may be expressed:

$$Nu = \frac{CDHT + CVHT}{CDHT} = 1 + \frac{CVHT}{CDHT}$$
Equation 3.37

or:-

$$Nu = \frac{h}{k/L} = \frac{hL}{k}$$

Equation 3.38

where h is the Heat Transfer Coefficient.

There are broadly four Nusselt thermal flow regimes:-

Regime	Nu Value
Conduction Only	1
Laminar Flow	1-10
Transitional Flow	10-100
Turbulent Flow	100-1000

A Hot Blast Iron Smelter Furnace admits a forced convection of compressed air from basal tuyeres. This forced air is typically at 1300°K. This blast may or may not be fortified with pure oxygen and or hydrocarbons. The high-pressure air blast has a buoying and separating effect upon the ore and fuel particles which increases their effective area and enhances both rates and yield of reaction.

A Hydrogen Direct Reduction furnace introduces a similar basal blast but of cold or hot hydrogen which, like the carbon of a traditional furnace, possesses both reducing and heating effects.

Under such conditions we may anticipate largely laminar flow, especially if the feedstock is pre-pelletised ore-flux-fuel composite of consistent and scientifically-designed "characteristic" diameters.

The Soon Analysis[3.2]

In 2021AD Soon, Zhang, Yang and Law published a method to model the isothermal melting of a packed bed subject to forced convection at a "high temperature". They based their model upon the fractional loss paradigm of Equation 3.33 where $\beta = 0$ for the Initial Mass (entirely solid) to $\beta = 1$ (complete liquefaction).

They idealised their treatment in terms of experimental ice spheres packed in a vertical tube and subjected to upwelling warm water. Soon and her colleagues did not relate their findings to irregular furnace feed fragments, but of course appropriate specialists could extend their work to such particles.

As I implied above, Soon Et Al studied particle ablation in terms of mass rather than volume in order to facilitate thermal discernments, and also as implied elsewhere they based the mechanics of the solids upon Newton's Second Law supplemented with a term to allow for the elastic collision of the time-wise ablating spheres. Also as above, they applied Navier-Stokes partial differential equation systems to the modelling of the moving flow, during the time that the flow was augmented by meltwater from the spheres. Reynolds Number, the Prantl Number and the Nusselt Number were all implicated in this important half-experimental and half-theoretical development.

Lee and Chung on Packed Reactor Beds[3.3]

Lee and Chung, who published the referenced paper in 2016AD, also examined packed spheres, but in the context of high-temperature Pebble Bed (Nuclear) Reactors (PBRs), specifically a reactor Pebble Bed Modular Reactor (PBMR) packed with 45000 spherical fuel pellets. As with Hot Blast furnaces, the picture is complicated by the fact that the particles are themselves sources of intense heat that complicates the trajectories and behaviors of forced convectional fluids.

For safety and convenience they simulated actual fuel pellet behavior using electrically-heated spheres in a vertical conduit through which water was collimated and forced.

According to the German Nuclear Safety Standards Commission (KTA) the appropriate range of applied Reynolds Number in the nuclear coolant water context is $100 < Re < 100000$ whilst the Volumetric Porosity $\varepsilon = 1 - V_{spheres}/V_{total}$ should lie in the range $0.36 < \varepsilon < 0.42$.

In such ranges the appropriate flavor of empirical formula for Nusselt Number is:-

$$Nu_d = 1.27 \left(\frac{Pr^{\frac{1}{3}}}{\varepsilon^{1.18}} \right) . Re_d^{0.36} + 0.033 \left(\frac{Pr^{\frac{2}{3}}}{\varepsilon^{1.07}} \right) . Re_d^{0.86}$$

Equation 3.39

Lee and Chung determined that Heat Transfer was greater at the base of the stack due to eddy currents in the fluid, whilst higher up the active column heat transfer was increased by increased porosities, specifically that the Heat Transfer Coefficient declined linearly from the wall to the center of the conduit.

These workers related this porosity effect to the findings of Ferng and Lin[3.4] that Heat Transfer differed as between Face Centered Cubic packed sphere stacks and Body Centered Cubic arrangements, where the mathematical porosity also differs between the packing geometries. Notwithstanding which, in both arrangements Heat Transfer was near to the KTA Nusselt Number of Equation 3.39

Mathematically speaking, as the number of packed spheres increases toward infinity, or indeed 45000, so the porosity converges to a finite limit value (0.25952).

I have not attempted further to rationalise the Equation 3.39 structure.

The mathematical void fractions in both the BCC and FCC cubic sphere packing arrangements yield porosities respectively of 0.32 and 0.26, markedly sparser than the KTA-approved porosities, which latter do not necessarily apply to close-packed spheres.

CHAPTER FOUR
THERMOCHEMICAL PRINCIPLES

Thermochemistry is the science of tracing the various heat transactions in forming the initial reagents, and of quantifying the total heat output (or consumption) of a balanced reaction. In practice, thermochemistry also involves Reaction Efficiencies and Thermodynamic Equilibrium.

Normal chemical reactions are reversable under the correct conditions of concentration, pressure and temperature.

<u>The States of Matter</u>

In the context of current and foreseeable iron smelting there are three states of Matter. Broadly speaking:-

(a)   Solid

A solid substance exhibits a rigid competency that resists deformation. Properly classified, all solids are crystals. Crystals have a geometrical 3D internal structure (lattice) that reflects organised attachments between their constituent atoms.

This 3D lattice, or more visually scaffold, is formed by surrendering heat to the environment.

At STP (Standard Temperature and Pressure) of 20°C and 1 Atmosphere (298.15 °K and 101.325 kPa) typical solids include iron, diamond, aspirin and quartz (sand). Glass is not a true solid, because on cooling it thickens gradually without surrendering Latent Heat of Fusion, and crystallising. Conversely, any heating will soften it gradually, and for example, if sheet glass is placed in a window it shall, over centuries, creep groundward. No crystal does this. Over two thousand years Roman glass has partially crystallised to a beautiful opaque iridescence. Glass is a "supercooled liquid".

A classical solid is a pure chemical, though some crystals contain impurities. No mixture is a solid though it may be a compacted aggregate of solids like concrete or furnace feed pellets.

Pure solids melt (turn to liquid) abruptly.

(b)   Liquid

Rare solids like Iodine sublimate (turn to gas) when heated, but most solids like Iron, Ice and Sand turn to liquid.

A liquid is a fluid that is not crystalline, and which will sit passively in a vessel under gravity though it may tend to evaporate according to its Vapor Pressure: Thus a liquid "seeks" its lowest level of potential energy in a gravity field. Liquids flow.

At the interface with a gas, liquids develop a meniscus, in which the planar surface curves slightly at the liquid's rim of containment, the zone of solid-liquid-gas triple contact.

Typical liquids at STP include Water, Mercury, Gasoline; and at elevated temperatures Molten Iron and Slag are liquids.

Liquids have little or no internal structure but constituent Molecules of distinct chemicals are held together by residual forces.

(c)   Gas

Gas molecules are entirely unattached to each other, but the molecules of gasses usually retain their chemical constitutions. A gas is therefore a fluid that is fugacious and occupies the whole of the volume available to it without ponding, except that immiscible gases may gravitationally segregate in cold, calm conditions.

Under both STP and viable furnace conditions the behavior of gases conforms to Kinetic Theory and in particular the Universal Gas Law.

Typical gases at STP include Air, Oxygen, Methane, Carbon Monoxide and Carbon Dioxide. Many gases are highly reactive chemically, but Inert Gases like Argon will only participate in compound chemistry under exceptional, artificial conditions.

When a gas forms it absorbs Heat of Vaporisation which, like Heat of Fusion, is yielded or absorbed without change of temperature, and is also fully recoverable upon condensation.

The Scheme of Thermal Transition

When a solid crystalline chemical ore like Hematite (Fe_2O_3) is heated from STP to working temperatures it passes through a number of *physical* energetic phases, in addition to the actual *chemical* reductive pathway process which engages Heat of Reaction. The general pattern for an STP gas (not metal ore) is:-

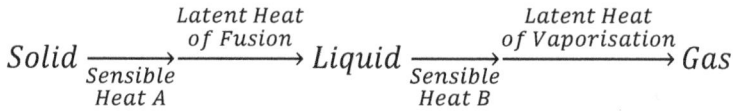

$$Solid \xrightarrow[\substack{Sensible \\ Heat\ A}]{\substack{Latent\ Heat \\ of\ Fusion}} Liquid \xrightarrow[\substack{Sensible \\ Heat\ B}]{\substack{Latent\ Heat \\ of\ Vaporisation}} Gas$$

Pathway 4.1

So called Sensible Heat is the heat acquired by a mass due to its temperature without Change of State.

To a linear approximation Sensible Heat is given by:-

$$\Delta H_{spx} = m s_{spx}.\Delta T = m s_{spx}.(T_2 - T_1)$$
Equation 4.1

where ΔH_{spx} is the Sensible Heat (J or KJ) added (or subtracted) due to heating (or cooling); m is the Mass of substance involved; ΔT is the Temperature Change; T_1 is the Initial Temperature; and T_2 is the Current or Higher Temperature of Interest.

s_{spx} is the Specific Heat of the Substance (in the relevant State), quasi-constant for small ΔT, but subject to slight variation in high-energy pathways. Under furnace conditions, s_{spx} is sometimes approximated by polynomial equations in T, for example the Shomate Equations.

The approximate Specific Heat of Solid Iron, $s_{sps}(Fe)$, or $C_p(Fe)$, is around 25.1 J/mol.°K

Figure 4.1 is a rough schematic of the behavior of absorbed heat during the heating of a typical solid.

The Latent Heats of Fusion and Vaporisation are different constants for any pure substance of a defined Chemical Formula.

For example, the Latent Heat of Fusion, $L_f(H_2)$ of Hydrogen is 0.05868 KJ/mol whereas its Latent Heat of Vaporisation $L_v(H_2)$ is 0.44936 KJ/mol.

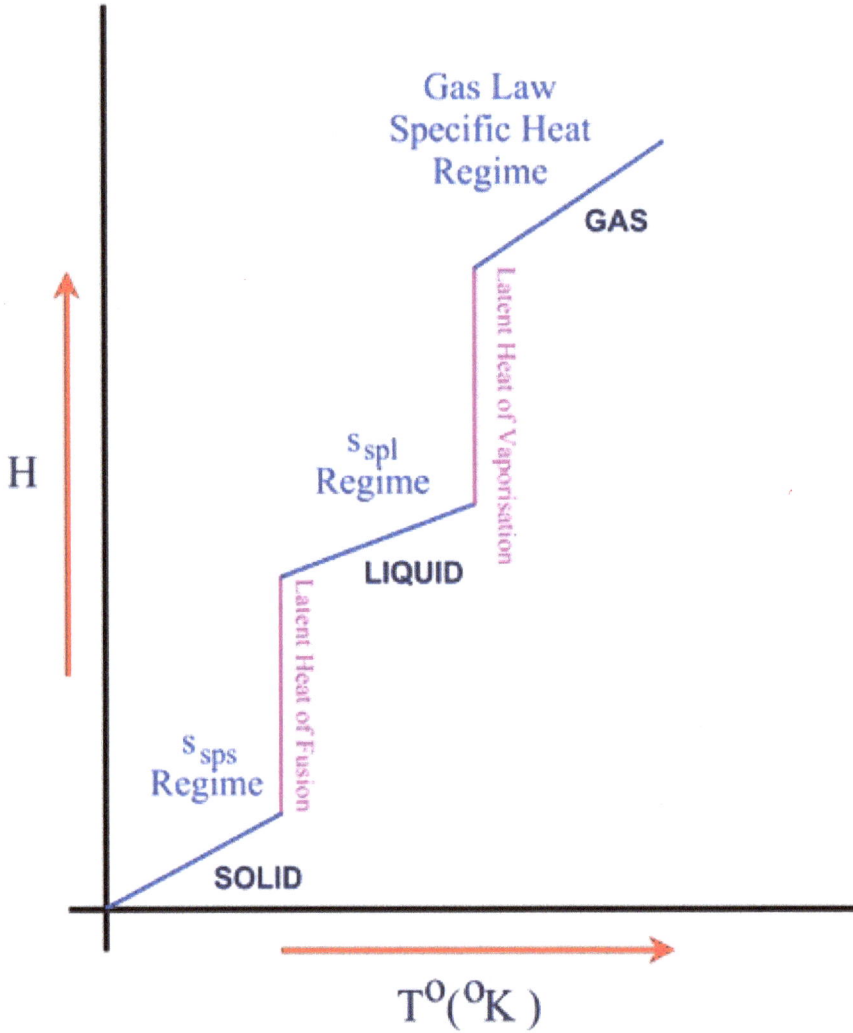

Figure 4.1
Schematic Diagram of Physical Heat Content H
for a Mass of Substance

The Sources and Sinks of Furnace Heat

When we attempt to list the inputs, processes and outputs of any chemical system, especially complex systems, it is inevitable that many factors are omitted by oversight or ignorance. Only experience and experiment may remedy some of these shortfalls.

Many of the most important economic factors in furnace use and design are almost irrelevant to the strictly thermochemical: Such as labor costs, maintenance costs, constructional materials, resource recovery including by-product sales, design and land acquisition and restitution. Payments for all this are unlikely to be compensated until many years after furnace installations, at any event in the case of iron smelting where margins may be as low as 3%.

So bear in mind that no list is or can be exhaustive.

Thermal Transputs for a Blast Furnace (BF)

These Hot Blast furnaces are fueled with coke (sometimes with supplementary coal or waste) and account for almost all 2024AD primary production of iron.

HEAT INPUTS	FURNACE GAINS	FURNACE LOSSES	HEAT OUTPUTS
Fuel Combustion Heat	Heats of Ore Reactions	Moisture Vaporisation	Heat in Hot Metal
Hot Blast Heat	Heats of Slag Reactions	Volatiles Vaporisation	Heat in Slag
			Heat in BF Gas
			Heat in Dust
			Heat in Coolant
			Other Losses

Table 4.1
Thermal Transputs for a Hot Blast Furnace

Thermal Transputs for a Hydrogen Direct Reduction Furnace (H2DRF)

Hydrogen DRF's are misnamed as they employ two distinct stages of reduction:-

(a) The Electrolysis of Water to provide Hydrogen Fuel
(together with oxygen which may or may not

be used in smelting)

(b)   The Actual Reduction of Iron Ore
using the Hydrogen

HEAT INPUTS	FURNACE GAINS	FURNACE LOSSES	HEAT OUTPUTS
Electricity	Heats of Ore Reactions	Moisture Vaporisation	Heat in Hot Metal
Water	Heats of Slag Reactions	Volatiles Vaporisation	Heat in Slag
	Heat of H_2 Combustion	Electrode Consumptions	Heat in Dust
	Heat of H_2-O_2 Combustion		Heat in Coolant
	Recovered Steam Heat		Steam
			Other Losses

Table 4.2
Thermal Transputs for a Hydrogen Direct Reduction Furnace

I have included water as an input because it is a feedstock of the electrolytic intermediary: It may or may not also be a means of generating electricity.

Thermal Transputs for a Molten Oxide
Electrolysis Furnace (MOEF)

MOEF's are experimental and are (2024AD) being prototyped and researched by Boston Metal[®].

HEAT INPUTS	FURNACE GAINS	FURNACE LOSSES	HEAT OUTPUTS
Electricity	Heats of Ore Reactions	Moisture Vaporisation	Heat in Hot Metal
	Heats of Slag Reactions	Volatiles Vaporisation	Heat in Slag
		Electrode Consumptions	Heat in Dust
			Heat in Coolant
			Oxygen
			Other Losses

Table 4.3
Thermal Transputs for a Molten Oxide Electrolysis Furnace

Entropy

Chemists and thermodynamicists often write about Entropy. It is important to remember that in common language all Entropy is Heat and all heat is Energy.

The distinction is that Entropy is a flavor of heat that inhabits specific theoretic objects or substances of a thermochemical system.

For example, Entropy of Formation ΔH_f^x , is the heat absorbed or surrendered when a specific amount of a pure chemical substance forms from its constituent elements. By convention, the Entropy of Formation of a chemical element like Carbon (C) or Oxygen (O_2) is zero. The subscript f stands for "formation" and the superscript x stands for a particular regime of Concentration, Pressure, Temperature, etcetera. A superscript Plimsoll Mark signifies STP , the regime of room conditions. A Plimsoll Mark is a struck roundel, very similar to the famous London Transport sign, $\ominus$: But where necessary I shall use the superscript abbreviation STP instead.

This rather begs the question of how Entropy of Formation is established in the first place. The classical method, if the substance in question is solid or liquid, is to burn a weighed sample in a Bomb Calorimeter. This is a heavy capsule of stainless steel about the size of a whisky tumbler, equipped with a robust screw lid.

A spatula to contain the sample is suspended from the inner base of the lid, which latter also supports an insulated electric terminal through which a spark is applied to ignite the sample when the apparatus is fired. The lid also has a gas-tight valve fitting through which, after you have screwed the lid carefully back into place, you feed a charge of "excess" oxygen at 20-30 atmospheres pressure.

The whole "bomb" is then immersed in an insulated and pre-calibrated Calorimeter filled with water and a finely-calibrated thermometer suspended in that.

You connect the terminal and press the firing button. You devoutly hope the apparatus does not explode!

Inevitably, when I showed lads how to do this forty years ago sometimes the contraption worked and sometimes nothing happened.

You log the apparatus temperature every thirty seconds for five minutes *before* you fire; then throughout the temperature rise; then for the first five or ten minutes of temperature decline.

After applying the Regnault-Pfaundler[4.1] Correction or some other suitable mathematical technique you determine the Corrected Temperature Rise ΔT in °K.

Now that we have ΔT to hand, the Total Quantity of Liberated Heat Q_{total} is given by:-

$$Q_{total} = q_{H2O} + q_{cal} = m_{H2O}c_{H2O}.\Delta T + C_{cal}.\Delta T$$
$$= \Delta T(m_{H2O}c_{H2O} + C_{cal})$$

Equation 4.2

where q_{H2O} is the Heat Taken Up by the Calorimeter Water; q_{cal} is the Heat Absorbed by the Calorimeter Apparatus (known from prior calibrations); m_{H2O} is the Mass of Water in the Calorimeter; c_{H2O} is the Specific Heat of Water (4.184 J/g°K); and C_{cal} is the Calibrated Heat Capacity of the Calorimeter.

Because the reaction is exothermic we need conventionally to switch the positive Q_{total} figure to a negative amount, i.e.:-

$$Q_r = Q_{totalneg} = -Q_{total}$$

Equation 4.3

where Q_r is the Total Heat Output of the Reaction (J/g).

The (Standard) Heat of Combustion, ΔH_c is then given by:-

$$\Delta H_c(reagent) = Q_r.M_{reagent}$$

Equation 4.4

where $M_{reagent}$ is the Molecular Weight of the Reagent in g/mol.

Therefore, ΔH_c(reagent) is in J/mol.

In practice, it is often more manageable to work in KiloJoules rather than Joules.

Gases' Heats of Combustion are assessed using a Flame Calorimeter in which a metered flow of excess oxygen is burnt in appropriate apparatus.

Heats of Combustion are Entropy Changes and accordingly are noted in terms of ΔH.

The Standard Enthalpy Change of a Reaction

For a reaction conducted at *Standard Temperature and Pressure (STP)* the Standard Enthalpy Change of a Reaction, $\Delta_r H^\ominus$, is given by:-

$$\Delta_r H^\ominus = \sum_{products} \lambda_j . \Delta_{f,j} H^\ominus - \sum_{reactants} \lambda_i . \Delta_{f,i} H^\ominus$$

Equation 4.5

or in our preferred idiom:-

$$\Delta_r H^{STP} = \sum_{products} \lambda_j . \Delta_{f,j} H^{STP} - \sum_{reactants} \lambda_i . \Delta_{f,i} H^{STP}$$

Equation 4.6

λ_i and λ_j are respectively the Stoichiometric Coefficients of the Reagents and the Products of the Reaction; $\Delta_{f,i} H^{STP}$ and $\Delta_{f,j} H^{STP}$ are respectively the Heats (Entropies) of Formation of the Reactants and the Products.

$$\Delta_{f,x} H^{STP} \equiv 0 \; by \; convention$$

Equation 4.7

for any element or compound substance x.

Hematite Reduction by Carbon

For example, the direct Reduction of Hematite using Carbon issues in Metallic Iron and Carbon Dioxide according to:-

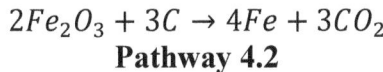

$$2Fe_2O_3 + 3C \rightarrow 4Fe + 3CO_2$$

Pathway 4.2

By expanding Equation 4.6 we may write:-

$$\Delta_r H_{PHem} = \left[\left(\lambda_{j=3} . \Delta H_{f,Fe} \right) + \left(\lambda_{j=4} . \Delta H_{f,CO2} \right) \right]$$
$$- \left[\left(\lambda_{i=1} . \Delta H_{f,Fe2O3} \right) + \left(\lambda_{i=2} . \Delta H_{f,C} \right) \right]$$

Equation 4.8

which by substitution for Stoichiometric Coefficients becomes:-

$$\Delta_r H_{PHem} = \left[\left(4 \times \Delta H_{f,Fe} \right) + \left(3 \times \Delta H_{f,CO2} \right) \right] \\ - \left[\left(2 \times \Delta H_{f,Fe2O3} \right) + \left(3 \times \Delta H_{f,C} \right) \right]$$
Equation 4.9

which by further substitution for Heats of Formation becomes:-

$$\Delta_r H_{PHem} = \left[(4 \times 0) + (3 \times -393.522) \right] \\ - \left[(2 \times -825.5) + (3 \times 0) \right]$$
Equation 4.10

in KJ/mol
Equation 4.10 simplifies to:-

$$\Delta_r H_{PHem} = \left[(3 \times -393.522) \right] - \left[(2 \times -825.5) \right] \\ = (-1180.566) - (-1651) \\ = 470.453 \; KJ/mol$$
Equation 4.11

So the Process Heat of Reaction for the Reduction of Hematite by Carbon is 470.453 KJ/mol.

To express the Process Heat of Reaction in terms of KJ/gram (effectively a heat of combustion) we need to divide $\Delta_r H_{PHem}$ by the Sum of the Coefficient-scaled Formula Weights of the Products, in this case:-

$$\Delta_r H_{PHemGrams} = \frac{\Delta_r H_{PHem}}{\left[\left(\lambda_{j=3} \cdot FW_{Fe} \right) + \left(\lambda_{j=4} \cdot FW_{CO2} \right) \right]} = \frac{\Delta_r H_{PHem}}{355.4} \\ = 1.32367473269555$$
Equation 4.12

For Endothermic (heat absorbing) reactions the ΔH values are conventionally positive.

Carbon Oxidation by Diatomic Oxygen

The Oxidation of Carbon using Oxygen forms Carbon Dioxide CO_2 and evolves Heat of Reaction ΔH_r according to:-

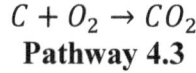

$$C + O_2 \rightarrow CO_2$$
Pathway 4.3

By expanding Equation 4.6 we may write:-

$$\Delta_r H_{PC} = \left[(\lambda_{j=3} \cdot \Delta H_{f,CO2}) + (0) \right] - \left[(\lambda_{i=1} \cdot \Delta H_{f,C}) + (\lambda_{i=2} \cdot \Delta H_{f,O2}) \right]$$
Equation 4.13

Note the absence of a second product.

By substitution for Stoichiometric Coefficients Equation 4.13 becomes:-

$$\Delta_r H_{PC} = \left[(1 \times \Delta H_{f,CO2}) + (0) \right] - \left[(1 \times \Delta H_{f,C}) + (1 \times \Delta H_{f,O2}) \right]$$
Equation 4.14

which by further substitution for Heats of Formation becomes:-

$$\Delta_r H_{PC} = \left[(1 \times -393.522) + (0) \right] - \left[(1 \times 0) + (1 \times 0) \right]$$
Equation 4.15

in KJ/mol
Equation 4.15 simplifies to:-

$$\Delta_r H_{PC} = \left[(1 \times -393.522) \right] - [0]$$
$$= -393.522 \; KJ/mol$$
Equation 4.16

So the Process Heat of Reaction for the Oxidation of Carbon by Oxygen is -393.522 KJ/mol.

To express the Process Heat of Reaction in terms of KJ/gram (this is the official heat of combustion of carbon) we need to divide $\Delta_r H_{PC}$ by the Sum of the Formula Weights of the Products, in this case:-

$$\Delta_r H_{PCGrams} = \frac{\Delta_r H_{PC}}{[(\lambda_{j=3}.\,FW_{CO2}) + (0)]} = \frac{\Delta_r H_{PC}}{44.01}$$
$$= -8.94164962508521$$
Equation 4.17

      The ΔH values are negative because the reaction is Exothermic (i.e. it gives out heat).

      For Endothermic (heat absorbing) reactions the ΔH values are conventionally positive.

Tabulation

Table 4.4 presents Heat of Reaction and or Heats of Combustion for several Chemical Reactions germane to Iron Smelting.

HEATS OF REACTION (COMBUSTION) ΔH_{rxn}

t °K	Reaction	v_1	ΔH_{form} KJ/mol	v_2	ΔH_{form} KJ/mol	= v_3	ΔH_{form} KJ/mol	v_4	ΔH_{form} KJ/mol	RAW HEAT OF REACTION KJ/mol ΔH_{rxn} KJ/mol	HEAT OF COMBUSTION Ideal Product Heat (ΔH_{rxn} KJ/gram)	Ideal Product Heat (ΔH_{rxn} KJ/Kg)
298	$3Fe_2O_3+CO=2Fe_3O_4+CO_2$	3	-825.5	1	-110.527	2	-1120.89	1	-393.522	-48.2833	-0.095219	-95.219154
1100	$Fe_3O_4+CO=3FeO+CO_2$	1	-1120.89	1	-110.527	3	-249.532	1	-393.522	89.3024	0.344078	344.077537
1100	$FeO+CO=Fe+CO_2$	1	-249.532	1	-110.527	1	12.4	1	-393.522	-21.0632	-0.210939	-210.938916
550	$Fe_3O_4+CO=3FeO+CO_2$	1	-1120.89	1	-110.527	3	-249.532	1	-393.522	89.3024	0.344078	344.077537
										109.2583 $\Sigma\Delta H_{rxn}$	0.381997	381.997004
1100	$1.5Fe_2O_3+0.333CO=Fe_3O_4+0.333CO_2$	1.5	-825.5	0.5	-110.527	1	-1120.89	0.5	-393.522	-24.14165	-0.095219	-95.219154
1100	$0.333Fe_3O_4+0.333CO=FeO+0.333CO_2$	0.33	-1120.89	0.33	-110.527	1	-249.532	0.33	-393.522	29.7674667	0.344078	344.077537
1100	$FeO+CO=Fe+CO_2$	1	-249.532	1	-110.527	1	12.4	1	-393.522	-21.0632	-0.210939	-210.938916
550	$0.333Fe_3O_4+0.333CO=FeO+0.333CO_2$	0.33	-1120.89	0.33	-110.527	1	-249.532	0.33	-393.522	29.7674667	0.344078	344.077537
										14.3300833 $\Sigma\Delta H_{rxn}$	0.381997	381.997004
1000	$Fe_2O_3+3CO=2Fe+3CO_2$	1	-825.5	3	-110.527	2	12.4	3	-393.522	1.3141	0.005392	5.391876
	$6C+3O_2=6CO$	6	0	3	0	6	-110.527	0	0	-663.1626	-3.945973	-3945.973060
	$6CO+3O_2=6CO_2$	6	-110.527	3	0	6	-393.522	0	0	-1697.9718	-6.430323	-6430.322998
	$6C+6O_2=6CO_2$	6	0	6	0	6	-393.522	0	0	-2361.1344	-8.941760	-8941.760302
	$C+O_2=CO_2$	1	0	1	0	1	-393.522	0	0	-393.5224	-8.941760	-8941.760302
										-5115.7912 $\Sigma\Delta H_{rxn}$	-28.259817	-28259.816662
	$2Fe_2O_3+3C=4Fe+3CO_2$	2	-825.5	3	0	4	0	3	-393.522	470.4328	1.323640	1323.639699
	$0.5Fe_2O_3+0.75C=Fe+0.75CO_2$	0.5	-825.5	0.75	0	1	0	0.75	-393.522	117.6082	1.323640	1323.639699
	$CaSiO_4$					1	-1630					
1273	$CaCO_3=CaO+CO_2$	1	-1207			1	-635	1	-393.522	178.4776	1.783234	1783.233503
1273	$CaO+SiO_2=CaSiO_3$	1	-635	1	-911	1	-89.61			1456.39	12.537577	12537.576832
	$3Fe_2O_3+H_2=2Fe_3O_4+H_2O$	3	-825.5	1	0	2	-1120.89	1	-241.826	-7.114	-0.014788	-14.787521
	$Fe_3O_4+H_2=3FeO+H_2O$	1	-1120.89	1	0	3	-249.532	1	-241.826	130.4717	0.558652	558.652145
	$FeO+H_2=Fe+H_2O$	1	-249.532	1	0	1	12.4	1	-241.826	20.1061	0.272218	272.217957
1950	$2C+O_2=2CO$	2	0	1	0	2	-110.527			-221.0542	-3.945973	-3945.973060
1400	$C+H_2O=CO+H_2$	1	0	1	-241.826	1	-110.527	1	0	131.2989	4.372826	4372.825642
										$\Sigma\Delta H_{rxn}$		
2173	$FeO+C=Fe+CO$	1	-249.532	1	0	1	0	1	-110.527	139.005	1.657681	1657.680928
	$C_8H_{18}+12.5O_2=8CO_2+9H_2O$	1	-250	12.5	0	8	-394	9	-286	-5476	-10.649269	-10649.268971
	$2C_8H_{18}+25O_2=16CO_2+18H_2O$	2	-250	25	0	16	-394	18	-286	-10952	-10.649269	-10649.268971
	$C+O_2=CO_2$	1	0	1	0	1	-393.522			-393.5224	-8.941760	-8941.760302
	$CO+0.5O_2=CO_2$	1	-110.527	0.5	0	1	-393.522			-282.9953	-6.430323	-6430.322998

Table 4.4
The Thermochemistry of
Selected Iron Smelting Reactions

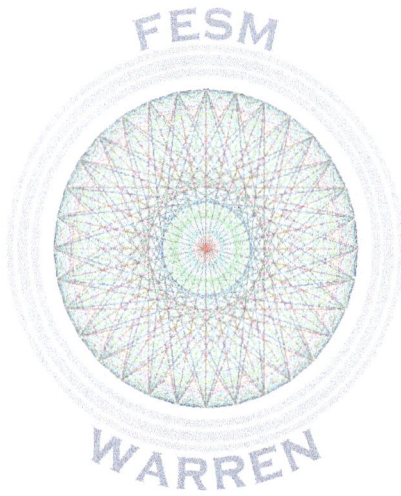

CHAPTER FIVE
THE PRINCIPLES OF
PROCESS ENERGY BALANCE

The Hot Blast Iron Furnace is a reasonably evolved and complex technology so we will examine Process Energy Balance with regard to it, but we should bear in mind that Energy Balance is applicable to any oven or furnace system and a good deal else beside.

The Law of the Conservation of Energy tells us:-

$$Energy\ In = Energy\ Out$$
Equation 5.1

Clear enough you might think, but when we actually attempt to list and quantify energy transactions things get complicated:-

$$\begin{array}{l} \dfrac{Ore}{Reaction\ Energy} + \dfrac{Flux}{Reaction\ Energy} + \dfrac{Carbon}{Oxidation\ Energy} \\[2ex] + \dfrac{Hot\ Blast}{Energy} \\[2ex] = \dfrac{Iron}{Melt\ Energy} + \dfrac{Slag}{Melt\ Energy} + \dfrac{Flue}{Gas\ Energy} \\[2ex] + \dfrac{Equipment}{Energy\ Losses} \end{array}$$
Equation 5.2

and even now our list is hardly exhaustive.

We can break-down Equation 5.2 into convenient parts, for example *by the mole*:-

$$\Delta r H_{PHem} + \Delta r H_{PHPC} + \dfrac{Blast}{Energy}$$
$$= \left(\Delta \dfrac{Sensible}{Heat} + \dfrac{Latent\ Heat}{Of\ Fusion} \right) + \dfrac{Flue}{Gas\ Energy}$$
Equation 5.3

whilst according to the Gas Law:-

$$\frac{Blast}{Heat} = nRT \times \frac{Partial\ Pressure}{of\ Oxygen}$$

Equation 5.4

where n is the Number of Moles of Oxygen (which may be dimensionalised as a mass); R is the Universal (Ideal) Gas Constant; and T is the Blast Temperature in Degrees Kelvin.

ΔrH_{PHem} and ΔrH_{PHPC} are respectively the Heats of Reaction of Hematite Reduction and Carbon Oxidation.

Whilst resolving all of these elements into even more closely-specified components we need to take care both to gather data and to identify, and if possible estimate the importance of, our known unknowns.

What it is We Want

We want to know the energetic cost of smelting one kilogram of liquid iron to a pourable state. In a simple model, arithmetic scaling can give us the energy requirement for one thousand or one million tonnes of metal, where a tonne is a thousand kilograms.

What we also need to account is a certain amount of hot gas. If we cannot sell it, we must try to use it to reclaim some of the heat generated by the primary process.

Limitations of System Modelling

We will never know in this life the most subtle and recondite of the iron smelting features we fail to recognise.

But we can identify *some* of the *grossest* omissions we commit.

For example:-

(A)   Flux and Slag Reactions and Behaviors
(B)   Quantification of Hot Blast Volumes
      and Rates
(C)   Equipment Heat Losses by Radiation,
               Convection and Conduction

This renders the Calculated Heat Demands of any system we examine a mere *minimum* of the energy the process needs.

Beyond this, we will make no attempt to document the wider finance and economics of iron smelting. For example, the physical and monetary costs of labor, mining or transport, or of capital costs, or of conveyance to the customer. It is these essentially social latter quantities that determine whether iron smelting is conducted here or there, in any particular juncture of history.

Specific Heat

Specific Heat is the amount of heat energy needed to raise the temperature of one Mole of a defined substance by one degree Kelvin. Therefore, specific heat is reconned in J/(mol.°K) so it has the dimensions $ML^2T^{-2}mol^{-1}\theta^{-1}$.

Sometimes J/(g.°K) is quoted, or J/(Kg.°K), in which case J/(mol.°K) must be scaled by an actual or representative Formula Weight.

We are taught in high school that Heat Content, ΔH, (or "enthalpy") relates linearly to Temperature Difference, ΔT, according to the equation:-

$$\Delta H = m.c_p.\Delta T$$
Equation 5.5

where m is the Mass of Substance and c_p is the Specific Heat.

Unfortunately this is a simplification. Specific Heat does not vary linearly with temperature. The amount and rates of variation vary with the state of the material, and gases obey their own special laws.

You can model these c_p vs, T variations in a number of ways but polynomial equations are usually good enough. For example you could use the Shomate Equation[5.1]:-

$$c_p = A + BT + CT^2 + DT^3 + \frac{E}{T^2}$$
Equation 5.6

where A, B, C, D and E are all Empirical Coefficients.
For example, *in the case of liquid iron only in the range 1809-3133.345°K only:-*

$$c_p = 46.024 - 0.0000000188T + 0.00000000609T^2$$
$$- 0.000000000664T^3 - \frac{0.00000000825}{T^2}$$

Equation 5.7

The c_p is the 46.02399998 J/(mol.°K) when T = 1900

Table 5.1 shows Gas and Iron Heat Capacities at various temperatures relevant to Hot Blast Iron Smelting.

For diatomic gases, Kinetic Theory suggests that the Molar Specific Heat is given by:-

$$c_p = \frac{5}{2}R \approx 29$$

Equation 5.8

The PSD(c_p,frac*R) shows this roughly to be true for H_2, N_2 and Ar (Argon) at low temperatures, but untenable for O_2 and high-temperature CO. For triatomic gases, c_p is supposed to be (8/6)R, but this is shown to be quite wrong for CO, CO_2 and H_2O (steam) at furnace temperatures.

Mixed Gas Heat Capacities

The Heat Capacity of a Mixture of Ideal Gases is given by:-

$$c_{p(mix)} = \sum_i c_{(p)i} \cdot f_i$$

Equation 5.9

where $c_{p(mix)}$ is the Equivalent Specific Heat of the Gas Mixture; $c_{(p)i}$ is the Specific Heat of the ith. component Gas and f_i is the (Mass) Fraction of the ith. Gas in the Mixture.

Knowledge of this $c_{p(mix)}$ value is itself of only partial use without accompanying knowledge of the Equivalent Mean Molecular Mass, $MW_{\mu(mix)}$, which as mol.T°K is given by:-

$$MW_{\mu(mix)} = \frac{1}{n}\sum_i MW_i \cdot \varphi_i$$

Equation 5.10

The Universal Gas Constant 8.314 J/mol

t Range °K	t °K	t/1000 °K	Formula	NIST Shomate Equation Coefficients					C_p (J/(mol.K))	frac*R (J/(mol.K))	PSD (Cp,frac*R)	Notes
				A	B	C	D	E				
298-1300	498	0.498	H_2	33.06618	-11.3634	11.43282	-2.772874	-0.158558	29.26077742	29.099	0.552881472	Gas Phase
100-700	498	0.498	O_2	31.32234	-20.2353	57.86644	-36.50624	-0.007374	31.05779143	29.099	6.30692441	Gas Phase
700-2000	1300	1.3	O_2	30.03235	8.772972	-3.98813	0.788313	-0.741599	35.99037651	29.099	19.14783112	Gas Phase
298-1300	498	0.498	N_2	28.98641	1.853978	-9.64746	16.63537	0.000117	29.57212226	29.099	1.599892827	Gas Phase
298-1300	1300	1.3	N_2	28.98641	1.853978	-9.64746	16.63537	0.000117	51.64035281	29.099	43.65065609	Gas Phase
298-6000	498	0.498	Ar	20.786	2.83E-07	-1.46E-07	1.09E-08	-3.66E-08	20.78599996	20.785	0.004810729	Gas Phase
298-6000	1300	1.3	Ar	20.786	2.83E-07	-1.46E-07	1.09E-08	-3.66E-08	20.78600012	20.785	0.004811519	Gas Phase
298-1300	498	0.498	CO	25.56759	6.09613	4.054656	-2.671301	0.131021	29.80741393	29.099	2.376636669	Gas Phase
1300-6000	498	0.498	CO	35.1507	1.300095	-0.20592	0.01355	-3.28278	22.51194895	29.099	-29.26024336	Gas Phase
298-1200	498	0.498	CO_2	24.99735	55.18696	-33.6914	7.948387	-0.136638	44.55558418	11.0853333	75.12021549	Gas Phase
500-1700	498	0.498	H_2O	30.092	6.832514	6.793435	-2.53448	0.082139	35.19756786	11.0853333	68.50539964	Gas Phase
298-1809	1800	1.8	Fe	23.97449	8.36775	0.000277	-0.000086	-0.000005	39.03683438			Solid Phase
1809-3133.345	1900	1.9	Fe	46.024	-1.88E-08	6.09E-09	-6.64E-10	-8.25E-09	46.02399998			Liquid Phase

Table 5.1
Gas and Iron Heat Capacities

where $MW_{\mu(mix)}$ is the Equivalent Molecular Weight of the Gas Mixture (dimensionally a Mass); MW_i is the Molecular Weight of the ith. component gas and φ_i is the Mass Fraction of the ith. component gas.

It is essential to specify the Temperature in Degrees Kelvin of the gas treated.

Table 5.2 presents arithmetic elaborations for the Equivalent Specific Heat of Hot Blast Air assumed to come in at 1300°K and Blast Furnace exhaust Gas (BFG), assumed to exit at 498°K.

Therefore the Heat Energy per Equivalent Mole of Mixed Gas $E_{mole(mix)}$ is given by:

$$E_{mole(mix)} = c_{p(mix)} \times MW_{\mu(mix)} = \sum_i c_{(p)i}.f_i \times \frac{1}{n} \sum_i MW_i.\varphi_i$$

Equation 5.11

which has the dimensions $ML^2T^{-2}mol^{-1}$.

Formula	AIR	1300					
	Mix Fraction (frac)	C_p at °K (J/mol.K))	C_p×frac			MW	MW×frac
N_2	0.7808	51.6403528	40.32079			28.0134	21.872863
O_2	0.2095	20.786	4.354667			31.9988	6.7037486
Ar	0.0098	20.7860001	0.203703			39.948	0.3914904
	1.0001		44.87916	$\Sigma(C_p\times frac)$		99.9602	28.968102 $\Sigma(MW\times frac)$
			14.95972	$\mu(C_p\times frac)$		33.32007	9.6560339 $\mu(MW\times frac)$

Formula	BFG	498					
	Mix Fraction (frac)	C_p at °K (J/mol.K))	Cp×frac			MW	MW×frac
N_2	0.5519	29.5721223	16.32085			28.0134	15.460595
O_2	0	31.0577914	0			31.9988	0
Ar	0	20.786	0			39.948	0
CO	0.2078	29.8074139	6.193981			28.0101	5.8204988
CO_2	0.2107	44.5555842	9.387862			44.0095	9.2728017
H_2	0.02076	29.2607774	0.607454			2.01588	0.0418497
H_2O	0	35.1975679	0			18.0153	0
	0.99116		32.51015	$\Sigma(C_p\times frac)$		192.011	30.595746 $\Sigma(MW\times frac)$
			4.644307	$\mu(C_p\times frac)$		27.43014	4.3708208 $\mu(MW\times frac)$

Table 5.2
Mixed Gas Heat Capacities

Selected Thermal Data

      Table 5.3 further prepares for simple modelling by conveniently assembling selected thermal data for water, aluminum and key grades of iron and iron products at specified temperatures, usually temperatures that are at or near state changes.

      Table 5.3 also presents the relevant Formula Weights. Thermal data include Molar and Kilogram Specific Heats, and Molar and Kilogram Latent Heats of Fusion.

      Latent Heat is the Energy per Unit Mass required to change the state of a substance.

Temporature of Water Fusion 273.2 °K

Formula	Nominal Formula Weight (g/mol)	Melting Point °C	Melting Point °K	Boiling Point °C	Boiling Point °K	Density (kg/m³)	Molar Specific Heat C_p (KJ/(mol·°K))	Molar Latent Heat of Fusion L (KJ/mol)	Specific Heat C_p (J/(kg·°K))	Latent Heat of Fusion L (KJ/kg)
H_2O	18.0153	0	273.17	100	373.17	0.999975	75.32737389	6.0171102	4181.3	334
Fe		1538					39.03683438	13.81		
Fe	55.845	1538	1811.17			6960.505500	46.02399998	13.81		
Fe	55.845	1538	1811.17	2861	3134.17	6980.000000	25.074405	13.81	449	247
Cast Iron	55.845	1204	1477.17			7150.000000	46.0548	15.6366	506	280
Wrought Iron	55.845	1482	1755.17			7874.000000	50.2416	13.793715	502.416	247
Steel	55.845	1371	1644.17			7850.000000	26.861445	15.07815	481	270
Stainless Steel	55.845	1510	1783.17			7750.000000	47.46825	15.6366	850	280
Al	26.981539	660.32	933.49	2519	2792.17	2375.000000	24.39131089	10.7	904	396

Table 5.3
Formula Weights and Selected Thermal Data

Reaction Balance

In our simple Reactions Model of the Hot Blast Iron Smelter we desire to calculate the basic complete Reduction Pathway for Hematite Iron Ore, and also the Oxidation Pathway that completes the participation of Carbon reacting with Hot Blast Oxygen gas as it burns to raise the process heat, and also as it reacts directly with Hematite to yield Iron and Carbon Dioxide.

That is:-

$$2Fe_2O_3 + 3C \rightarrow 4Fe + 3CO_2$$

Pathway 5.1

and:-

$$C + O_2 \rightarrow CO_2$$

Pathway 5.2

Equations for the (scaled) Heats of Formation, ΔH_f* in KJ/mol, are associated with these pathways in terms of:-

$$\Delta H_f{}^* = [(\lambda_3 FW_3 + \lambda_4 FW_4) - (\lambda_1 FW_1 + \lambda_2 FW_2)]$$

Equation 5.12

where ΔHF^* is a Scaled Heat of Formation; λ_3 and λ_4 are respective Stochiometric Coefficients for the First and Second Chemical Products (where the second product is present); and FW_3 and FW_4 are the respective Formula Weights for the First and Second Reagents. Subscripts 1 and 2 denote the First and Second Reagents.

Technically, it is only necessary to establish the thermal behavior of the chemical products, but it is good practice to compute a balance of formula weights for the products against the reagents: The scaled sum of the formula weights for each side should be identical.

You can check such balances on Table 5.4

Another striking thing about Table 5.4 is that whilst ΔH_{red} for Pathway 5.1 is +470.434 KJ/mol, the ΔH_{oxy} for Pathway 5.2 is -393.522

If a Heat of Reaction is *positive* the Reaction is Endothemic, or in other words it has to *absorb* heat from its environment. If the Heat of Reaction is *negative* then the Reaction is

Exothermic, which means it *gives out heat*. Note: An <u>opposite</u> convention is often seen in literature.

Accordingly, to establish a grand balance of zero on all sides we must note the difference ΔH_{red}-ΔH_{oxy} as +76.912 KJ/mol and to support the Hematite Reduction Reaction this heat has got to be found from somewhere, presumably the burning of carbon.

The appropriate Balance Constant, κ_{bal}, is:-

$$\kappa_{bal} = \frac{|\Delta H_{red}|}{|\Delta H_{oxy}|} \equiv \left|\frac{\Delta H_{red}}{\Delta H_{oxy}}\right| = \left|\frac{+470.434}{-393.522}\right| = 1.195445236$$

Equation 5.13

REDUCTION X

	Chemical Formula	Stoich. Coeff. λ_k	Molar Weight FW_k (g/mol)	Heat of Formation ΔH_f (KJ/mol)	Scaled Molar Weight FW_k^* (g/mol)	Scaled Heat of Formation ΔH_f^* (KJ/mol)	
Scale Factor κ_x 1 $2Fe_2O_3+3C=4Fe+3CO_2$							
Reagents							
Reduction Reagent a1	Fe_2O_3	2	159.688	-825.5	319.376	-1651	
Reduction Reagent a2	C	3	12.0107	0	36.0321	0	
Products							
Reduction Product a3	Fe	4	55.845	0	223.38	0	
Reduction Product a4	CO_2	3	44.0095	-393.522	132.0285	-1180.566	
					355.4081		Σ_{reg}
					355.4085		Σ_{pro}
						470.434	ΔH_{red}

COMBUSTION Y

	Chemical Formula	Stoich. Coeff.	Molar Weight	Heat of Formation	Scaled Molar Weight	Scaled Heat of Formation	
Scale Factor κ_y 1 $C+O_2=CO_2$							
Reagents							
Combustion Reagent b1	C	1	12.0107	0	12.0107	0	
Combustion Reagent b2	O_2	1	31.9988	0	31.9988	0	
Products							
Combustion Product b3	CO_2	1	44.0095	-393.522	44.0095	-393.522	
Combustion Product b4	nul						
					44.0095		Σ_{reg}
					44.0095		Σ_{pro}
						-393.522	ΔH_{oxy}

NET HEAT per formulas weight (g)						76.912	ΔH_{net}
Scale Factor κ_{bal}	1.195445236						
Balance Heat						0	ΔH_{bal}

Table 5.4
Carbon-based Hot Blast Iron Furnace
Reaction Balance

<u>Simple Balance Modelling</u>

Heat per Gram and Heat per Kilogram

We saw from Table 5.4 that ΔH_{red} for Pathway 5.1 is +470.434 KJ/mol, the ΔH_{oxy} for Pathway 5.2 is -393.522. These are Heat Quantities or "Enthalpies" *per simplest balanced equation.*

To estimate Heat in real life we should re-express it as heat *per molar mass of products,* Σ_{pro}. This will lead directly to Heat per Gram of output, Σ_{prog}, or, multiplied by 10^3, Heat per Kilogram, Σ_{proKg}. For the case of the first Fe-C reaction:-

$$\Sigma_{progFe-C} = \left| \frac{\Sigma_{proFe-C}}{\Sigma FW_k^*} \right| = \left| \frac{+470.434}{355.4} \right| = 1.32367473269555$$
Equation 5.14

which means that Absorbed Heat due to Carbon reduction of Hematite is 1.32367473269555 Joules per Reagent Gram. Clearly, this scales to about 1324 Joules per Kilogram.

By the same token, the Carbon-Oxygen Reaction generates Evolved Heat given by:-

$$\Sigma_{progC-O2} = \left| \frac{\Sigma_{proC-O2}}{\Sigma FW_k^*} \right| = \left| \frac{-393.522}{44.01} \right| = 8.94164962508521$$
Equation 5.15

which is around 8.942 Joules per Gram of Reagents (or Products) giving about 8942 Joules per Kilogram.

Data Assembly for Modelling

Table 5.5 presents relevant Thermal Data and also selected Molar Equivalent Masses (grams) to prepare elementary model outputs for the Hot Blast Iron Furnace case.

Note that several features are neglected, especially flux-slag behaviors and furnace carcass losses.

Carbon Heat Contributions:-
To Complete the Hematite Reaction

$$\Delta H_{assi} = 76.912$$
Equation 5.16

Carbon Heat Contributions:-
Sensible Heat Required per Mole of Liquid Iron

$$\Delta H_{FE} = \Delta h_{solFe} + \Delta h_{liqFe} + L_{fusFe} = 3581235.47381156$$
Equation 5.17

Carbon Heat Contributions:-
To Heat Inherent in Exit Iron and Exit Blast Furnace Gas

$$\Delta H_{rais} = \left| \frac{\Delta H_{Fe} + \Delta h_{gasBFG}}{\Delta H_{cMole}} \right| = 9610.00211882888$$
Equation 5.18

Carbon Heat Contributions:-
Total Carbon Heat Contribution by Oxidation

$$\Delta H_{cOxid} = \Delta H_{assi} + \Delta H_{rais} = 9686.91411882888$$
Equation 5.19

Carbon Heat Contributions:-
Balance Heat due to Carbon

$$\Delta H_{BFbal} = \Delta H_{Fe} + \Delta h_{gasBFG} - \Delta h_{gasBlast}$$
Equation 5.20

Kilogram Demands:-
Kilograms of Iron Required

We decide we want to have one metric tonne of iron at the end of the process. Hence:-

$$Fe_{Kg} = 1000$$
Equation 5.21

Kilogram Demands:-
Formula Moles of Iron per Kilogram

$$mols_{FeReq} = \frac{1000}{\lambda_{Fe}.FW_{Fe}} = 4.476676515355$$
Equation 5.22

Kilogram Demands:-
Calculated Carbon Demand (Kg)

$$C_{Kg} = Fe_{Kg}.mols_{FeReq}.C_{KgMolFe} = 1519.87944310144$$
Equation 5.23

Kilogram Demands:-
Calculated Oxygen Demand (Kg)

$$O2_{Kg} = Fe_{Kg}.mols_{FeReq}.O2_{KgMolFe} = 4049.24927925508$$
Equation 5.24

Please refer to Gurprit Singh[5.2] for a comprehensive blast furnace computer model developed by a steel production professional.

Description	Symbol	Equation	Value	Units
Melting Point of Iron	T_{MPFe}		1811.15	°K
Ambient Temperature (taken as 0°C)	T_{amb}		273.15	°K
Required Temperature of Molten Iron	T_{meltFe}		1900	°K
Required Temperature of the Hot Blast	T_{Blast}		1300	°K
Temperature of Blast Furnace Gas at Exit	T_{BFG}		498	°K
Temperature Elevation of Solid Iron	ΔT_{Fe}	$T_{MPFe}\text{-}T_{amb}$	1538	°K
Temperature Elevation of Solid Iron	ΔT_{FeSol}	$T_{MPFe}\text{-}T_{amb}$	1538	°K
Temperature Elevation of Liquid Iron	ΔT_{FeLiq}	$T_{meltFe}\text{-}T_{MPFe}$	88.85	°K
Temperature Elevation of Hot Blast	ΔT_{Blast}	$T_{Blast}\text{-}T_{amb}$	1026.85	°K
Temperature Elevation of Blast Furnace Gas	ΔT_{BFG}	$T_{BFG}\text{-}T_{amb}$	224.85	°K
Specific Heat of Hot Blast Gas (Equivalent)	c_{pBlast}		44.87916	J/(mol.°K)
Specific Heat of BFG (Equivalent)	c_{BFG}		32.51015	J/(mol.°K)
Specific Heat of Solid Iron (Shomate)	c_{pFeSol}		39.03683	J/(mol.°K)
Specific Heat of Liquid Iron (Shomate)	c_{pFeLiq}		46.024	J/(mol.°K)
Molar Equivalent Mass of Hot Blast Gas	m_{Blast}		33.32007	g
Molar Equivalent Mass of BFG	m_{BFG}		27.43014	g
Molar Equivalent Mass of Solid Iron	m_{FeSol}		55.845	g
Molar Equivalent Mass of Liquid Iron	m_{FeLiq}		55.845	g
Sensible Heat for Hot Blast Gas	$\Delta h_{gasBlast}$	$c_{pBlast}{\cdot}m_{Blast}{\cdot}\Delta T_{Blast}$	1535527	J
Sensible Heat for BFG	$\Delta h_{gasBFGt}$	$c_{BFG}{\cdot}m_{BFG}{\cdot}\Delta T_{BFG}$	200511.8	J
Sensible Heat for Solid Iron	Δh_{solFe}	$c_{pFeSol}{\cdot}m_{FeSol}{\cdot}\Delta T_{FeSol}$	3352858	J
Sensible Heat for Liquid Iron	Δh_{liqFe}	$c_{pFeLiq}{\cdot}m_{FeLiq}{\cdot}\Delta T_{FeLiq}$	228363.2	J
Latent Heat of Fusion of Iron	L_{fusFe}		13.81	KJ/mol
Molar Heat Difference for Carbon Oxidation	ΔH_{cMole}		-393.522	J/ForMole
Mass per Mole for Carbon	$mols_C$		28.26735	g/mol
Mass per Mole for Diaatomic Oxygen	$mols_{O2}$		28.26735	g/mol
Mass Heat Difference for Carbon per Iron Mole	$C_{KgMolFe}$		0.339511	Kg/mol
Mass Heat Difference for Diatomic Oxygen per Iron Mole	$O2_{KgMolFe}$		0.904521	Kg/mol

Table 5.5
Hot Blast Iron Furnace
Gathered Key Data

Simple Balance Modelling in Spreadsheet Form

Table 5.6 presents my developed Simple Balance Model in EXCEL® spreadsheet form:-

Item	Temp. (T_r) (°K)	Symbol	Value	Inherent Heat ΔH (J/mol)	Molar Specific Heat $c_p(x)$ (J/(mol.°K))	Mole Equivalent Mass (m_x) (g)	Temp. Differ. $(T_r-T°)$ (°K)	Mole Equivalent Sensible Heat (solid) (J/mol)	Mole Equivalent Sensible Heat (liquid) (J/mol)	Mole Equivalent Sensible Heat (gas) (J/mol)	Mole Equivalent Latent Heat Of Fusion (J/mol)
Ambient Temperature	273.15 T°						0				
Melting Point of Iron (°C)	1538										
Melting Point of Iron (°K)	1811.2 T_{MPFe}						1538				
Hot Blast				1535527.37	44.87915727	33.3200667	1026.85			1535527.371	
BFG				200511.78	32.51015022	27.43014	224.85			200511.78	
Solid Iron				3352858.48	39.03683438	55.845	1538				
Liquid Iron				228363.183	46.02399998	55.845	88.85		228363.1833		13.81
Net Heat Input per Mole of Carbon [1]		ΔH_{NFeC}	-393.522								
Moles of Carbon Input		$mols_C$	24.61594046								
Kilograms of Carbon per Mole of Iron Output		C_{kgMoFe}	0.295654676								
Moles of Oxygen Input		$mols_{O2}$	24.61594046								
Kilograms of Oxygen per Mole of Iron Output		$O2_{kgMoFe}$	0.787680555								
Carbon Heat Contribution to Complete Hematite Reaction		ΔH_{BSI}	76.912								
Carbon Heat Contribution to Raise Iron and BFG to Exit States [5]		ΔH_{RIO}	9610.00212								
Total Carbon Heat Contribution by Oxidation		ΔH_{COXE}	9686.91412								
Balance Heat due to Carbon [1]		ΔH_{BFSE}		2246219.88					13.81		
Heat Input per Mole Equivalent of Hot Blast [2]	1300	ΔH_{Blast}		1535527.37		33.3200667	1026.85			44.87915727	
Heat Required per Mole of Liquid Iron [3]	1900	ΔH_{Fe}		3581235.47			1626.85	39.036834	46.02399998		13.81
Heat Inhered by a Mole Equivalent of Blast Furnace Gas at Exit [4]	498	ΔH_{BFG}		200511.78		27.43014	224.85			32.51015022	
Kilograms of Iron Required		Fe_{kg}	1000								
Formula Moles of Iron per Kilogram		$mols_{FeperKg}$	4.476676515								
Calculated Carbon Demand (kg)		C_{kg}	1323.550345								
Calculated Oxygen Demand (kg)		$O2_{kg}$	3526.191044								

[1] $\Delta H_{Fe} + \Delta H_{BFGEXC} - \Delta H_{HotBlast}$

[2] $m_{Blast} \cdot C_{pBlast} \cdot \Delta T_{Blast}$

[3] $m_{FeLiq}((C_{pFeSol} \cdot (T_{MPFe}-T_{amb})+(C_{pFeLiq} \cdot (T_{MPFe}-T_{MPFe}))+L_{Fe}$

[4] $m_{BFG} \cdot C_{pBFG} \cdot \Delta T_{BFG}$

[5] $ABS[(\Delta H_{Fe}+\Delta H_{BFGEXC})/\Delta H_{COXE}]$

Table 5.6
A Simple Blast Furnace Energy Balance Model
Tabulated

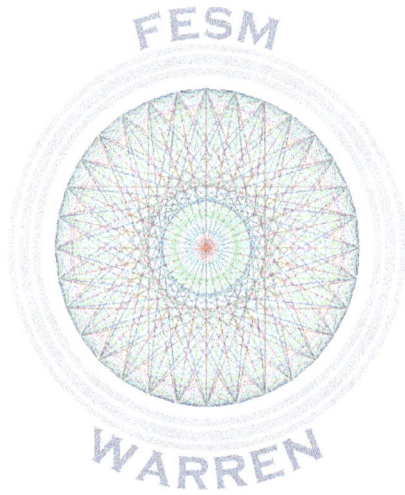

CHAPTER SIX
AN IMPRESSION OF ACTIVATION ENERGY

Activation Energy is a thermodynamic construct which allows for the stability of unreactive mixtures, even dry gunpowder.

If you will, Activation Energy is a "hump" of excess energy that a reaction must surmount in order to progress to the making of products.

Activation Energy is intimately related to a "spark" of intense temperature that a smidgin of the mixed reagents must endure in order to start work.

At an intuitive level, if you want your heap of gunpowder to explode (or to be pedantic, rapidly burn) you have to strike a match and apply the burning end to the powder. Otherwise, the gunpowder will just sit there indefinitely until it eventually dampens and decays.

In the context of iron smelting, we can mix iron ore of any kind with coke or charcoal or form it into pellets or mounds in a storage hopper and nothing will happen. How convenient if it would immediately and calmly start reducing itself to fine steel!

Unfortunately not. We have to supply it with excess free oxygen and set a match to it.

The process of starting a blast furnace is called "Blowing-In". Before a blast furnace can be started it must first of all be carefully dried by experts otherwise the refractory bricks that line the structure are liable to fracture and ruin the facility. Then furnace men (or today women) construct a carefully arranged charge of dry coke above dry timber and ignite that with the help of accelerant if necessary. The blast is gradually increased. Further calibrated coke charges are added when appropriate, and further oxygen admitted until the furnace fire becomes sustainable. Finally, the first small charges of ore and reduction coke are introduced.

The same principle applies to furnaces fueled by hydrogen or any other chemical, though electric furnaces whether of traditional or plasma design conform to their own rules.

Some of you will object that heaps of coal, etcetera, can combust spontaneously. This phenomenon is deceptive. What is actually happening is that the coal has become damp, and that iron sulfide (pyrite) within the coal has become partly hydrolysed to dilute sulfuric acid. Assisted by microbial activity, this metal-acid interaction generates enough internal heat to activate smoldering of the heap center, which can degenerate to naked fire, as it did in the bunkers of numerous nineteenth-century steamer ships with tragic results. Dry coal will not burn unless carefully activated and persistently aeriated.

The Wikipedia article about "Activation Energy" is characteristically excellent[6.1]. If you lack access to the Internet please consult your librarian.

Figure 6.1 from the Wikipedia article is by Jerry Crimson Mann, Tutmosic and Fvasconcellos[6.2] and illustrates the decomposition of the carbohydrate food chemical glucose ($C_6H_{12}O_6$) with oxygen to water and carbon dioxide with the release of chemical energy. The picture illustrates the big hump of Activation Energy necessary without the intervention of an enzyme (a biological catalyst) and the lower hump with an enzyme. In the context of this reaction the catalyst involved is called insulin and is generated naturally in the pancreatic gland, except in diabetics such as myself, for whom it must be made synthetically and injected through the skin.

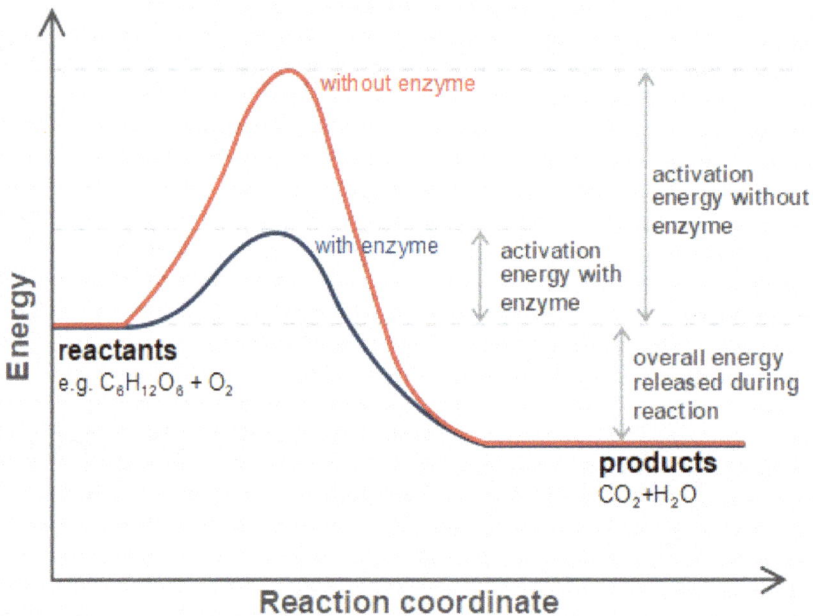

Diagram: Mann Et Al

Figure 6.1
Qualitative Energetics of the
Metabolism of Glucose

Figure 6.2 also from Wikipedia "Activation Energy" defines the thermodynamical symbolism conventional in discussions of Reaction Energy and Activation Energy:-

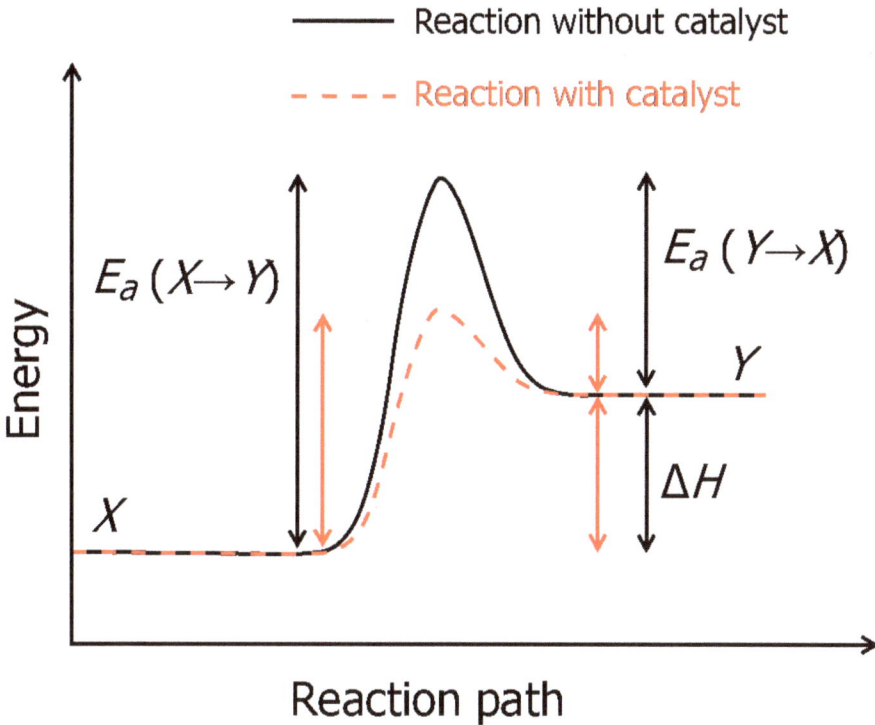

Diagram: Anonymous

Figure 6.2
Thermodynamic Symbolism attaching to
Activation and Reaction Energies

In Figure 6.2 X→Y is the Reaction Forward Pathway and Y→X is the Reaction Backward Pathway. ΔH is the Heat (or Enthalpy) of the Reaction in J/mol.

At Thermodynamic Equilibrium ΔH is zero and the reaction terminates until one or more of the products or reagents are added or removed. (c.f. Le Chatelier's Principle).

$E_a(X \rightarrow Y)$ is the Activation Energy of the Forward Reaction and $E_a(Y \rightarrow X)$ is the Activation Energy of the Backward Reaction

The red lines illustrate the effects of catalysis.

The *principles* can be extended to iron smelting or any other chemical reaction.

The Arrhenius and Eyring Equations

The Number of Molecular species in a Mole of Substance is, by Avogadro's Number (Avogadro Constant), N_A, equal to $6.02214076 \times 10^{23}$ mol^{-1}. Meanwhile, K_B, the Boltzmann Constant is 1.380649×10^{-23} J/°K.

The Universal Gas Law, R, is given by:-

$$R = N_A K_B = 8.31446261815324 \text{ J/(mol.°K)}$$
Equation 6.1

This simple relationship is essential, not only in discussions of gas conditions and reaction heats, but also in electrochemistry.

The Arrhenius Equation

The Arrhenius Equation was published by Svante Arrhenius in 1889AD. It is given as:-

$$k = A . exp \left(\frac{-E_a}{RT} \right)$$
Equation 6.2

where:-

$k =$ Rate Constant (Hz $\equiv$ T^{-1})
       This is the Frequency of Molecular Collisions Resulting in a Reaction
$A =$ Frequency Factor
$E_a =$ Molar Activation Energy
$R =$ The Universal Gas Constant
$T =$ Absolute Temperature (°K)

The Eyring-Polanyi Equation

The Eyring Equation was given by Henry Eyring in 1935AD. It was simultaneously arrived at by Meredith Gwynne Evans and Michael Polanyi.

$$k = \frac{\kappa . k_B . T}{h} . exp\left(\frac{-\Delta G_a}{RT}\right)$$

Equation 6.3

where:-

k = Rate Constant (Hz $\equiv$ T^{-1})
    This is the Frequency of Molecular Collisions Resulting in a Reaction
κ = Transmission Coefficient
k_B = Boltzmann Constant
    = 1.380649×10^{-23} J/$^{\circ}$K
T = Absolute Temperature ($^{\circ}$K)
h = Planck Constant (J/Hz $\equiv$ ML^{2}T^{-1})
    = $6.62607015 \times 10^{-34}$ J/Hz
ΔG_a = Gibbs Energy of Activation
E_a = Molar Activation Energy
R = The Universal Gas Constant

The Transmission Coefficient κ is essentially the amount of energy conveyed from one side of the reaction to the other and thus analogous to the Heat of Reaction ΔH.

It can be seen that the Eyring Equation is a sophistication of the Arrhenius Equation and that equivalation allows us to write:-

$$A . exp\left(\frac{-E_a}{RT}\right) = \frac{\kappa . k_B . T}{h} . exp\left(\frac{-\Delta G_a}{RT}\right)$$

Equation 6.4

In 1873AD Josiah Willard Gibbs enunciated his concept of "available energy" as:-

"The greatest amount of mechanical work which can be obtained from a given quantity of a certain substance in a given initial state, without increasing its total volume or allowing heat to pass to or from external bodies, except such as at the close of the processes are left in their initial condition."

Since his day this concept has been called the Gibbs Free Energy of a closed system at a given pressure and temperature and is quantified as:-

$$G(p,T) = U + pV - TS = H - TS$$
Equation 6.5

where:-

G(p.T) = Gibbs Free Energy at
Pressure p and Temperature T
U = System Internal Energy
p = System Pressure
V = System Volume
T = Absolute Temperature
S = System Entropy
H = System Enthalpy

In principle, the system in question could be a balanced molar equation.

Therefore, in terms of system energy *changes*, in say Joules:-

$$\Delta G = \Delta H - T\Delta S$$
Equation 6.6

Knowing Equation 6.6 we can move forward to recast Equation 6.4 as:-

$$k = \frac{\kappa.k_B.T}{h} \times exp\left(\frac{\Delta S_a}{R}\right) \times exp\left(\frac{-\Delta H_a}{RT}\right)$$
Equation 6.7

And Equation 6.7 may re-expressed by taking logarithms of its terms as:-

$$\ln\left(\frac{k}{T}\right) = \frac{-\Delta H_a}{R} \times \frac{1}{T} + \left[\ln\left(\frac{\kappa.k_B}{h}\right) + \frac{\Delta S_a}{R}\right]$$
Equation 6.8

Because Equation 6.8 is a first-degree polynomial equation of form $y = \alpha + \beta x$ (i.e. a straight line equation), and k_B, h, κ, k and R are all actual or assumed constants, linear regression or some other suitable mathematical technique can be used to establish the value of Activation Enthalpy, ΔH_a and Activation Entropy ΔS_a.

α is the Intercept of the said Straight-Line Equation on the y-axis (that is to say, when x = 0): It is equivalent to the square-bracketed expression:-

$$\alpha = \left[\ln\left(\frac{\kappa.k_B}{h}\right) + \frac{\Delta S_a}{R} \right]$$

Equation 6.9

so we may write:-

$$\alpha = \ln(C_1.\kappa) + C_2.\Delta S_a$$

Equation 6.10

where C_1 and C_2 are further constants easily calculated from elementary constants of known values. κ is unity if we can be assured that the reaction drives to entire production.

Meanwhile, dx/dy = β = Line Slope is given by:-

$$\beta = \frac{-\Delta H_a}{R} = C_2.-\Delta H_a$$

Equation 6.11

where $(x,y) \equiv [1/T, \ln(k/T)]$

This elementary mathematical treatment is contingent upon the performance of a series of experiments upon the system in which the temperature is varied at constant volume.

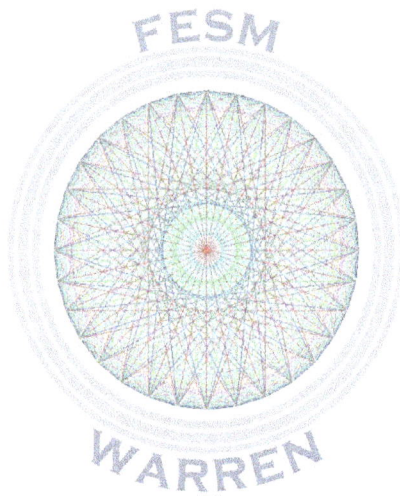

CHAPTER SEVEN
THE PRINCIPLES OF ELECTROLYTIC SCIENCE

Electrolysis is the art and science of using an electric current to break apart liquid chemicals dissolved or molten in an Electrolytic Cell, basically a glass or ceramic bath to which current is admitted via two electrodes, a "negative" Cathode which is the source of flowing Electrons, and a "positive" Anode which is their sink. Different, simpler, chemicals emerge in an electrically-neutral state at either or both of the electrodes.

Figure 7.1 is a sketch of an idealised glass cell for splitting water (H_2O) into its component Hydrogen and Oxygen gases. Because at Standard Temperature and Pressure both gases are physically Ideal you get twice the volume of Hydrogen as you get Oxygen even though you get only 2.01588 kilos of H_2 for every 15.9994 kilograms of O.

Note that you have to use Direct Current (DC) for electrolysis. This is industrially inconvenient, but great if you are using batteries to supply current in a laboratory, as did the pioneers of electrical science.

Large industrial electrolysis installations either have to generate DC *ab initio* using Dynamos (a trick that incurs its own penalties) or else they have to convert Alternating Current (AC) to DC using a heavy electronics device called a Rectifier, which inevitably wastes some of the current as heat.

Pure water is a Dielectric (insulator) and will not conduct electricity, which needs Ions, or in other words, electrically-charged atoms to move through liquid (other than for instance mercury, which conforms to being a metal). Notwithstanding, a trace of acid, alkali or salt will give enough ionisation to render water conductive and electrically decomposable.

For managerial reasons it is a good idea to keep the Electrolyte surface at an even level across the electrolyser: Closed arms would of course hold twice as much hydrogen volume as oxygen unbalancing the feedstock pool.

Current I is effectively the amount of Charge, Q, passed in an Electric Circuit every second, assuming that the current is steady.

When the great English scientist and lecturer Michael Faraday was working at The Royal Institution in Mayfair, London, in the Eighteen-Twenties and Eighteen-Thirties electrical theory was barely understood and Faraday reached his conclusions by experiment, based upon the prior work of his patron and mentor, Sir Humphry

Davy. Humphry Davy, a Cornishman, the famous safety lamp inventor, had used electrolysis to isolate the alkali metals. In fact, Davy had used the electrolysis of molten salts to discover Potassium and Sodium in 1807, and Calcium, Strontium, Barium and Magnesium one year later.

Both men are venerated as heroes in my country, and Faraday appears on the back of the £20 note (treasury bill) as it was printed in the Nineteen-Nineties.

We now know that electricity is the flow of atom-orbital sub-atomic particles called Electrons through a suitable medium, usually a metal or a salty liquid. An individual Electron a negatively-charged Lepton and is truly tiny: It has a uniform Electron Mass, m_e, of $9.1093837015 \times 10^{-31}$ Kilograms and a standard Charge e_q of $-1.602176634 \times 10^{-19}$ Coulombs.

Faraday's Laws and the Passage of Charge

"The mass of a substance deposited at any electrode is directly proportional to the amount of charge passed."

To put that algebraically as Faraday's First Law:-

$$m \propto Q$$
Proportionality 7.1

To be more exact:-

$$m = ZQ$$
Equation 7.1

where Z is the Electro-chemical Equivalent (ECE) of the substance.

Faraday understood the concept of Atomic Weight and the related Molecular Weight of chemical compounds, and that there was a definite relationship between Atomic or Molecular Weights and the amount of Charge passed.

So Faraday said:-

"If the same amount of electricity is passed through different electrolytes, the masses of ions deposited at the electrodes are directly proportional to their chemical equivalents."

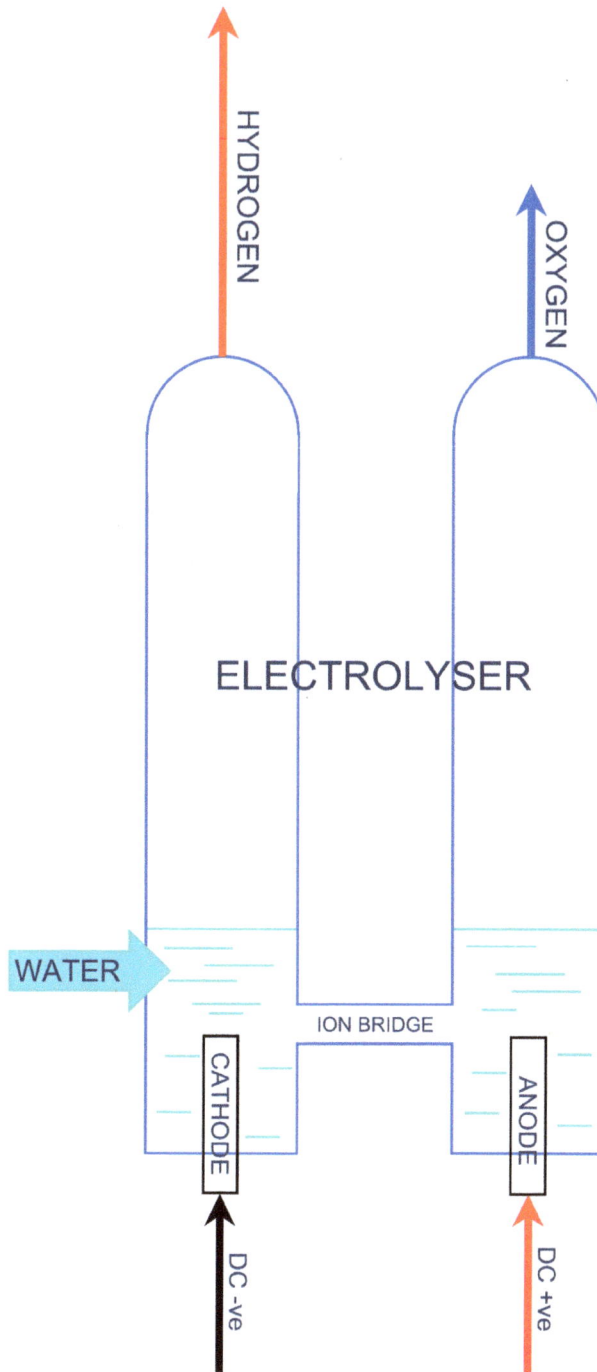

Figure 7.1
The Electrolysis of Water

Today we call this Faraday's Second Law of Electrolysis.

In algebra it is:-

$$E_i = \frac{molar\ mass}{valency} = \frac{M}{v}$$
Equation 7.2

where E_i is the Equivalent Weight of Species i [M^{+1}] measured in grams or Kilograms; Molar Mass is the Mass [M^{+1}] the Formula Weight of a Mole of Substance read as grams or Kilograms; and v is the Valency of the Ionised Reagent.

The *systeme international d'unites* (SI) convention prefers Electric Current (I), measured in Amperes, "amps", as the basic dimension, so that Q then becomes:-

$$Q = It = [I^{+1}T^{+1}]$$
Equation 7.3

where t is Time passed in seconds.

In 1865AD Johann Josef Loschmidt used the Universal Gas Law, combined with arguments relating to James Clerk Maxwell's Kinetic Mean Free Path Theory, to furnish an estimate of the Number Density, n_0, of (any) gas at 0°C and one atmosphere pressure as approximately 1.81×10^{24} m^{-3} (i.e. molecules per cubic meter). My transposition of Loschmidt's Equation is:-

$$N_A = n_0 \times \frac{RT_0}{p_0}$$
Equation 7.4

Knowing the Temperature, T_0, at 0°C to be 273.15°K; the Pressure, p_0, to be 1 atmosphere = 101325 Pa; n_0 (Loschmidt) to be 1.81×10^{24} m^{-3}; and R, the Universal Gas Constant to be 8.31446261815324 J/(mol.°K) we can transpose Loschmidt's formulation to yield an estimate of Avogadro's Number, N_A, as:-

$$N_A = n_0 \times \frac{RT_0}{p_0} = 1.81 \times 10^{24} \times \frac{8.31446261815324 \times 273.15}{101325}$$
$$= 4.05692848764756 \times 10^{22}$$
Equation 7.5

The accepted modern estimate of Avogadro's Number is $6.02214076 \times 10^{23}$ mol^{-1} so the Loschmidt Formulation is only 6.74% of the "true" figure, but the principle was sound and I am sure that, given the circumstances you will forgive the Austrian genius the odd 93% here or there!

Maxwell and others refined the Loschmidt figure to be one of the most important constants in science and industry.

In 1897AD, Manchester physicist Joseph John Thomson recognised that the Cathode Rays that he shone through a vacuum tube at Cambridge University were actually streams of tiny matter particles because they could be manipulated using charged electrode plates and electromagnets. This was impossible with light or any other electromagnetic radiation. By balancing the electric charge and the magnetic flux he could bend and re-straighten the beam. This enabled Thomson to calculate the Charge to Mass Ratio of the Electron.

Thomson won a Nobel Prize, and the British King knighted him and admitted him to The Order of Merit, that confers an even greater honor but no title.

In 1909AD Robert Millican and Harvey Fletcher of the University of Chicago set out to measure the Charge on the Electron, and indeed establish whether it was Elementary, by experiment.

They too favored a Balance of Known Forces approach. They used electrified plates to levitate frictionally-charged droplets of watchmaker's oil in a sealed air chamber. By balancing the known electric field force against the gravitational force and air drag on the falling droplets Millican and Fletcher could bring the individual droplets to a stand-still. They measured this effect for many droplets and by taking the Lowest Common Multiple of the stasis force for the set of droplets they could estimate the Charge on the Electron as 1.5924×10^{-19} Coulombs, within 0.6% of the modern value.

The computation of the Mass of the Electron was now a matter of simple arithmetic.

Millican picked up his own Nobel Prize.

So having to hand now the Charge on the Electron and Avogadro's Number we are able to declare that for any electrolytic reaction:-

$$F = N_A . e_q$$
Equation 7.6

F is Faraday's Constant and has a value near to 96485.33212331 Coulombs/mol.

(Wikipedia gives 96485.3321233100184).

There are Avogadro's Number of electrons in a mole of electrons. Faraday's Constant is the Number of Coulombs of charge represented by that mole of electrons.

The Mass of Electrolytic Product

An electrolysis may produce several products at both the cathode and the anode. On the whole, one or more electropositive elements like metals or hydrogen are attracted to the negative cathode where they take up donated electrons to become electrically-neutral deposits. On the other hand electronegative elements like oxygen or halogens form at the anode. Where precious metals or other exotica are present in a melt or an aqueous solution they may drop to beneath the anode as an "anode slime".

As with thermal smelting, real electrolytic processes tend to move through a complex cascade of intermediate reactions and half-reactions until the final Hess's Law outcome is arrived at. This is especially true of electrolyses mediated by third chemicals, such as the electrolysis of alumina involving the active participation of cryolite. In thermal chemistry, these sub-processes are influenced by pathway Heats of Reaction. In electrochemistry, the stages of reaction are influenced by respective Electrochemical Potentials (J/mol).

At both electrodes, any gases formed tend to bubble out of the electrolyte and escape, unless they are captured for sale, or for use in a further stage of the chemical process.

Most designs of electrolyser that involve high-temperature melt usually arrange for the product metal to rest in a liquid state upon a conductive cathodic bath made of graphite or something. The whole gaseous atmosphere above the melt is then produced by probe anodes partially submerged in the molten electrolyte.

The Mass of the ith. Product, m_i, can be calculated using:-

$$m_i = \frac{\xi QM}{n_e F} = \frac{EIt}{F} = \frac{MIt}{zF}$$

Equation 7.7

where:-

m_i = Mass of the ith. Product
(grams or Kilograms)
ξ = Stoichiometric Factor
Q = Charge passed
M = Molar Mass
n_e = Number of Moles of Electrons involved
F = Faraday Constant
E = Equivalent Weight (grams or Kilograms)
I = Electric Current (amperes) [I^{+1}]
t = Duration of the (steady)
Electric Current (seconds)
z = Electrons Transferred per Ion

Water Electrolysis[7.1,7.2,7.3]

In the electrolysis of water the Molar Mass of diatomic hydrogen H_2 is 2.01588 (grams) and of (monoatomic) Oxygen 15.9994 grams.

In acidified water (PEM) the Ionised fluid experiences the following Half-Reactions (Kumar and Lim):-

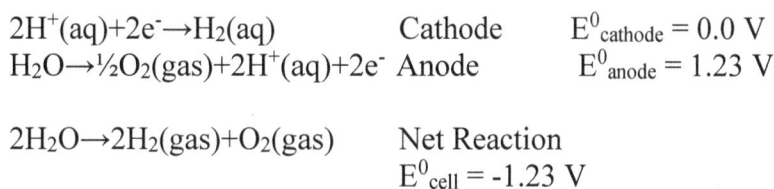

$2H^+(aq)+2e^-\rightarrow H_2(aq)$ Cathode $E^0_{cathode}$ = 0.0 V
$H_2O\rightarrow\frac{1}{2}O_2(gas)+2H^+(aq)+2e^-$ Anode E^0_{anode} = 1.23 V

$2H_2O\rightarrow 2H_2(gas)+O_2(gas)$ Net Reaction
E^0_{cell} = -1.23 V

Half-Reactions 7.1a

And in alkaline water these half-reactions:-

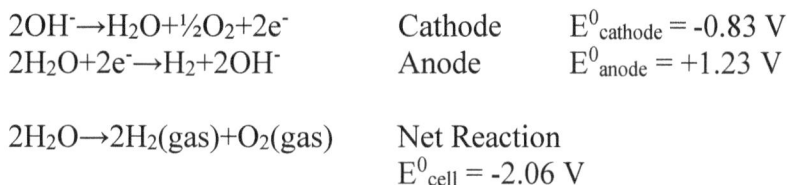

$2OH^-\rightarrow H_2O+\frac{1}{2}O_2+2e^-$ Cathode $E^0_{cathode}$ = -0.83 V
$2H_2O+2e^-\rightarrow H_2+2OH^-$ Anode E^0_{anode} = +1.23 V

$2H_2O\rightarrow 2H_2(gas)+O_2(gas)$ Net Reaction
E^0_{cell} = -2.06 V

Half-Reactions 7.1b

In general, the Cell Potential E^0_{cell} is calculated from:-

$$E^0{}_{cell} = E^0{}_{cathode} - E^0{}_{anode}$$
Equation 7.8

For standard water the Standard Anodic Electrode Potential, E_{rev}, is +1.23V. From the point of view of practical water electrolysis this is too low an estimate of potential difference (voltage) because of the existence of Overpotential, E_{ov}. This gives rise to an Operant Cell Potential, E_{op}, given by:-

$$E_{op} = E_{rev} + E_{ov}$$
Equation 7.9

For water the Overpotential is around 0.25 volts so that E_{op} is 1.48 volts.

In the context of water, the Valency of Hydrogen is 2 and that of Oxygen is 1. Accordingly E_{H2} is Molar Mass H_20/ Hydrogen Valency = 18.01528/2 = 9.00764 whilst E_O is Molar Mass H_2O / Oxygen Valency = 18.01528/1 = 18.01528

Thermodynamics of Water Electrolysis

Electrolysis is ultimately controlled by Gibbs Free Energy of Reaction, ΔG_r, and Enthalpy of Reaction, ΔH_r, as we discussed in the earlier elementary thermodynamics material.

Relevant basic equations include:-

$$\Delta G_r = n_e . F . E_{rev}$$
Equation 7.10

and:-

$$\Delta H_r = \Delta G_r + T . \Delta S_r$$
Equation 7.11

The electrolysis of water (or anything else) generates significant waste heat. According to Kumar and Himabindu[7.4] the ergo-material summary of water electrolysis is:-

$$1H_2O = \Delta H_{elec} + \Delta H_{heat} + H_2 + \tfrac{1}{2}O_2$$
Equation 7.12

where ΔH_{elec} is the Electricity Consumption, 237.2 KJ/mol, and ΔH_{heat} is the Heat Output, 48.6 KJ/mol. Accordingly, 17% of the energy input to the water electrolysis is waste heat which may assist efficiency of reduction a little, but is mostly to be removed by coolant systems.

Allow that the Total Input Energy, ΔH_{data}, is:-

$$\Delta H_{data} = \Delta H_{elec} + \Delta H_{heat} = 285.8 \; KJ/mol$$
Equation 7.13

and given that the Molecular Weight of Diatomic Hydrogen, MW_{H2} is 2.01588 we may specify the moles of Evolved Hydrogen, $mols_{H2}$, as:-

$$mols_{H2} = \frac{\Delta H_{elec}}{MW_{H2}}$$
Equation 7.14

Further given that $n_e = 2$; $F = 96485.33212331$ and $V_{rev} = 1.23$ we may calculate:-

$$\Delta G_r = n_e . F . V_{rev} = 237353.917023343 \; J/mol$$
Equation 7.15

and:-

$$\Delta H_{dataJmol} = 1000 \times \Delta H_{data} = 285800 \; J/mol$$
Equation 7.16

Therefore the Cell Efficiency, ξ_{cell}, is given by:

$$\xi_{cell} = \frac{\Delta G_r}{\Delta H_{dataJmol}} = 10^{-3} . \frac{n_e . F . V_{rev}}{\Delta H_{data}} = 0.830489562712885$$
Equation 7.17

Confirming our findings by simple arithmetic above.

The quoted Minimum Required Cell Voltage (Thermo-Neutral Voltage), V_{TN}, was 1.48. This can now be computed and confirmed using:-

$$V_{TN} = 10^3 \frac{\Delta H_{data}}{n_e.F} = 1.48105413387986$$

Equation 7.18

The Percentage Specific Defect between the quoted voltage of 1.48 and the computed V_{TN} is -0.071225262153016

The V_{TN} incorporates the Heat Energy element of the pathway to allow for Overvoltage.

The Use of Gibbs Free Energy in the Calculation of Minimum Required Cell Voltage and Thermo-Neutral Voltage[7.4]

The Minimum Theoretical Electrodes' Potential difference, V_{rev}, is given by:-

$$V_{rev} = \frac{\Delta G}{n_e.F}$$

Equation 7.19

where ΔG is the Change of Gibbs Free Energy of Reaction; n_e is the Number of Electrons Passed; and F is the Faraday Constant.

Accordingly the Cell Efficiency, ξ_{cell} is:-

$$\xi_{cell} = \frac{V_{rev}}{V_{TN}} = 0.830489562712885$$

Equation 7.20

which further tallies with our estimate, based upon Equation 7.12, that 17% of the applied current is "wasted" as heat.

Not only can Operant Voltage, V_{TN}, by related to Gibbs Free Energy Change, ΔG, but also to Reaction Entropy Change, ΔS, as specified below:-

$$V_T = \frac{T.\Delta S}{n_e F}$$

Equation 7.21

where V_T is the Cell Voltage at Temperature T (°K); ΔS is the Change in Gibbs Free Energy, the Reaction Entropy Change; n_e is the Number of Electrons Passed and F is the Faraday Constant.
And:-

$$\Delta S_{rev} = \frac{\Delta G}{T}$$
Equation 7.22

whilst:-

$$\Delta S_{TN} = \frac{\Delta H_{dataJmol}}{T}$$
Equation 7.23

Thus for V_{rev}:-

$$V_{rev} = \frac{T}{n_e.F} \times \frac{\Delta G}{T} = \frac{\Delta G}{n_e.F} = 1.23$$
Equation 7.24

and for V_{TN}:-

$$V_{TN} = \frac{T}{n_e.F} \times \frac{\Delta H_{dataJmol}}{T} = \frac{\Delta H_{dataJmol}}{n_e.F} = 1.48105413387986$$
Equation 7.25

Energy Equivalent of One Kilogram of Hydrogen: I
Power for One Kilogram is Known

Consider the case that we know how much Power to apply and for how long to gather m_{H2} grams of hydrogen, using the Applied Voltage as $V_{TN} = 1.48$
The Energy is 39.4 KWh.
Efficiency $\xi_{100} = 1.00$ *given* $V_{TN} = 1.48$
Then the Available Energy, E, is:-

$$E = \xi_{100}Pt = 1.00 \times 39.4 \times 1000 \times (60 \times 60)$$
$$= 1.4184 \times 10^8 \text{ } Joules$$
Equation 7.26

and the Power applied is:-

$$P = \xi_{100} \times 39.4 \times 1000 = 39400 \; Watts$$
Equation 7.27

The Electrolytic Current, I, is then:-

$$I = \frac{P}{V} = \frac{39400}{1.48} = 26621.6216216216 \; Amperes$$
Equation 7.28

Time, t, is 3600 seconds and Valency z = 2. The Passed Charge, Q, is therefore:-

$$Q = It = 95837837.8378378 \; Coulombs$$
Equation 7.29

Therefore the Moles of Electrons Passed, n_e, is:-

$$n_e = \frac{Q}{F} = 993.289194624478$$
Equation 7.30

We know that the Molecular Weight of Diatomic Hydrogen, MW_{H2} is 2.01588 (grams) and as stated the Valency is 2, so the Moles of H_2 released, $mols_{H2}$, are:-

$$mols_{H2} = \frac{n_e}{z} = 496.644597312239$$
Equation 7.31

Accordingly, the Mass of Hydrogen liberated in grams is:-

$$m_{H2} = mols_{H2} \times MW_{H2} = 1001.1759108298$$
Equation 7.32

We may gain the same information by direct application of Faraday's Second Law as:-

$$m_{H2} = \frac{Q.MW_{H2}}{z.F} = \frac{I.t.MW_{H2}}{z.F} = 1001.1759108298$$

Equation 7.33

and since we want the answer in Kilograms:-

$$m_{H2} = 10^{-3} \times \frac{Q.MW_{H2}}{z.F} = 10^{-3} \times \frac{I.t.MW_{H2}}{z.F}$$
$$= 1.0011759108298$$
Equation 7.34

Error is roughly one part in one thousand.

Energy Equivalent of One Kilogram of Hydrogen: II
Amount of Hydrogen (and Oxygen) Liberated by One KWh[7.5]

We know from the terms of the question that Power, P, is 1000 Watts, and it is implied that the Duration of Power, t, is 60×60 = 3600 seconds.

It is accordingly the case that the Available Energy at entry, E, is 3600000 Joules.

The Cell Voltage, V, is 1.48

Therefore the Current passed, I, is:-

$$I = \frac{P}{V} = 675.675675675676 \; Amperes$$
Equation 7.35

It follows that Charge passed, Q, is:-

$$Q = I.t = 675.675675675676 \times 3600$$
$$= 2432432.43243243 \; Coulombs$$
Equation 7.36

I have included all available digits at 15-figures so that you can check the work, not only better to comprehend the material, but also to assure that I have done my sums correctly!

As I remark in one of my other books, the trailing repeated digits are redundant and afford no further information.

Now we can compute the Number of Electrons passed, n_e, as:-

$$n_e = \frac{Q}{F} = 25.2103856503674 \; moles \; of \; electrons$$
Equation 7.37

Accordingly, the Moles of Hydrogen liberated, $mols_{H2}$, is:-

$$mols_{H2} = \frac{n_e}{z} = 12.6051928251837$$
Equation 7.38

and therefore the Moles of Oxygen, $mols_{O2}$, is:-

$$mols_{O2} = \frac{mols_{H2}}{z} = 6.30259641259186$$
Equation 7.39

The Mass of Water Divided, m_{H2O}, is:-

$$m_{H2O} = mols_{H2}(MW_{H2} + MW_O) = 227.086078199676 \ grams$$
Equation 7.40

where MW_O, the Molecular Weight of *Monatomic* Oxygen is 15.9994

The next task is to establish the number of moles of hydrogen in one kilogram of Hydrogen, $mols_{1KgH2}$. This is given by:-

$$mols_{1KgH2} = \frac{1000}{MW_{H2}} = 496.061273488501$$

Equation 7.41

To compute the Number of Kilowatt-Hours to form One Kilogram of Hydrogen, KWh_{1KgH2} we may write:-

$$KWh_{1KgH2} = \frac{mols_{1KgH2}}{mols_{H2}} = 10^3 \frac{z}{MW_{h2}n_e} = 10^3 \frac{F.z}{MW_{h2}Q}$$
$$= 10^3 \frac{z}{MW_{h2}.I.t}$$
Equation 7.42

The value of KWh_{1KgH2} is 39.3537235303079 Kilowatt-Hours, which is the amount of electricity required to evolve One Kilogram of Hydrogen.

The Smelting of Aluminum

The Hall-Héroult Process[7.6]

During the Nineteenth-Century Aluminum was rated a precious metal despite being one of the most common metals in the Earth's Crust and indeed the Earth's Mantle.

That was because being so electropositive as an element, the metal was exceedingly expensive to smelt by traditional means. Aluminum element was first isolated in 1824AD by the Danish scientist Hans Christian Ørsted who reacted Aluminum Chloride $AlCl_3$ with potassium amalgam. Potassium and sodium are even more electronegative than aluminum, and so should be able to thermochemically reduce aluminum from its ore.

Davy had isolated sodium and potassium as early as 1807AD by the electrolysis of their hydroxides. But this depended upon the destructive supply of electric current from zinc-acid cells. So the whole aluminum production process was horrendously wasteful and expensive: Hence the price.

The Danish king ordered his crown cast of aluminum, perhaps a back-handed tribute to his brilliant subject Ørsted.

Aluminum was more precious than Gold and it is said that the French king, presumably Napoleon III, awarded medals of aluminum and for sure he commissioned a dinner service made of the metal. In Britain, the 1893AD sculpture of the winged archer Anteros that graces Piccadilly Circus in Mayfair, London was cast of it. By then, steam-driven dynamos could provide the current for alkali metal production, preceding the reduction of Aluminum.

But the economic production of the metal had to await the American Charles Martin Hall and his French contemporary Paul (Louis-Toussaint) Héroult.

Both men were 22-year-olds when they simultaneously discovered the electrolytic process for smelting aluminum which ever since bore their names.

The Hall-Héroult Process involves placing a charge of Aluminum Oxide (Alumina) Al_2O_3 in a bath built from refractory ceramic, together with an active flux called Cryolite Na_3AlF_6. Cryolite was itself a rare and expensive mineral, but is now made synthetically and is much cheaper.

The bath is lined with graphite carbon which forms a cathode to which metallic ions migrate, whilst the anode above is also graphite and actively takes part in the reduction reaction. Oxygen gas is evolved.

An immense electric current is applied, 100-300 KA at 5 volts, though my post-war Lowry and Cavell[7.7] specifies 40000 Amperes at 4.5 Volts.

Besides the purely electrochemical reduction of the melt this discharge raises the temperature of the mix to some 1010°C or about 1283°K. The actual alumina reduction is endothermic so external heat has to be applied beyond the ionisation of the liquid. This extra heat is provided by $I^2\Omega$ electrolyte resistance and also cell current heating.

(Electricians usually write I^2R, but we prefer to avoid confusion with R, the Universal Gas Constant).

The primary reaction is:-

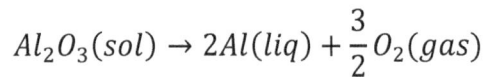

$$Al_2O_3(sol) \rightarrow 2Al(liq) + \frac{3}{2}O_2(gas)$$
Pathway 7.1

You can see that Aluminum is trivalent, so that z = 3 in subsequent calculations of Aluminum yield.

In the case of the Hall-Héroult Process it is also the case that the electrodes are consumed according to:-

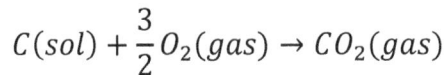

$$C(sol) + \frac{3}{2}O_2(gas) \rightarrow CO_2(gas)$$
Pathway 7.2

Therefore, it is the furnacemen's practice to float coke on the surface of the molten aluminium product in order to conserve the expensive prebaked anodes, which in any event need frequent or indeed continuous replacement. At the working temperature cryolite is lighter than aluminum, so it tends to float. Alumina is periodically added to the bath to maintain a percentage ratio to the cryolite of around 2% to 6%.

The net endothermic reaction is given by:-

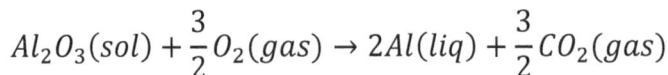

$$Al_2O_3(sol) + \frac{3}{2}O_2(gas) \rightarrow 2Al(liq) + \frac{3}{2}CO_2(gas)$$
Pathway 7.3

In the costing of this Process the anode-cathode interdistance (in meters) is critical, because it determines the heating resistance.

Aluminum Metal Yield By Faraday's Second Law

The Number of Passing Electrons, n_e, in an electrolysis is given by:-

$$n_e = \frac{Q}{z.F} = z.N_A.q_e$$

Equation 7.43

where:-

Q	Charge (Coulombs)
z	Valency (depends upon participating Al atoms)
F	Faraday Constant
N_A	Avogadro's Number (Avogadro Constant)
q_e	The Charge on the Electron

In the context of the Hall-Héroult Reaction $z = 3$ whilst F, N_A and q_e are all universal constants respectively 96485.33212331 (964853.321233100184) Coulombs per mol, 6.02214076×10^{23} and $1.602176634\times10^{-19}$ Coulombs.

Faraday's Second Law specifies:-

$$m_x = n_e mols_x AT_x = \frac{Q.mols_x.AT_x}{zF} = \frac{I.t.AT_x}{zF}$$

Equation 7.44

where:-

m_x	Deposited Mass of Element x (grams)
$mols_x$	Deposited Moles of Element x
AT_x	Atomic Weight of Element x (grams $\equiv M^{+1}$)

For example, a small aluminum smelter has the penstock gate power of 82MW whilst the applied cell voltage is set to 4.5V. The Atomic Weight of Aluminium is 26.9815384.

How many tonnes of aluminum can it smelt in one hour of operation?

First of all note that 82MW is 82000000 Watts and that 1 hour is 60×60 = 3600 seconds.

Further noting that:-

$$P = IV$$
Equation 7.45

so that:-

$$I = \frac{P}{V} = \frac{82000000}{4.5} = 18222222.2222222 \; Amperes$$
Equation 7.46

In practice, this huge current will be divided in a large number of electrically-parallel channels to assist management. In the case of our example smelter there are eighty reduction cells, so the indicated Cell Current is 227777.778 amperes per cell, and this division enables several cells at a time to be closed for maintenance or in response to market demand.

We may invoke Faraday's Second Law to compute the Mass produced as:-

$$m_x = \frac{I.t.AT_x}{zF} = \frac{18222222.2222222 \times 3600 \times 26.9815384}{3 \times 96485.33212331}$$
$$= 6.11488095115474 \times 10^6$$
Equation 7.47

This result is in grams, so you must multiply by 10^{-6} to render it to metric tonnes:-

$$m_x = 10^{-6}.\frac{I.t.AT_x}{zF}$$
$$= 10^{-6}.\frac{18222222.2222222 \times 3600 \times 26.9815384}{3 \times 96485.33212331}$$
$$= 6.11488095115474$$
Equation 7.48

Concluding that the smelter produces 6.11488095115474 metric tonnes of aluminum per hour.

Note that I have neglected heating losses and side-reactions, as well as a good deal else.

Realistic hydroelectric aluminum smelter Process Efficiencies ξ range from 0.84 to 0.95 so that the real tonnage produced is around 5.8 at the higher limit.

Will these 5.8 tonnes of Aluminium fit on a lorry (road truck)?

Well a standard European 44-tonne truck will surely bear the load or in theory six hours of production but as a guide what is the size of a cube of aluminium?

First we need to know that the Density, ρ, of a cool, solid lump of aluminum is 2699 kg/m^3 whilst 5.8 metric tonnes (m) is 5800 kilograms.

The Volume (in cubic meters) of 5.8 tonnes of aluminum is:-

$$V = \frac{m}{\rho} = 2.14894405335309$$

Equation 7.49

so that the Cubic Side Length, l, is given by:-

$$l = \sqrt[3]{2.14894405335309} = 1.29045154822112$$

Equation 7.50

Since a cube of 5.8 tonnes of aluminum has a side of about 1.3 meters it will easily fit on a truck, and if the truck is at least eight meters long six hours' worth should engage one truck to the customer.

One of the well-recollected summer's evenings of my boyhood found me contemplating the tidal rapids at Coalasnacon in Scotland. At this point the sea fiord, Loch Leven, is some 58 meters wide.

As I watched a swarm of jellyfish helplessly glide landward I was astonished to see a small steamship, a Clyde puffer, shoot effortlessly toward the head of the inlet. I marvelled at both the seamanship and the courage of the helmsman, and indeed the other man aboard. As a twelve-year-old I had no idea why they took such a risk, but as an old man I know that they were taking supplies to the aluminum smelter at Kinlochleven, which as its Gaelic name *Ceann Loch Lìobhann* implies stands on the level patch of till at the loch head.

The skipper must have closed his regulator as only the lightest wisps of tawny smoke rose from the funnel, and the little ship glided silently with the tide. Somehow the still, steeped Highland air seemed to drift with the little ship.

The smelter at Kinlochleven commenced production of aluminum in 1907AD and closed in 2000AD. It employed 700 in a village whose current population is about 760.

By the way, Clyde Puffers[7.9] were so called because they exhausted uncondensed steam to the atmosphere like railway locomotives and were correspondingly inefficient, but in sober hands were reliable and easily-maintained by a crew of two or three. All puffers were built on the Forth and Clyde Canal, either at Kirkintilloch or Maryhill, suburbs of Glasgow, and had to be less than sixty meters long in order to negotiate the canal locks, and, if required, enter the Clyde river and steam seaward. In later days, some of the seagoing puffers were fitted with condensing engines or even diesel.

Many of the communities of the Scottish Western Isles depended upon the beachable puffers for their supplies and the small ships probably prevented many islands from becoming uninhabited as their people might have otherwise been obliged to migrate to North America or Glasgow. The puffers have won a mystique, not to say romance, of their own and have entered British folklore.

The cost of electric power in Britain has always been extremely high. Until this century almost all was generated by burning coal or oil, though efforts are underway to develop cost-effective wind power and a viable form of nuclear power supply. In the early Twentieth-Century the Government required a sovereign, strategic aluminum source, perhaps involving Irish bauxite, a source of alumina. Very cheap hydroelectric power was obtainable in the Scottish Highlands and North Wales, provided the necessary high-level reservoirs could be built, and they quickly were. Amongst the bonuses were that the sites were far away from potential Continental enemies; that in remote places, spies could be quickly identified; and that, in the Second World War, these smelters were out of Nazi bomber range and their location in narrow ravines in any event made their destruction by aerial bombing almost impossible.

Photo: JM Briscoe (adapted by the author)[7.8]

Figure 7.1
The Clyde Puffer *Eilean Eisdeal* liveried as the "Vital Spark"
Possibly for a Television Production

My Late Mother was directed to work in the laboratories of several chemical companies during the Second World War. Due to her scientific skills she was not allowed to join the Royal Navy which was her desire, later fulfilled. At one stage she worked at the Booth Aluminium Company assisting the development of aviation Duralumin, a light but tough alloy which is around 95% Al (aluminum), 4% Cu (copper), 0.5% Mn (manganese) and 0.5% Mg (magnesium). The magnesium was extracted from seawater near Maryport in Cumberland. Lowry and Cavell remarks that duralumin is as "strong as boiler steel" but only one third its weight.

CHAPTER EIGHT
THE IRON SMELTING BLAST FURNACE

The Iron smelting Blast Furnace is a tower oven, a fusiform structure made of a steel carcass lined with refractory bricks and having at its base a refractory receptacle (hearth) to receive liquid metal and any slag or other waste which may float atop.

According to the Britannica[8.1] a typical blast furnace has a diameter, d, of ten meters and a height, h, of 32 meters. This is a very approximate statement, and since the girth of a blast furnace varies along its height one might assume that the compilers mean the greatest width, at the bosh.

Taking the Britannica dimensions at face value the Radius, r, is five meters and the height/radius ratio, s, is 6.4

Modelling the internal blast furnace as a right cylinder the Volumetric Capacity of such a structure, V, is given by:-

$$V = \pi.r^2.h = \pi.r^2.r.s = 2513.274 \; cubic \; meters$$
Equation 8.1

This value of volume is around half of the actual capacity of a modern furnace which is about 5000 m^3, whilst the largest (2024) iron blast furnace is 6000 m^3 volumetric capacity.

Transposition of Equation 8.1 gives:-

$$r = \sqrt[3]{\frac{V}{\pi s}}$$
Equation 8.2

Assuming similarity, for a 5000 m^3 furnace the indicated height is h = s×r = 40.25 meters and the indicated diameter d = 2×r = 12.58 meters. For the 6000m^2 furnace h = 42.77 meters and d = 13.37 meters.

I stress that these are approximate figures because a blast furnace is not a right cylinder: It is a series of stacked frustra, and in any case the reactors vary enormously in size and design.

An Iron Smelting Blast Furnace[8.2] is designed to operate at 1950°K to produce liquid iron, slag (mainly calcium and silicon oxides), and Blast Furnace Gas.

The slag and gas may be regarded as waste products but the exhaust gas is a mixture of Nitrogen (N_2), Carbon Monoxide (CO), Carbon Dioxide (CO_2) and Hydrogen (H_2). Carbon Monoxide and Hydrogen participate in the chemical reduction of the Iron Ore, but the remaining Blast Furnace Gas (BFG) is fed into Cowper Stoves and burnt in order to recover some of the expensive waste heat by heating the Hot Blast, which is basically atmospheric air pumped into the Blast Furnace at 1200°K to 1550°K in order to oxidise the carbon fuel and to augment the speed and thoroughness of iron production.

The slag may itself be saleable as a chemical feedstock or an agricultural fertiliser, whilst the coke fuel used by the furnace is made by roasting coal, which also produces saleable waste.

A key advantage of the Blast Furnace as opposed to some other smelting technologies is that it outputs *liquid* metal relatively free of gaseous contaminants which can vitiate the quality of the metal product, and necessitate expensive post-processing.

Besides coke, which is about 86.7% carbon and thus a high-class material, some modern furnaces are supplemented with up to 3% natural gas (Methane: CH_4) or even petroleum and coal tar. Use of hydrocarbon wastes such as these is poor practice.

No iron ore decomposes to iron spontaneously. Six of the nine intermediate furnace reactions involving iron are endothermic (that is they need to absorb environmental heat) as are all of the six strongly-endothermic slag reactions.[8.3]

Accordingly, heat must be supplied to the process, principally by the combustion of coke.

The main Fuel Semi-Combustion Pathway is:-

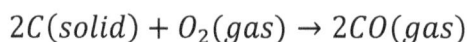

$$2C(solid) + O_2(gas) \rightarrow 2CO(gas)$$
Pathway 8.1

which yields Carbon Monoxide, a combustible and a powerful reducing agent, especially when hot. The Heat of Reaction, ΔH_r, is +209.2 KJ/mol, a respectable amount of heat *output*, because the positive figure indicates an Exothermic Reaction.

The Main Hematite-Carbon Monoxide Reaction for hematite ore is:-

$$Fe_2O_3(solid) + 3CO(gas) \rightarrow 2Fe + 3CO_2$$
Pathway 8.2

Essentially an extension of Pathway 8.1 the ΔH_r for this is -40.56 KJ/mol indicating a modest endothermic demand.

The Hematite-Carbon Gross Reaction is:-

$$2Fe_2O_3(solid) + 3CO(gas) \rightarrow 4Fe + 3CO_2$$
Pathway 8.3

which according to my calculations has a ΔH_r of -785.64 KJ/mol indicating a strongly Endothermic Reaction that absorbs heat that must be provided from adequate external sources such as Pathway 8.1

Other reactions evolve hydrogen H_2 so that we may summarise the reducing agencies as Carbon Monoxide 60%, Carbon element 35% and Hydrogen 5%.

Furnace Anatomy

The Blast Furnace includes at its base a sump into which liquid iron collects with the slag floating on top. Holes in the wall of this sump exist or can be drilled to allow iron and slag products to be tapped under gravity at their respective base levels. Above the sump is a widened, bulbous volume of the tower called the bosh.

This sump is often termed "the hearth" though no actual combustion takes place within it.

It is in the bosh that the major reactive activity takes place, the carbon being oxidised, and the iron ore reduced by complex cascades of reactions between Hematite (Fe_2O_3) and its derived Wüstite (FeO) with the powerful reducing gas Carbon Monoxide (CO), itself a product of the partial combustion of carbon.

Above the bosh is the cooler shaft of the furnace where dehydration and low-temperature changes take place.

At the base of the furnace, but above the level of interference by ponding liquids, are arranged a ring of gas input ports called tuyeres (pronounced "twee-ers") through which the oxidising hot blast is admitted: Compressed air sometimes supplemented with hydrocarbons.

At the top of the furnace, Blast Furnace Gas is abstracted to fuel the Cowper Stoves. This material is a mixture of nitrogen (from the compressed air), unconsumed carbon monoxide, carbon dioxide waste and hydrogen. The CO and the H_2 are the only flammables and research is current to separate and vent the unburnable nitrogen and carbon dioxide, because it is not usually practicable to sell them.

The rôle of the external Cowper Stoves is to heat the Hot Blast.

Figure 8.1, which I have adapted from Girprit Singh[8.4] shows a schematic section of a typical iron smelting blast furnace with isotherms.

All figures are Degrees Kelvin. You can see that the isotherms, which are three-dimensional, describe a graded series of subconical frustra from 2273°K at the main locus of reactions in the basal raceway to a surprisingly cool 473°K near to the BFG exhaust portal.

This latter confirms that heat transfer in the Cowper Stove is not a matter of sensible heat: It is rather a matter of the actual combustion of waste gas which implies a further consumption of ambient air.

Ironworks Conformation

It has always been the case that no blast furnace can be viewed as a chemical reactor in isolation, because it requires ancillary systems to load fuel, flux and ore; to furnish the blast; and to abstract liquid iron and slag.

The production of *liquid* iron is the great strength of the Blast Furnace, because the physical homogeneity of this product minimises the contamination by extraneous foreign matter, especially vesicular gas. Therefore, the most expensive and inconvenient efforts to improve the purity and fabric of the iron are abridged.

Figure 8.2 is a schematic representation of a complete ironworks (without steel-making and steel semi-fabrication post-smelting factories)[8.5]

In best practice Cowper Stoves are provided in sets of three to allow for repair and maintenance.

Solid charges are conveyed to the head of the structure by cableway trolleys that raise ore, coke and limestone in the respective proportions 3.062, 1 and 0.186

Slag is conveyed hot to a dump ready for landscaping or supply to customers using ground-based railway tubs, and iron is shunted in its liquid state to steel plants using thermally-insulated railway torpedo cars.

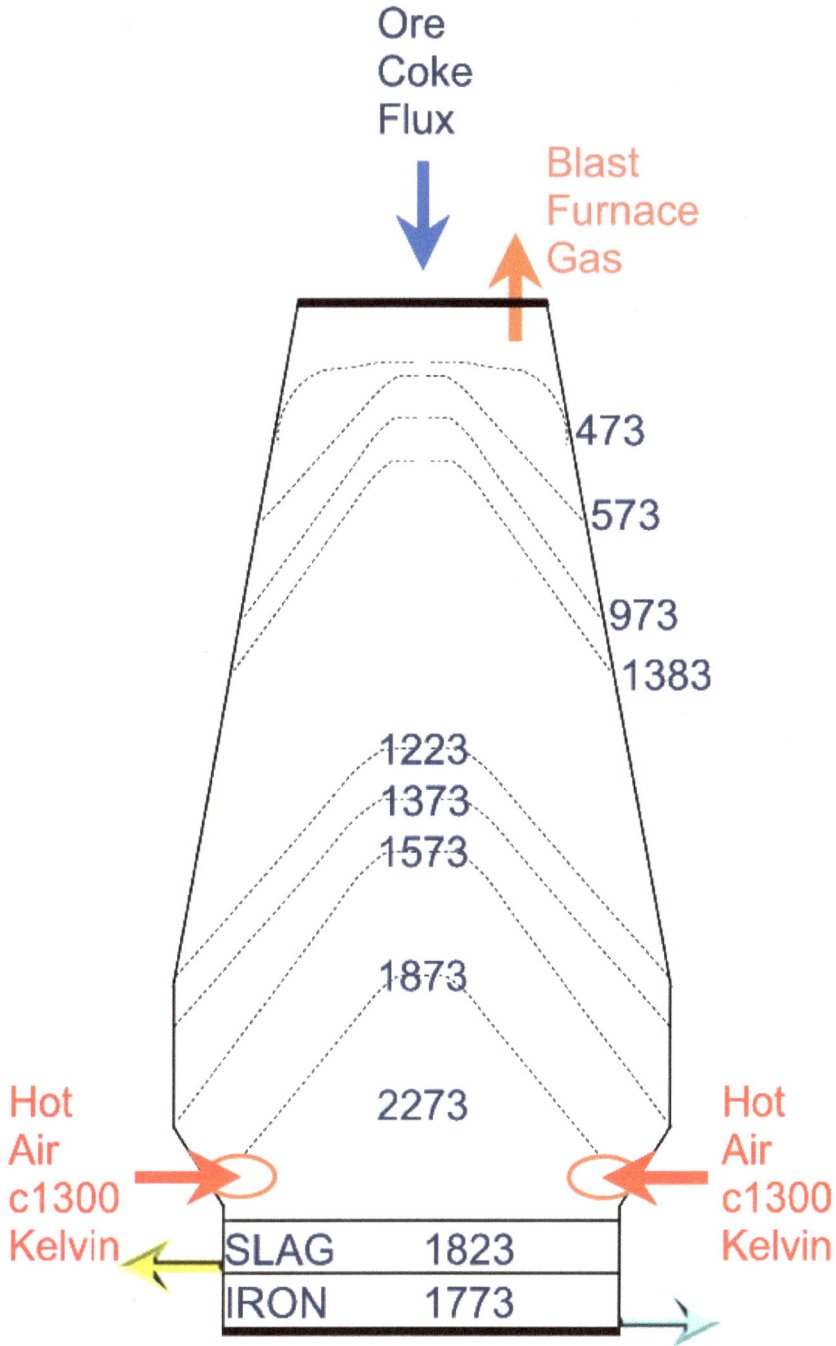

Figure 8.1
Schematic Section of a
Typical Iron Smelting Blast Furnace

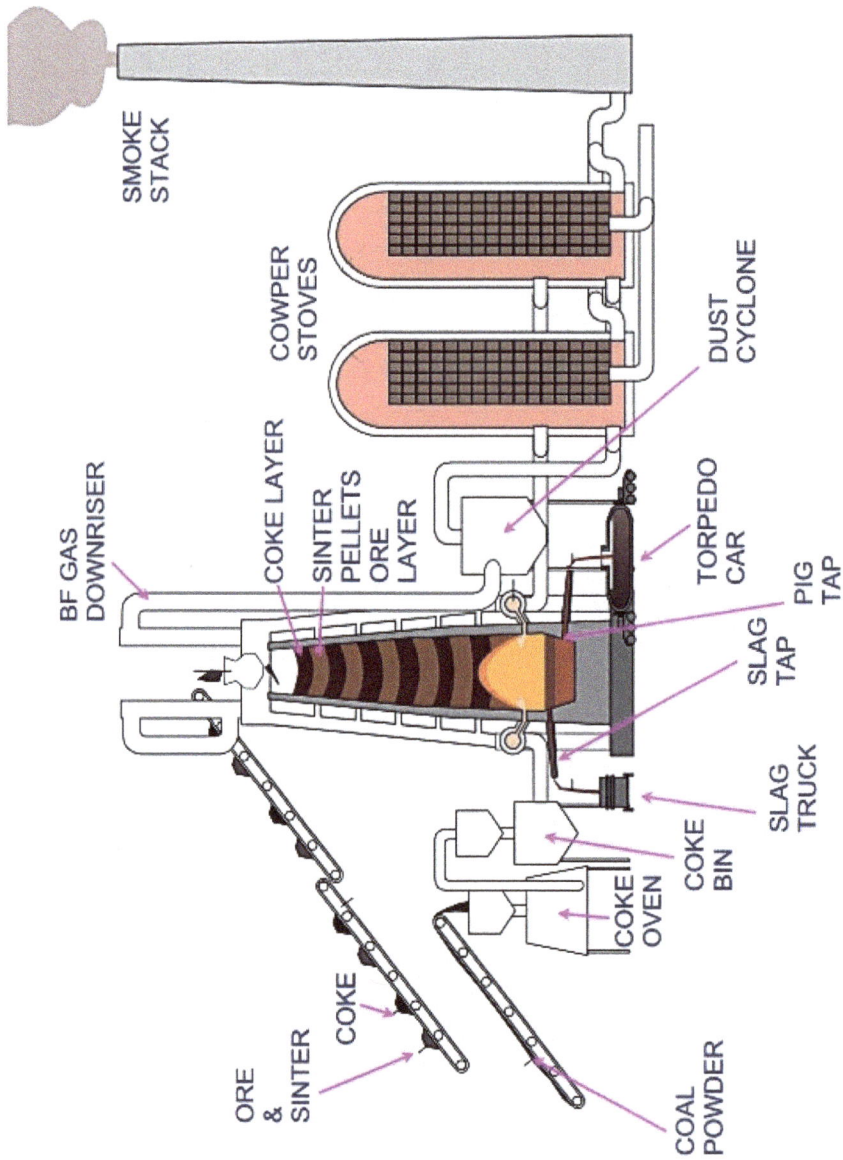

Diagram: Tosaka[8.5] (adapted by the author)

**Figure 8.2
A Blast Furnace
amidst its
Service Installations**

Tosaka's diagram also shows provision for the addition of sinter and raw coal which are fed in some ironworks, as well as dust traps to prevent clogging of the Cowper Stoves' checkerwork. Finally, there is a tall smokestack to discharge the spent gases.

Charge Ratios[8.7,8.8]

Charge ratios vary enormously according to the type and quality of the input, and the modernity and other design aspects of the furnace system.

Table 8.1 presents a selection of ore-coke-flux-air charge ratios from various authorities in regard to various built furnaces or traditions.

Weirdly, only two contributors remember to include air in their advice, even though smelting with fossil fuel would be impossible without it.

Equally strange on the part of some is to neglect whether they write gravimetrically of volumetrically, or to specify the type and condition of the materials.

On the whole, 7 of Ore to 2 of Coke to 1/50 of Flux to 10 of Air (tonnes) is probably about right.

Porosity

As aforenoted, the permeability of the stacked melt is critical to the oxygenation and reduction mechanisms of a blast furnace. It is especially the case when the mix is semi-molten that the mass may congeal and stifle the fire. Flux can in part mitigate this by lowering the viscosity of slushes so that slag and liquid metal drains freely.

But the porosity of the stacked materials still has to be optimised by proper screening and layering.

The use of pre-prepared feed pellets can be of great assistance. The mix of ore, coke and flux can be optimised from the chemical viewpoint as can the fabric and friability of the charge, and, crucially, the size distribution and three-dimensional geometry of the pellet.

The pellets are made spherical or cylindrical. V_x is the Volume of State x. Porosity, ε, is defined as the ratio:-

$$\varepsilon = \frac{V_{gas}}{V_{gas} + V_{liquid} + V_{soild}} = 1 - \left(V_{gas} + V_{liquid} + V_{soild}\right)$$

Equation 8.3

L-Furnace — Meng, Shao and Zou

Proportions				Ore	Coke	Flux	Air	
Ore	Coke	Flux	Air					
182	2	91		7	2	0.019231		
91	7	13		364	104	1		
13	13	1		3.5	1	0.009615		104 recip

Algoma No.1 Furnace — Meng, Shao and Zou

Proportions				Ore	Coke	Flux	Air	
Ore	Coke	Flux	Air					
44	2	22		11	3	0.083333		
22	2	11		132	36	1		
11	11	1		3.67	1	0.041667		24 recip

Mean — Meng, Shao and Zou

Ore	Coke	Flux	Air	
3.5	1	0.009615		
3.67	1	0.041667		
Mean	3.58	1	0.025641	39 recip

Epa.gov — 12.5 Iron And Steel Production
https://www3.epa.gov/ttnchie1/ap42/ch12/final/c12s05.pdf   Nov-86

Tons				Ratios				
Ore	Coke	Flux	Air	Ore	Coke	Flux	Air	
1.4	0.65	0.25	2	2.15	1	0.384615	3.08	2.6 recip

For 1 ton of Iron

Brain and Lamb — "How Iron and Steel Work"
Marshall Brain and Robert Lamb
https://science.howstuffworks.com/iron3.htm

Tons				Ratios				
Ore	Coke	Flux	Air	Ore	Coke	Flux	Air	
2	1	0.5	5	2	1	0.5	5	2 recip

World Steel Organisation
https://worldsteel.org/steel-topics/raw-materials/

Tonnes				Ratios			
Ore	Coke	Flux	Air	Ore	Coke	Flux	Air
1.6	0.45			3.56	1		

Table 8.1
Ore-Coke-Flux-Air Charge Ratios

Porosity is maximised for *geometrically-congruent convex hulls* by the sphere.

Allow that Sphere Radius, r, is unity. ε_{sphere} is Porosity due to Sphere Containment in 3D Cartesian Space; V_{sphere} is the Volume of a Sphere and V_{cube} is the volume of the Containing Cube:-

$$\varepsilon_{sphere} = 1 - \frac{V_{sphere}}{V_{cube}} = 1 - \left(\frac{4}{3}\pi r^3 \times \frac{1}{(2r)^3}\right) = 1 - \frac{\pi}{6} \approx 0.4764$$

Equation 8.4

So a solid charge that is exclusively uniform spheres leaves more than half the stack free for gas permeation.

As with Charge Ratios, stack Porosities vary from furnace to furnace, but also they vary dynamically in time and its analog, charge position within the stack.

Blast Furnace Feedstocks

Classical Coke from Coking Coal

Coke[8.9] is made by anoxically roasting suitable "coking" coal in specialist coking ovens (Pyrolysis), or using suitable by-product from town-gas retorts, which latter is almost universally obsolete.

Coke is a grey, vitreous, hard, light, highly-vesicular substance clean to handle. Metallurgical coke is at least 82% carbon the remainder being mineral ash. Coking coal is, correspondingly low in ash, moisture, gas and volatile hydrocarbons.

The Heat of Combustion of Coke is 19 MJ/Kg[8.9] and its Mass Density is 0.85 tonnes/m^3, and therefore coke floats on water.

Furnace coke functions both as an exothermic charge heating fuel, and as a reducing agent for the iron ore.

It is crucial that metallurgical coke has sufficient crushing strength to maintain the integrity of the furnace burden.

High-Reactivity Coke

This is metallurgical coke coated with iron oxides or calcium oxides in order to improve its characteristics as a chemical reduction agent.

Classical Lump Ore

This is Iron Ore in pebble or cobble form, screened or unscreened, as received from the mine. It is of random morphology, and hence its porosity characteristics are widely variable, thus conditioning its speed and completeness of reaction.

Most iron ore used today is Hematite (Fe_2O_3) with a density of 5300 kg/m^3; Magnetite ($Fe^{2+}Fe_2^{3+}O_4$) with a density of 5175 kg/m^3; or Limonite ($FeO(OH).nH_2O$) with a density of 3500 kg/m^3. In former times, Siderite "blackband" ($FeCO_3$), a coal stratum seat earth was highly important, especially as it was laminated with coal, and upon calcining presented an ore 70% to 80% iron. Pure siderite is of density 3960 kg/m^3, but is a low-grade feed with less than half its mass iron.

Lump ore should not exceed 30 mm across and 3 mm is often desirable.

Sinter

Sinter is iron-oxides-rich dusty waste material including actual furnace dust and dusts from steel fettling and other processing sources. It can include "blue iron" dusts from natural deposits and miscellaneous highly-ferruginous materials.

After compaction, it is often dumped back into the furnace.

Classical Anthracite

Anthracite is a natural coal, often altered tectonically, and is 86% to 97% carbon. It is the traditional maritime steam-raising coal with a heat of combustion of 26-33 MJ/Kg and a density of 1350 kg/m^3, and therefore sinks in water.

Where it is abundant it is occasionally used as a coke supplement but is falling out of favor.

Pulverised Coal Injectant (PCI)

This is finely comminuted coal dust of the correct quality sometimes blown into furnaces to augment heating and reduction. As a material it is highly reminiscent of the comminuted coal blown into the former coal-burning power stations of Britain. Like other furnace coals and cokes it should be low in ash and volatiles.

Limestone

Limestone ($CaCO_3$) is the common blast furnace flux. A flux is designed to lower the melting temperature of the burden, to purify the iron, and lower the viscosity of slag and iron liquids.

Formerly, Flourite (CaF_2) was used as a *steel* melting flux, especially in wartime Britain.

Pelletised Ore Mix

This is a pre-prepared optimised mixture of coke, ore and flux formed into spheres or sub-cylindrical pellets of consistent size and geometry. POM has found great favor in recent decades because of its consistent, predictable, characteristics and calculable porosity.

Hot Compressed Air

This is the definitional Hot Blast Furnace feed, and will be discussed in detail later. It has both heating and (indirectly) reductive functions.

Oxygen

This is, like Hot Compressed Air, definitionally an oxidising agent, but paradoxically it has been found to promote reaction of iron ore in a quite drastic fashion.

Methane

Methane gas (CH_4) (sometimes known as "natural gas" or "firedamp") is sometimes injected into the lower levels of furnaces to promote heating and reduction, both of which can be significantly assisted as long the methane is restricted to around 2% of the gaseous input.

At high concentrations, methane precipitates carbon soot which impedes blast furnace processes.

Re-cycled Blast Furnace Gas

BFG is highly reducing, especially if it can be scrubbed for carbon dioxide. Excess BFG spared from Cowper Stove duties can be recycled to the furnace.

Petroleum

Hydrocarbon liquids such as heptane (C_7H_{16}) or octane (C_8H_{18}) are sometimes added to fluid injectants. Such additives are frowned upon in Europe, not least because they are expensive and highly-taxed!

Coal Tar Liqors

These include moist and highly-variable liquid distillate products of coal coking and gasification. They may be rated as benzene (C_6H_6) or even phenol (C_6H_5OH). Essentially these are noxious waste and affect the melt in dubious and unpredictable ways, but may be desired in markets where there is little demand for aromatic chemicals, or where the alternative is environmental dumping.

Operating Conditions[8.9]

In his 1967AD paper Iwao Muchi lists the parameters of two contemporary blast furnaces, A and D from which many thermal, charge and product characteristics may be inferred.

By aggregating Muchi's figures for Sinter, Ore and Metal, pellet; as "Ore": And also Limestone and Silica as "Flux" we may average the results for Furnaces A and D to arrive at the charge ratios Ore:Coke:Flux as 32311:10553:1964.5 or 3.062:1:0.186

I have rendered everything in SI.

Table 8.2 represents a synopsis giving the Muchi data for the mean of Furnace A and Furnace D values only.

The Standard Product, σ, in kilograms is given by:-

$$\sigma = \frac{Pig\ Iron\ Production}{Charge\ Frequency} = \frac{37.5578704}{0.00183449} = 20473.19$$

Equation 8.5

From which the Fractional Yield of Iron, y_{Fe}, is:-

$$y_{Fe} = \frac{\sigma}{All\ "Ore"} = \frac{20473.19}{32311.00} = 0.633629$$
Equation 8.6

The Energy Balance

Elaborate models of heat accounting for the blast furnace have been developed by thermodynamic furnacemen over the post-war period[8.3,8.4,8.10], and these were anticipated by the empirics of industrial craftsmen through the centuries. Intercomparison of real furnace behaviors with theoretical models has been essential in all investigations.

Ayush Bhattacharya and Sadhasivam Muthusamy[8.3] provide the synopsis of Table 8.4

Table 8.2
Operating Characteristics of Two Blast Furnaces

VARIABLE	UNITS	FURNACE A Mass (Kg)	Volume (m³[bed])	Diameter (meters)	FURNACE D Mass (Kg)	Volume (m³[bed])	Diameter (meters)	MEANS Mass (Kg)	Volume (m³[bed])	Diameter (meters)	
Blast											
Volume Rate at STP	m³/s	50.45			49.36666667			49.9083333			
Temperature	°K	1366.15			1195.15			1280.65			
Pressure	Pascal	189170.2785			176029.3675			182599.823			
Humidity at STP	kg/m³	0.0225			0.0355			0.029			
Injected Oil at STP	kg/m³	0.04193			0			0.020965			
Charge											
Sinter		21000	12.22	0.0185	26284	14.934	0.0175	23642	13.577	0.018	
Ore		12400	4.98	0.0218	4738	2.051	0.0211	8569	3.5155	0.02145	32311.00 All "Ore"
Metal, pellet		200	0.1	0	0	0	0	100	0.05	0	
Limestone		1700	1.04	0.0331	2026	1.213	0.0185	1863	1.1265	0.0258	1964.50 All "Flux"
Mn Ore		450	0.24	0	100	0.387	0.0219	275	0.3135	0.01095	
Silica		0	0	0	203	0.119	0.0151	101.5	0.0595	0.00755	
Coke		10100	21.1	0.0053	11006	15.951	0.0563	10553	18.5255	0.0308	10553.00 All "Coke"
Sum		45850	39.68	0.0787	44357	34.655	0.1504	45103.5	37.1675	0.11455	
Charge Frequency	s⁻¹	0.001875			0.001793981			0.00183449			20473.19 Fe Product./Charge Freq.
											0.633629 Yield
BFG (Tar Gas)											
Temperature	°K	482.15			482.15			482.15			
Pressure	Pascals	39226.6			58447.634			48837.117			
CO, CO₂ and H₂	vol%	23.6, 18.4, 4.4			23.6, 18.6, 2.7			23.6, 18.5, 3.55			
Pig Iron											
Production	kg/s	40.45138889			34.66435185			37.5578704			
Temperature	°K	1723.15			1650.15			1686.65			
Si, Mn, P, S	wt%	0.58, 0.184, 0.038			0.76, 0.122, 0.28, 1			0.67, 0.153, 0.16			
Frequency of Tapping	s⁻¹	0.0000925926			0.000810185			8.6806E-05			
Slag											
Production	Kg/s	11.57407407			11.9212963			11.7476852			
CaO, MgO, SiO₂, Al₂O₃, and S	wt%	39.54, 33.5, 15.9, 4.24, 0.98			43.4, 34.8, 14.4, 6.29, 0.77			41.47, 34.15, 15.15, 5.265, 0.875			

VARIABLE	UNITS	MEANS			
Blast		Mass (Kg)	Volume (m³(bed))	Diameter (meters)	
Volume Rate at STP	m³/s	49.9083333			
Temperature	°K	1280.65			
Pressure	Pascal	182599.823			
Humidity at STP	kg/m³	0.029			
Injected Oil at STP	kg/m³	0.020965			
Charge					
Sinter		23642	13.577	0.018	
Ore		8569	3.5155	0.02145	32311.00 All "Ore"
Metal, pellet		100	0.05	0	
Limestone		1863	1.1265	0.0258	1964.50 All "Flux"
Mn Ore		275	0.3135	0.01095	
Silica		101.5	0.0595	0.00755	
Coke		10553	18.5255	0.0308	10553.00 All "Coke"
Sum		45103.5	37.1675	0.11455	
Charge Frequency	s⁻¹	0.00183449			20473.19 Fe Product./Charge Freq.
					0.633629 Yield
BFG (Tor Gas)					
Temperature	°K	482.15			
Pressure	Pascals	48837.117			
CO, CO₂ and H₂	vol%	23.6	18.5	3.55	
Pig Iron					
Production	kg/s	37.5578704			
Temperature	°K	1686.65			
Si, Mn, P, S	wt%	0.67	0.88	0.153	0.159
Frequency of Tapping	s⁻¹	8.6806E-05			
Slag					
Production	Kg/s	11.7476852			
CaO, MgO, SiO₂, Al₂O₃ and S	wt%	41.47	5.265	34.15	15.15
					0.875

Table 8.3
Averaged Operating Characteristics of
Two Blast Furnaces

HEAT INPUTS
Sources

	ΔH (Kcals/THM)	ΔH (KJ/THM)	Percent
Energy from Fuel	3533420	14783829.28	85.92
Energy from Combustion Air	528378	2210733.55	12.85
Energy from Slag Production	50708	212162.27	1.23
Total Heat Input	4112506	17206725.10	100.00

HEAT OUTPUTS
Sinks

	ΔH (Kcals/THM)	ΔH (KJ/THM)	Percent
Sensible Heat of Hot Metal	696730	2915118.32	16.94
Sensible Heat in Volatile Matter	3091	12932.74	0.08
Heat Needed to Vaporise Moisture	18929	79198.94	0.46
Latent Heat in BF Gas	1480740	6195416.16	36.01
Sensible Heat in BF Gas	88809	371576.86	2.16
Total Heat in Dust	39084	163527.46	0.95
Heat Carried away by Water	170143	711878.31	4.14
Sensible Heat in Slag	153299	641403.02	3.73
Summation of BF Reactions	359261	1503148.02	8.74
Calculated Heat Output	3010086	12594199.82	73.19
Other Heat Losses	1102420	4612525.28	26.81
Total Heat Output	4112506	17206725.10	100.00

Table 8.4
Heat Energy Balance Sheet

You can see from Table 8.4 that whilst Fuel provides around 86% of BF heat input a substantial amount is contributed (12.85%) by the hot air blast admitted at the base of the furnace through tuyeres just above the maximum level of the slag pond. This admitted air would have the ambient temperature of 273 to 300°K if it were not for the combustive superheating by the Cowper Stoves.

The Hot Blast typically enters at between 1500 to 1550°K. It should be remembered that the Hot Blast is *mixed gases*, and accordingly its heat content at STP *per unit volume* is restricted in comparison to molten metal, or indeed boiling water.

This volumetric leanness of the blast is partially compensated by admitting the blast at high pressure[8.10].

Table 8.5 presents the stoichiometric characteristics of the component gases for both air and Blast Furnace Gas (BFG):-

Formula	AIR Mix Fraction (frac)	1300 C_p at °K (J/mol.K))	C_p×frac		MW	MW×frac	
N_2	0.7808	51.6403528	40.32079		28.0134	21.872863	
O_2	0.2095	20.786	4.354667		31.9988	6.7037486	
Ar	0.0098	20.7860001	0.203703		39.948	0.3914904	
	1.0001		44.87916	$\Sigma(C_p$×frac)	99.9602	28.968102	Σ(MW×frac)
			14.95972	$\mu(C_p$×frac)	33.32007	9.6560339	μ(MW×frac)

Formula	BFG Mix Fraction (frac)	498 C_p at °K (J/mol.K))	Cp×frac		MW	MW×frac	
N_2	0.5519	29.5721223	16.32085		28.0134	15.460595	
O_2	0	31.0577914	0		31.9988	0	
Ar	0	20.786	0		39.948	0	
CO	0.2078	29.8074139	6.193981		28.0101	5.8204988	
CO_2	0.2107	44.5555842	9.387862		44.0095	9.2728017	
H_2	0.02076	29.2607774	0.607454		2.01588	0.0418497	
H_2O	0	35.1975679	0		18.0153	0	
	0.99116		32.51015	$\Sigma(C_p$×frac)	192.011	30.595746	Σ(MW×frac)
			4.644307	$\mu(C_p$×frac)	27.43014	4.3708208	μ(MW×frac)

Table 8.5
Stoichiometric Characteristics of Gases

At admission, the Hot Blast has a pressure of between 2.5 and 4.2 kgForce/cm^2, for which the mean is 3.35 kgForce/cm^2. Since one kilogramforce per centimeter squared is equivalent to 98066.5 Pascals, the respective Pascal values of these pressures is 245166.25, 411879.3 and 328522.775

The Mean Molecular Weight of Air, MW_{air}, is 33.32007 (grams), and the Molecular Weight of Diatomic Oxygen, MW_{O2}, is 31.9988. Meanwhile, the Fraction of Oxygen, f_{O2}, in air is 0.2095

Therefore, the Molar Proportion of Oxygen in Air, $Prop_{O2}$, is:-

$$Prop_{O2} = MW_{O2}.f_{O2} = 6.7037486$$
Equation 8.7

and the Moles of Air per Kilogram, mol_{air}, is given by:-

$$mol_{air} = \frac{1000}{MW_{air}} = 30.011941751623$$
Equation 8.8

and the *effective* Moles of Oxygen per Kilogram, mol_{O2}, is given by:-

$$mol_{O2} = \frac{1000}{Prop_{O2}} = 149.170271689484$$
Equation 8.9

Given that the Mean Hot Blast Pressure is p_μ = 328522.775 pascals, and V = 1 m^3, R = 8.314 J/(mol.°K) and T = 1300 °K we may invoke the Universal Gas Law to write:-

$$E_{blast} = p_\mu.V = mol_{air}.R.T = 328522.775 \, Joules$$
Equation 8.10

where E_{blast} is the Imparted Energy per cubic meter of hot blast air.

So at 1300°K, and 328522.775 pascals of pressure, one cubic meter of hot air contains 328.5 KJ of energy. The Pressure 328522.775 pascals is about 3.24 atmospheres.

We may transpose the Universal Gas Law to derive the Number of Cubic Meters of Blast, V_{blast}, required to impart One Kilogram of Oxygen in the following terms:-

$$V_{blast} = \frac{mol_{O2}.RT}{p_\mu} = 4.90761144482079$$

Equation 8.11

where V_{blast} is 4.907 Cubic Meters of Blast per Kilogram of Oxygen.

The Hot Blast in History

In 1824AD the Wilsonstown Ironworks at Forth in Central Scotland possessed two blast furnaces. Developing the work of Thomas Botfield, who invented the Iron Furnace Hot Blast a little earlier, Dr James Beaumont Neilson fed heated compressed air to the work's furnaces.

Dramatic gains in the speed and completeness of iron smelting were manifest.

Let I Lowthian Bell[8.13] put it in his own way:-

"In the year 1828, J. B. Neilson patented an 'improved application of air to produce heat in fires, forges, and furnaces, where bellows and other blowing apparatus are required'. This discovery consisted, as is well known, in heating the air before it is propelled into the furnace; and although, from the title of the patent, Neilson and his colleagues appear to have seen it generally employed in all furnaces driven by compressed air, its use has, practically, been exclusively confined to those employed in smelting the ores of iron.

In 1834, Monsieur Dufrenoy was sent over to this country [Britain], by the Director-General of Mines of France, to report to the authorities at Paris on an invention, which at the time was truly described as one revolutionizing, in

Scotland at all events, where it was first put into practice, the art of making iron.

This gentleman in a report gave good reasons apparently for this statement, by quoting the experience of the owners of Clyde Iron Works, which was as follows [please see Table 8.6].

From this it would appear that heating the air with 5 cwt. of coal had saved 47 cwt. of fuel in the furnace, and 8 cwt. similarly applied had been followed with an economy of 93 cwt., or above 69 per cent.

Besides this advantage, the make was increased by more than one-third, and a blowing engine, which only supplied three furnaces with cold blast, was equal to four when the air was heated.

The iron trade hesitated somewhat in crediting the heat generated from 8 cwt. of fuel burnt outside the furnace, should be able to perform the duty of a very much larger weight burnt inside."

One Imperial Hundredweight (cwt.) is 50.8023 Kilograms.

Twenty hundredweights make one Imperial Long Ton, which latter is therefore 1016.046 kilograms or 1.016046 tonnes.

The mean ambient outdoor temperature in Scotland is roughly 9°C or 282.15°K given that 0°C is 273.15°K.

To convert °F to °K use $°K=(5/9)\times(°F-32)+273.15$

Table 8.6 displays the Dufrenoy tabulation that compares the Clyde Iron Works Cold and Hot Blast consumption accounts for 1829, 1831 and 1833AD, both in its original nineteenth-century form and in my SI renditions:-

For the year ...	1829	1831	1833
Temperature of blast ...	Cold.	450°F	612°F
Coal used per ton of iron .	As coke.	As coke.	In raw state.
For fusion, cwt ...	133	86	40
heating air, raw coal.	nil.	5	8
blowing engines	20	7	11
	153	98	59
Cwt. limestone per ton of iron ...	10½	9	7

Year	1829	1831	1833
Blast Temperature (°K)	282.2	505.4	595.4
Coal Used Per Tonne of Iron	Coke	Coke	Coal
For fusion, Kilograms ...	6865.1	4439.1	2064.7
heating air, raw coal.	0.0	258.1	412.9
blowing engines	1032.3	361.3	567.8
Total Process Fuel (Kgs.)	7897.5	5058.5	3045.4
As Percent of 1829 Figure	100	64.1	38.6
Kilograms Limestone per Tonne of Iron	542.0	464.6	361.3
As Percent of 1829 Figure	100	85.7	66.7

Table 8.6
The Hot Blast Revolution at
The Clyde Iron Works

Hot Gas Reactivity and Kinetic Theory[8.11]

Kinetic Theory concerns the likelihood of two chemical molecules reacting when they collide during random thermal agitation. By and large it is expected (in the technical, stochastic sense) that the likelihood of reaction rises with temperature. This applies to the Oxygen (O_2) gas which is around 21% of hot air, and its target solid Hematite (Fe_2O_3) in the iron ore. It also applies to any intermediate compounds like Carbon Monoxide gas (CO) and Wüstite (FeO) liquid.

Gas-Solid relations are however related in a complex way to gas diffusivity and solid comminution.

Rates of Reaction relate to Temperature via the Arrhenius Equation, promulgated by Svante Arrhenius in 1889AD; and by the Eyring-Polanyi Equation proposed in 1935AD.

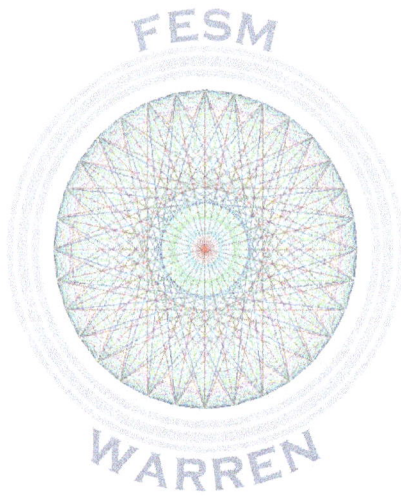

CHAPTER NINE
GAS DENSITY AND HEATING VALUES

In 2016AD Tatsuro Ariyama[9.1] and his colleagues published their "Total Model of Mass and Energy Balance in Iron Making Process" for both Mass and Energy Flows in a Conventional Blast Furnace (BF) and a Blast Oxygen Assisted Blast Furnace (OBF). We will develop this model, and also the slag-related treatments advanced by Fang and his group[9.4].

Some Convenient Functions

Percentage Specific Defect

As per our usual convention for comparing the mutual agreement of two numbers we shall define the Percentage Specific Defect (PSD) as:-

$$PSD(x,y) = 100 \left(\frac{x-y}{x}\right)$$
Equation 9.1

For example, if x is 95 and y is 94 the PSD(95,94) is:-

$$PSD(95,94) = 100 \left(\frac{95-94}{95}\right) = 100 \frac{1}{95} = 1.05263157894737$$

So the PSD of 95 and 94 is about +1.053%.

PSD can be either positive or negative depending upon the relative size of the number arguments.

Shomate Molar Specific Heat Equation

NIST[9.2] offers a suite of polynomial Shomate Equations for selected thermodynamic parameters for a wide range of pure chemicals and specified temperature ranges.

Of immediate interest is the Shomate Equation for Molar Specific Heat (at Constant Pressure), C_p, in J/(mol.°K) or in alternative notation $Jmol^{-1}K^{-1}$

The relevant function is:-

$$Cp(c,t) = c_1 \cdot t^0 + c_2 \cdot t^1 + c_3 \cdot t^2 + c_4 \cdot t^3 + \frac{c_5}{t^2}$$

Equation 9.2

where:-

$$t = \frac{T^\circ K}{1000}$$

Equation 9.3

and c_i are Empirical Coefficients.

$T^\circ K$ is the Temperature in Degrees Kelvin of the material under consideration.

The Heating Value of a Combustible Gas

It is evidently useful to be able to compute the Heat of Combustion ΔH_c of some fuel, typically a gas, in terms of Mass Heat of Combustion $\Delta H(m.x)_c$ in KJ.Kg^{-1}; Volumetric Heat of Combustion $\Delta H(V.x)_c$ in KJ/m^{-3}; or Evolved Heat H_x in KJ or GJ.

Allow that V_{BFG} is the Volume of Blast Furnace Gas and that x is some Gas Species (e.g. Carbon Monoxide, Hydrogen, Steam).

f_x is the Volume Fraction of Gas x in the BFG considered.

Then the Volume of x, V_x, in Cubic Meters at STP is given by:-

$$V_x = f_x \cdot V_{BFG}$$

Equation 9.4

Also the (Mean for Mixtures) Molecular Weight of Any Gas, MW_x, in grams may be calculated from the chemical formula or formulas and the respective formulae where n is the Number of Gas Species in the Mixture according to:-

$$MW_x = \frac{\sum_{i=1}^{n} f_i \cdot MW_i}{n}$$

Equation 9.5

Where a single chemical is in question, say Carbon Monoxide (CO), then it is sufficient to write:-

$$MW_{CO} = 28.01$$
Equation 9.6

The Liters per Mole of Any Gas at Standard Temperature and Pressure (STP) is noted as MV and is in units of liters per mole. It has the constant value 22.4

Accordingly, the Molar Density of Gas x is ρ_x in grams/liter is yielded by:-

$$\rho_x = \frac{MW_x}{MV}$$
Equation 9.7

Hence:-

$$m_x = \rho_x \times MV = MW_x$$
Equation 9.8

where m_x is the number of Grams of Gas in 22.4 liters, equivalent to the Molecular Weight of the Gas or Gas Mixture.

Now by the Universal Gas Law and at the Standard Temperature and Pressure taken to be $T_x = 293.15°K$ and $p_x = 101325$ Pascals (NIST Standard Atmosphere), and noting that the Universal Gas Constant R is 8.31446261815324 $Jmol^{-1}K^{-1}$:-

$$D_x = \frac{m_x . p_x}{1000. R. T_x}$$
Equation 9.9

D_x is the calculated Gas Mass Density at T_x in $Kg.m^{-3}$

For the case of x is Carbon Monoxide, the calculated D_x is 1.16440922550048 Kg/m^3 whilst the Toolbox[9.3] value is 1.165 Kg/m3. The PSD($D_x,D_{xToolbox}$) is -0.050735985818733 showing an error of about a twentieth of a percentage point.

$\Delta H(m.x)_c$, Gas Heat of Combustion (Mass) ($KJ.Kg^{-1}$) is taken from tables, and given that, $\Delta H(m.x)_c$, the Gas Heat of Combustion (Volume) ($KJ.m^{-3}$) is given by:-

$$\Delta H(V.x)_c = D_x . \Delta H(m.x)_c$$
Equation 9.10

Therefore the Evolved Heat due to Combustible Gas x, H_x, measured in Joules, or more practically in KJ or even GJ is computed using:-

$$H_x = \Delta H(V.x)_c . V_x$$
Equation 9.11

Table 9.1 presents these data for Ariyama Et Al Exported Conventional Blast Furnace BFG where the Volume of Mixed Gas (CO and H_2 neglecting incombustibles) was 1096 cubic meters:-

Correspondingly, Table 9.2 presents these data for Ariyama Et Al Stove-routed Conventional Blast Furnace BFG where the Volume of Mixed Gas (CO and H2 neglecting incombustibles) was 654 cubic meters

The Sensible Heat of Blast Gases

"Sensible" Heat is physical heat content due to the Specific Heat of the substance and its rise or fall in temperature. The concept can be extended to include Latent Heat of state change, but Latent Heat is not relevant to gases involved in iron smelting.

The equation for Sensible Heat is:-

$$Sensible\ Heat = m.\,C_p.\,\Delta T$$
$$= Mass \times Specific\ Heat \times Temperature\ Change$$
Equation 9.12

In the Ariyama context, the Conventional Blast Furnace Volume of Blast, $V0_{Blast}$ is 1318 m^3 at NIST STP whilst the Fraction of gas x in that total volume is f_x.

Property	Symbol	Units	Carbon Monooxide (CO)	Diatomic Hydrogen (H_2)
Volume Of Blast Furnace Gas	V_{BFG}	m^3	1096	1096
Fraction of BFG	f_x	none	0.232	0.021
Volume of Gas at STP	V_x	m^3	254.272	23.016
(Mean) Molecular Weight of Gas	MW_x	(grams)	28.01	2.016
Liters per Mole of Any Gas at STP	MV	L/mol	22.4	22.4
Molar Density of Gas at STP	ρ_x	g/L	1.250446429	0.09
Molar Mass of Gas	m_x	(grams)	28.01	2.016
Gas Pressure	p_x	Pascals	101325	101325
Gas Temperature (NIST standard)	T_x	°K	293.15	293.15
Universal Gas Constant	R	$Jmol^{-1}K^{-1}$	8.314462618	8.31446262
Gas Mass Density at T_x (calculated)	D_x	$Kg.m^{-3}$	1.164409226	0.08380753
Gas Mass Density at T_x (Toolbox)	$D_{xToolbox}$	$Kg.m^{-3}$	1.165	0.0899
$PSD(D_x, D_{xToolbox})$		none	-0.050735986	-7.2695936
Gas Heat of Combustion (mass)	$\Delta H(m.x)_c$	$KJ.Kg^{-1}$	10104	141584
Gas Heat of Combustion (volume)	$\Delta H(V.x)_c$	$KJ.m^{-3}$	11765.19081	11865.8057
Evolved Heat from Gas Combustion	H_x	KJ	2991558.599	273103.385
Evolved Heat from Gas Combustion	H_x	GJ	2.991558599	0.27310339
Total BFG Combustion Heat	$\Sigma Heat_{BFG}$	KJ	3264661.984	
Total BFG Combustion Heat	$\Sigma Heat_{GJBFG}$	GJ	3.264661984	

Table 9.1
Conventional BF Exported BFG
Thermometric Data and Results

Property	Symbol	Units	Carbon Monooxide (CO)	Diatomic Hydrogen (H_2)
Volume Of Blast Furnace Gas	V_{BFG}	m^3	654	654
Fraction of BFG	f_x	none	0.232	0.021
Volume of Gas at STP	V_x	m^3	151.728	13.734
(Mean) Molecular Weight of Gas	MW_x	(grams)	28.01	2.016
Liters per Mole of Any Gas at STP	MV	L/mol	22.4	22.4
Molar Density of Gas at STP	ρ_x	g/L	1.250446429	0.09
Molar Mass of Gas	m_x	(grams)	28.01	2.016
Gas Pressure	p_x	Pascals	101325	101325
Gas Temperature (NIST standard)	T_x	°K	293.15	293.15
Universal Gas Constant	R	$Jmol^{-1}K^{-1}$	8.314462618	8.31446262
Gas Mass Density at T_x (calculated)	D_x	$Kg.m^{-3}$	1.164409226	0.08380753
Gas Mass Density at T_x (Toolbox)	$D_{xToolbox}$	$Kg.m^{-3}$	1.165	0.0899
PSD($D_x,D_{xToolbox}$)		none	-0.050735986	-7.2695936
Gas Heat of Combustion (mass)	$\Delta H(m.x)_c$	$KJ.Kg^{-1}$	10104	141584
Gas Heat of Combustion (volume)	$\Delta H(V.x)_c$	$KJ.m^{-3}$	11765.19081	11865.8057
Evolved Heat from Gas Combustion	H_x	KJ	1785108.872	162964.976
Evolved Heat from Gas Combustion	H_x	GJ	1.785108872	0.16296498
Total BFG Combustion Heat	$\Sigma Heat_{BFG}$	KJ	1948073.848	
Total BFG Combustion Heat	$\Sigma Heat_{GJBFG}$	GJ	1.948073848	

Table 9.2
Conventional BF Stove BFG
Thermometric Data and Results

The (Mean) Molecular Weight of Gas x is MW_x (in grams) and T is the Gas Temperature in Degrees Kelvin whilst t = T/1000

The Shomate Operative Temperature Range for $x = N_2$ is 500 - 2000°K and the Toolbox value of Specific Heat (mass) is 1.244 $KJKg^{-1}K^{-1}$ for N_2 in this range.

Given n appropriate Shomate coefficients c_i we have:-

$$C_p(c,t) = 34.7644440707435$$
Equation 9.13

as the Molar Specific Heat of Gas x (N_2 in this instance) measured in $Jmol^{-1}K^{-1}$

The Moles per Kilogram of Gas x, mol_{Kg}, in $mol.Kg^{-1}$ is given by:-

$$mol_{Kg} = \frac{1000}{MW_x}$$
Equation 9.14

whilst the Volume of Blast Gas x at STP, $V0_x$, in m^3 is:-

$$V0_x = f_x.V0_{Blast}$$
Equation 9.15

The Molar Density of Gas x at STP, ρ_x, is:-

$$\rho_x = \frac{MW_x}{MV}$$
Equation 9.16

The Mass of Gas x in Kilograms is:-

$$m_x = \rho_x.V0_x$$
Equation 9.17

so that calculated Mass Specific Heat of Gas x, C_{pxKg}, in $J.kg^{-1}.K^{-1}$ is:-

$$C_{pxKg} = C_p(c,t).mol_{Kg}$$
Equation 9.18

For N_2 C_{pxKg} is 1240.99338426408
The Toolbox value for $C_{pxKgToolbox}$ is 1244 $J.kg^{-1}.K^{-1}$
Therefore PSD(C_{pxKg},$C_{pxKgToolbox}$) is -0.242274920594
In the case of Nitrogen x = N_2; and the Sensible Heat of the Gas Component x of the Blast, $Heat_x$, is given by:-

$$Heat_x = m_x.C_{pxKg}.T$$
Equation 9.19

Table 9.3 presents data and results for Ariyama Et Al Hot Blast gases where Nitrogen, Oxygen and Argon are all accounted for. The Ariyama blast is not perfectly of atmospheric composition, being 0.713 N_2 and anomalously 0.266 O_2. I assumed that the difference, 0.019, was all Argon though of course ambient air includes significant Carbon Dioxide and Water Vapor.

Property	Symbol	Units	Diatomic Nitrogen (N_2)	Diatomic Oxygen (O_2)	Argon (Ar)
Volume Of Blast at STP	VO_{Blast}	m³	1318	1318	1318
Fraction of Blast	f_x	none	0.713	0.266	0.019
(Mean) Molecular Weight of Gas	MW_x	(grams)	28.0134	31.9988	39.948
Shomate Range of Validity (°K)			500 - 2000	700 - 2000	298 - 6000
Liters per Mole of Any Gas at STP	MV	L/mol	22.4	22.4	22.4
Molar Density of Gas at STP	ρ_x	g/L	1.250598214	1.428517857	1.783392857
Molar Mass of Gas	m_x	(grams)	28.0134	31.9988	39.948
Gas Temperature	T_x	°K	1473.15	1473.15	1473.15
Gas Temperature (×0.001)	t_x	°K/1000	1.47315	1.47315	1.47315
C_p Shomate Co-efficient #1	c_1	none	19.50583	30.03235	20.786
C_p Shomate Co-efficient #2	c_2	none	19.88705	8.772972	2.825911E-07
C_p Shomate Co-efficient #3	c_3	none	-8.598535	-3.988133	-1.464191E-07
C_p Shomate Co-efficient #4	c_4	none	1.369784	0.788313	1.092131E-08
C_p Shomate Co-efficient #5	c_5	none	0.527601	-0.741599	-3.661371E-08
Gas Specific Heat	$C_p(c,t)$	Jmol⁻¹K⁻¹	34.76444407	36.4798263	20.78600012
Moles per Kilogram of Gas	mol_{Kg}	mol.kg⁻¹	35.69720205	31.25117192	25.0325423
Volume of Blast Gas at STP	VO_x	m³	939.734	350.588	25.042
Mass of Gas	m_x	Kg	1175.229662	500.8212185	44.65972393
Specific Heat of Gas (mass) (Calculated)	C_{pxKg}	J.Kg⁻¹.K⁻¹	1240.993384	1140.037323	520.3264273
Specific Heat of Gas (mass) (Toolbox)	$C_{pxKgToolbox}$	J.Kg⁻¹.K⁻¹	1244	1143	520
PSD(C_{pxKg},$C_{pxKgToolbox}$)			-0.242274921	-0.259875414	0.062735093
Sensible Heat of Gas Component	$Heat_x$	Joules	2148518911	841102183.5	34232521.4
Total Blast Sensible Heat	$\Sigma Heat_{Blast}$	Joules	3023853616		
Total Blast Sensible Heat	$\Sigma Heat_{GJBlast}$	GJ	3.023853616		

Table 9.3
Conventional BF Hot Blast
Sensible Heat

Iron Heat Content

Specific Heat is the heat inherent in a hot body by virtue of the random jiggling of its component atoms whilst the substance remains in its current state, in this case liquid iron.

Latent Heat of Fusion is absorbed as the substance changes state *without change of temperature* (i.e. as the substance melts). When the substance resolidifies the mass yields Latent Heat to the environment, again without change of temperature. Mixtures like slag are impure substances and have a slurred melting point. Pure chemicals like iron or magnesia (MgO) have abrupt melting points. Latent Heat is intimately involved with crystal formation in solid substances, and the crystallisation of melts can be exploited to separate and purify substances.

In this book we define Sensible Heat to include both Heat due to Specific Heat, and also Heat added or subtracted due to Latent Heat *if the body of material changes state*.

Sensible Heat is a purely physical phenomenon and explicitly excludes Heats of Formation and the related Heats of Combustion. If needed those must be calculated and added separately.

Allow x to represent any Pure Substance (in this subsection x is Iron [Fe]) then the Change in Sensible Heat due to Specific Heat, ΔH_s, is given by:-

$$\Delta H_s = m_x. C_{pxKg}. (T - T^{\ominus})$$
Equation 9.20

where m_x is the Mass of the Substance in Kilograms; C_{pxKg} is the (Range Mean) Specific Heat of the Substance in Kilograms; T is the current Substance Temperature in Degrees Kelvin; and $T^{\ominus}$ is the Standard Temperature, usually the internationally-accepted contemporary room temperature. I use the NIST convention for $T^{\ominus}$ which is 293.15°K or 20°C.

(In Victorian times rooms in France and Britain were much colder and the standard for laboratory calibration was then 15°C, as I discovered as a teenaged technician at Aberdeen University who "inherited" nineteenth-century glassware, but not quite in Victoria's reign!).

Latent Heat, L_x, arises from a process that takes place at the Melting Point only and so change of temperature is not involved. But Latent Heat is a function of Mass so the Total Latent Heat liberated or absorbed is given by:-

$$\Delta H_L = m_x. L_x$$
Equation 9.21

Thus the Total Sensible Heat of the mass, ΔH_{sens}, is given by:-

$$\Delta H_{sens} = \Delta H_s + \Delta H_L = m_x . C_{pxKg} . (T - T^\Theta) + m_x . L_x$$
$$= m_x \left[L_x + C_{pxKg} . (T - T^\Theta) \right]$$

Equation 9.22

Computing Specific Heat

The Specific Heat of liquid Iron at 1809 or 1900°K is not the same as its Specific Heat at room temperature, however room temperature is defined.

To compute C_p at an arbitrary temperature within experimental range we need to use a Shomate Equation, basically an algebraic polynomial of the form:-

$$C_{pxT} = \sum_{i=1}^{n} \kappa_i . 10^{e\kappa_i} . T^{eT_i}$$

Equation 9.23

where C_{pxT} is the Specific Heat at Temperature T in J/(mol.°K); κ_i is the ith. Empirical Constant considered as having one digit before the decimal point; $e\kappa_i$ is the Exponent of the (Denery) Empirical Constant; and eT_i is the Temperature Exponent. n is the Number of Coefficients of the Shomate Equation.

The 3×15 relevant values are presented in Table 9.4 in the "C_p Shomate Coefficient #i" rows.

Using the Shomate Equation for the Specific Heat of Iron in the Range 1809 - 3133.345 degrees Kelvin I computed 46.024 J/(mol.°K). NIST gives this statistic as 46.02. Accordingly, the PSD(PSD(Cp_{calc},Cp_{NIST}) is 0.008691073, less than one part in ten thousand, corroborating the arithmetic accuracy so far.

To convert Molar Specific Heat in J/(mol.°K) to Mass Specific Heat in J/(Kg.°K) use:-

$$Cp_{mass} = \frac{1000}{FW_x} . Cp_{molar}$$

Equation 9.24

where FW_x is the Formular Mass of the pure Substance, or the Mean Weighted Formula Mass of a Mixture.

IRON SENSIBLE HEAT

			Fe	c exp.	T exp.
Mass of Liquid Iron (m_{Fe}) (Kg)	1000				
Temperature T (°K)	1900	Range (°K) 1809 - 3133.345			
Temperature Argument t	1.9				
Substance			Fe	c exp.	T exp.
Cp Shomate Co-efficient #1			46.024	0	0
Cp Shomate Co-efficient #2			-1.884667	-8	1
Cp Shomate Co-efficient #3			6.09475	-9	2
Cp Shomate Co-efficient #4			-6.640301	-10	3
Cp Shomate Co-efficient #5			-8.246121	-9	-2
Molecular Weight (grams/mol)			55.845		
Moles per Kilogram			17.9067061		
Cp (J/mol.°K)			46.024	46.02	
Cp (J/Kg.°K)			824.138239		
PSD(Cp_{calc}, Cp_{NIST})			0.00869107		
L_{Fe} (J/mol) (fusion)			13.81		
L_{Fe} (J/Kg) (fusion)			247.291611		
Heat due to Specific Heat (ΔH_s) (Joules)			1324266530		
Heat due to Latent Heat (ΔH_L) (Joules)			247291.611		
Total Heat (ΔH_{sens}) (Joules)			1324513822		

Table 9.4
Iron Heat Content

Slag Makeup and Specific Heats

.Slag is indeed a mixture and successfully to assess its thermal behavior we need knowledge of the proportions of its normative components (usually metal oxides) and their individual specific heats at the appropriate temperature.

Fang and his colleagues[9.4] provided a normative analysis of five components of nine samples which I reproduce at the head of Table 9.5 under the title SLAG COMPOSITION (mass%).

I totalled and averaged the sample proportions and checked that their means summed to 100%.

I then reduced the mean values to fractions which totalled unity.

For the five oxides there were between n = four and n = five Shomate Coefficients subject to the form:-

$$Cpx_{gram°K} = \sum_{i=1}^{n} \kappa x_i \cdot 10^{e\kappa_i} \cdot T^{eT_i}$$

Equation 9.25

analogously to Equation 9.23

The five Percentage Specific Defects, PSD(Cp$_{calc}$,Cp$_{NIST}$), ranged from -0.508 for TiO$_2$ to -4.949 for SiO$_2$, except for Al$_2$O$_3$ for which PSD anomalously exceeded -487

For Al$_2$O$_3$ I adopted the specific heat 225.413961 J/(Kg.°K)

The mix-averaged Specific Heat of Slag, Cp$_{slag}$, in J/(kg.°K) is then:-

$$Cp_{slag} = \sum_{i=1}^{n} f_i \cdot Cp_i$$

Equation 9.26

where f$_i$ is the Mass Fraction of Oxide i and Cp$_i$ is the Specific Heat of Oxide i at the relevant temperature.

With regard to Latent Heat the relevant mathematics is:-

$$H_{L,i} = \sum_{i=1}^{n} f_i L_i$$

Equation 9.27

where H$_{L,i}$ is the Total Latent Heat due to the Mixture and L$_i$ is the Latent Heat contributed by Substance i.

Summatively, the Contained Heat due to Specific Heat, ΔH$_s$, in Joules is given by:-

$$\Delta H_s = m_{slag} Cp_{slag} (T - T^{\Theta})$$

Equation 9.28

whilst the Contained Heat due to Latent Heat, ΔH$_L$, in Joules is rendered by:-

$$\Delta H_L = m_{slag} L_{slag} = m_{slag} \sum_{i=1}^{n} f_i L_i$$
$$\textbf{Equation 9.29}$$

The Total Sensible Heat in the Slag Body, ΔH_{sens}, in Joules is yielded by:-

$$\Delta H_{sens} = \Delta H_s + \Delta H_L$$
$$\textbf{Equation 9.30}$$

Heat Input from Fuel Combustion

In the 5.3.1(a) Total Model of Ariyama and Sato, for every 1000 kilograms of Iron output, 446 kilograms of Coke is burnt, which I have taken to be Grade I Metallurgical Coke, together with 64 Kilograms of Coal, which I have assumed Anthracite.

I rate the Heat of Combustion of Coke, H_{cCoke} as 29305.274 KJ/Kg and the Heat of Combustion of Coal (Anthracite), H_{cCoal} as 32500 KJ/Kg

Therefore the Modelled Yield of Coke Heat, H_{coke}, is:-

$$H_{coke} = m_{coke}. H_{cCoke} = 13070152.2$$
$$\textbf{Equation 9.31}$$

where m_{coke} is the Mass of Coke Tipped into the Blast Furnace and the Modelled Yield of Coal Heat, H_{coal}, is:-

$$H_{coal} = m_{coal}. H_{cCoal} = 2080000$$
$$\textbf{Equation 9.32}$$

Both H_{coke} and H_{coal} are in KiloJoules (KJ).

Therefore, the Available Combustive Heat Energy, H_{avail}, is:-

$$H_{avail} = H_{coke} + H_{coal} + H_{blast}$$
$$= 13070152.2 + 2080000 + 3023853.616 \, KJ$$
$$\textbf{Equation 9.33}$$

or the H_{avail} = 18.17400582 GigaJoules (GJ)

Similarly, the Output Heat documented by our calculations, H_{traced}, is:-

$$H_{traced} = H_{BFGsens} + H_{BFGfuel} + \Delta H_{Fe} + \Delta H_{slag}$$
$$= 0 + 3.264661984 + 1.324513822 + 0.121305335 \ GJ$$
$$= 0 + 3264661.984 + 1324513.822 + 121305.335 \ KJ$$
$$= 4710481.14 \ KJ$$

Equation 9.34

The Heat Balance, H_{bal}, is:-

$$H_{bal} = H_{avail} - H_{traced}$$

Equation 9.35

so that the Apparent Efficiency of the Furnace System, ε_{BF}, is:-

$$\varepsilon_{BF} = \frac{H_{traced}}{H_{avail} + H_{traced}} = 0.205837306$$

Equation 9.36

This efficiency value implies that a mere 20.58% of the supplied energy is "usefully" employed, the other 80% presumably wasted as lost heat.

Also notice that nether Ariyama and Sato, or myself, have seen fit to compute the sensible heat of the Blast Furnace Gas (BFG or Top Gas): It is relatively cool and its sensible heat contribution can probably be neglected.

Also, you can see from Table 9.6 that the contribution of slag to the sensible heat is tiny.

Ariyama Model Thermal Content Summary

Table 9.6 presents a conspectus of the thermogenic inputs and outputs of an Ariyama model Conventional Blast Furnace in a manner that is readily scalable for practical furnaces.

	CaO	SiO2	MgO	Al2O3	Al2O3	TiO2	Total	Liquid Temperature (°K)
Sample #1	30.03	27.3	8.5	14.17	14.17	20	100	1432
Sample #2	31.08	28.25	8.5	14.17	14.17	18	100	1430
Sample #3	32.13	29.2	8.5	14.17	14.17	16	100	1427
Sample #4	33.17	30.16	8.5	14.17	14.17	14	100	1421
Sample #5	34.22	31.11	8.5	14.17	14.17	12	100	1413
Sample #6	30.03	27.3	6.48	16.19	16.19	20	100	1440
Sample #7	30.03	27.3	7.56	15.11	15.11	20	100	1436
Sample #8	30.03	27.3	9.33	13.34	13.34	20	100	1428
Sample #9	30.03	27.3	10.08	12.59	12.59	20	100	1424
Total	280.75	255.22	75.95	128.08	128.08	160	900	
Mean	31.1944444	28.3577778	8.43888889	14.2311111	14.2311111	17.7777778	100	
Mean Fraction	0.31194444	0.28357778	0.08438889	0.14231111	0.14231111	0.17777778	1	

SPECIFIC HEATS

	1800 Range (°K)	CaO	SiO2	MgO	Al2O3	Al2O3	TiO2
Substance							
Temperature T (°K)		298 - 2845	298 - 1996	298 - 3098	298 - 1200	1200 - 2327	298 - 2130
Slag Mass (Kg)	82.956						

Substance	CaO	c exp.	T exp.	SiO2	c exp.	T exp.	MgO	c exp.	T exp.	Al2O3 (298-1200)	c exp.	T exp.	Al2O3 (1200-2327)	c exp.	T exp.	TiO2	c exp.	T exp.
Cp Shomate Co-efficient #1	1.048	0	0	1.332	0	0	1.516	0	0	-0.1541	0	0	-7.724	0	0	0.975	0	0
Cp Shomate Co-efficient #2	-2.046	4	-2	-5.903	4	-2	-1.541	4	-2	-2.973	4	-2	6.461	4	1	-4.217	4	-4
Cp Shomate Co-efficient #3	-2.388	0	-0.5	-3.999	0	-0.5	-7.349	0	-0.5	-4.899	0	-0.5	2.587	6	-2	5.045	6	-3
Cp Shomate Co-efficient #4	1.836	6	-3	8.181	6	-3	1.45	6	-3	-1.099	4	-3	-1.496	4	-1			
Cp Shomate Co-efficient #5										69.33	3	0	6.56	2	-0.5			
Molecular Weight (grams/mol)	56.077			60.0848			40.3044			101.9613			101.9613			79.866		
Moles per Kilogram	17.832623			16.6431444			24.8111869			9.8076427			9.8076427			12.5209726		
Cp (J/gram.°K)	0.9857143			1.22092631			1.33802872			-6.409E+12			0.22541396			0.96284962		
Cp (J/Kg.°K)	985.7143			1220.92631			1338.02872			-6.409E+15			225.413961			962.849623		
Cp (J/mol.°K)	55.2759008	57.63		73.359113	76.99		53.9284448	54.94		-6.409E+18			22.9835005	135		76.898948	77.29	
PSD(Cp_calc-Cp_test)	-4.2588165			-4.9494696			-1.8757358						-487.37789			-0.5085271		
I_i (KJ/mol) (fusion)	53.78						77.4			1093			1093			875		
I_i (KJ/Kg) (fusion)	959.038465			159.8			1920.38586											
I_i.Cp_i	307.4881			346.227569			112.914757						32.0789112			171.173266		
I_i.L_i	299.166721			45.3157289			162.059229						155.546044			155.555556		
Cp_slag (J/Kg.°K)	969.882604																	
L_slag (J/Kg)	817.643279																	
Heat due to Specific Heat (ΔH_s) (Joules)	121237506																	
Heat due to Latent Heat (ΔH_L) (Joules)	67828.4159																	
Total Heat (ΔH_tom) (Joules)	121305335																	

Table 9.5
Slag Makeup and Specific Heats

THE ARIYAMA MODEL CONVENTIONAL BLAST FURNACE
For One Tonne (1000 Kilograms) of Iron Hearthed

	Coke	Coal	Iron	Slag	
Slag Temperature (°K)	1800	1800	1900	1800	
Modelled Mass of Substance (kg) (Ariyama)	446	64	1000		
Expectation of Fraction	1	0.143497758	2.24215247	0.186	
Mass of Substance (Kg)	446	64	1000	82.956	
Mean Specific Heat (J/gram.°K) (Tang Et Al)				0.9698826	
Mean Specific Heat (J/Kg.°K)			824.138239	969.8826	
Mean Latent Heat (J/Kg)			247.291611	817.64328	
Modelled Heat Content (ΔH_x) (Joules)			1324513822	121305335	
Modelled Heat Content (ΔH_x) (GJ)			1.32451382	0.1213053	

Add Hot Blast	3.02385362	GJ
Less BFG Sensible Heat	0	GJ
Less BFG Fuel Value	3.26466198	GJ

Modelled Coke Fuel Mass (Kg)	446
Heat of Combustion of Coke (KJ/Kg)	29305.27
Modelled Coke Heat Yield (KJ)	13070152
Modelled Coke Heat Yield (GJ)	13.07015

Modelled Coal Fuel Mass (Kg)	64
Heat of Combustion of Coal (KJ/Kg)	32500
Modelled Coal Heat Yield (KJ)	2080000
Modelled Coal Heat Yield (GJ)	2.08

HEAT IN (GJ)

Modelled Coke Heat Yield	13.07015	
Modelled Coal Heat Yield	2.08	
Hot Blast	3.023854	
	18.17401	Sum

HEAT OUT (GJ)

BFG Sensible Heat	0	
BFG Fuel Value	3.264662	
Modelled Iron Heat Content	1.324514	
Modelled Slag Heat Content	0.121305	
	4.710481	Sum

SUMMATIVE RESULTS (GJ)

Balance	13.46352
Heat Efficiency Fraction	0.205837

Table 9.6
Thermal Content Summary

CHAPTER TEN
THE THERMAL POWER OF A BLAST FURNACE

We think of the iron smelting blast furnace as a fiery splendor. I have throughout my adult life harbored the prejudice that the rateable power of a blast furnace is some 55MW±10MW. Where I got that notion I do not know, nor when.

I suppose I compared it with known and familiar things like the Rolls-Rolls-Royce Merlin engine at 1MW or The Flying Scotsman railway locomotive at two thousand horsepower or 1.47 MW

We remember viewing the nocturnal canvases of the likes of Turner, or the 1801AD painting "Coalbrookdale by Night" by Philip Jacob De Loutherbourg, that celebrated the almost biblical visions of the early Industrial Revolution and the Romantically-set furnaces and forges the pride of England before Gothicism soured our appreciation of the Works of Man.

We remember also the would-be revolutionaries of Pentrich who sought ordnance to steal, and in an age of few and feeble lights followed the glow in the night sky to The Butterley Company furnaces.

But a well-designed and well-managed modern furnace intends to minimise the fuel spent and the heat raised and the major part of the fire is confined to the raceway and its immediate burden, whilst the rest of the massive structure is in comparison cool.

We do not wish to make assertions without evidence, so first a word about the basics:-

The Heat Value of Coke

The major fuel of the blast furnace is coke. For sure, a few percent of the charge in a modern furnace may be Pulverised Coal Injectate (PCI) and in Asian furnaces methane and liquid petroleum is a few percent more; or even, in India, coal tar or other fluid flammable refuse.

But we will focus on coke for the meantime, and try to figure its Heat of Combustion in Joules per kilogram (J/Kg) as this can be scaled to Megawatts per Tonne. It will also be convenient, rather than calculating from first principles, to know how many kilograms of prime metallurgical coke are needed to produce one tonne of molten iron. This latter figure is known to engineers as the Coke Rate, and varies according to the furnace and the skill of its design and operation.

There are of course economies of scale and it would be silly to think that the eighteenth-century furnaces in Butterley or Coalbrookdale could be matched with a small modern furnace producing a thousand tonnes per annum, or even that the latter could compare with a modern Chinese or Korean blast furnace outputting ten thousand tonnes in a year.

The purpose of all this is to provide a "benchmark" against which proposed or experimental non-hydrocarbon iron smelting furnaces may be assessed.

Table 10.1 reproduces the findings of the LECO Corporation regarding the Heat Value of Metallurgical Coke[10.1]. They applied ASTM Method ASTM D5865 using the calorimeter AC600, a product of their firm. Benzoic Acid was used in five firings for calibration purposes and then five samples of coke were burnt.

I have computed the Mean Gross Heat of Combustion of Metallurgical Coke arising from these experiments to be 29306204.4 J/Kg or 29.3MJ/Kg

Substance Name	Mass (grams)	Mass (Kg)	Gross Heat (BTU/lb)	Gross Heat (J/kg)
Benzoic Acid	1.0711	0.001071	11370	26446620
Benzoic Acid	0.9026	0.000903	11373	26453598
Benzoic Acid	0.9914	0.000991	11376	26460576
Benzoic Acid	0.9864	0.000986	11381	26472206
Benzoic Acid	0.9859	0.000986	11375	26458250
Mean	0.98748	0.000987	11375	26458250

Substance Name	Mass (grams)	Mass (Kg)	Gross Heat (BTU/lb)	Gross Heat (J/kg)
Metallurgical Coke	0.7507	0.000751	12586	29275036
Metallurgical Coke	0.7514	0.000751	12615	29342490
Metallurgical Coke	0.7516	0.000752	12604	29316904
Metallurgical Coke	0.7577	0.000758	12605	29319230
Metallurgical Coke	0.7534	0.000753	12587	29277362
Mean	0.75296	0.000753	12599.4	29306204.4

Table 10.1
The LECO Experiment for the
Heat Value of Metallurgical Coke

Figure 10.1
"Coalbrookdale by Night"
By Philip Jakob de Loutherbourg (1801)
(The Science Museum, London)

The Energy Requirements of Steel Production Processes

A useful list of process energy requirements is given by Fruehan and his colleagues in the US Government publication "Theoretical Minimum Energies to Produce Steel"[10.2].

Based upon Table. E-1 of their executive summary, this tabulation extended by me is presented as Table 10.2

Our focus is upon the initial part of the steel mill process "Liquid Hot Metal (5%C)" which demands a Mean Actual Requirement of Energy of 13.5 GJ/tonne per tonne of Liquid Hot Metal.

Clearly, if we know the Heat Value of Coke and the Process Energy Demand *per tonne of liquid iron* we can assess the *theoretical* Coke Rate. But it behoves us to rely upon the empirical coke rates produced by furnacemen and other experts, as long as a consensus seems tenable. Note that nothing is said about the hot blast, or BFG, or heat losses from the structure carcass.

The European Steel Association

The European Steel Association[10.3] publishes annual statistics for the 57 surviving iron blast furnaces in Europe. By their definition of "Europe" they include the UK but exclude Russia, Ukraine and the Caucasian states.

Their synoptic table of European hot iron production in Tonnes per Annum (tpy) of Tonnes of Hot Metal (thm) is reproduced in Table 10.3

I have extended the ESA data to include:-

Total and Mean Output Tonnes of Hot Metal
Tonnes per Blast Furnace per Year
Kilograms per BF per Year
Kilograms per BF per Second
Mean BF Energy Requirement (GJ/year)
    liquid iron only
Mean BF Energy Requirement (J/s = W)
    liquid iron only
Metallurgical Coke ΔH_c (J/kg)
The Modern Coke Rate (Kg/thm)
The Traditional Coke Rate (kg/thm)

Process	Actual Requirements (GJ/tonne Product) MIN	Actual Requirements (GJ/tonne Product) MAX	Actual Requirements (GJ/tonne Product) MEAN	Actual Requirements (J/Kg Product) MEAN	Absolute Minimum (GJ/tonne Product)	Percent Difference	Practical Minimum (GJ/tonne Product)	Percent Difference
Liquid Hot Metal (5%C)	13	14	13.5	13500000	9.8	27.41	10.4	5.77
Liquid Steel (BOF)	10.5	11.5	11	11000000	7.9	28.18	8.2	3.66
Liquid Steel (EAF)	2.1	2.4	2.25	2250000	1.3	42.22	1.6	18.75
Hot Rolling Flat	2	2.4	2.2	2200000	0.03	98.64	0.9	96.67
Cold Rolling Flat	1	1.4	1.2	1200000	0.02	98.33	0.02	0.00
18-8 Stainless Melting				0	1.2		1.5	20.00

Table 10.2
Steel Mill Energetic Process Costs
(Fruehan, Fortini, Paxton and Brindle)

Nation	Location	Hot Metal Capacity (tonnes *10³/year)	Finished Steel Capacity (tonnes *10³/year)	Number of Furnaces
Austria	DONAWITZ (Leoben)	1370	1570	2
Austria	LINZ	430	6000	3
Belgium	GHENT	4430	5000	2
Chechia	OSTRAVA – BF only	3200		3
Chechia	TRINEC	2100	2400	2
Finland	RAAHE	2400	2600	2
France	DUNKERQUE	6800	6750	3
France	FOS-SUR-MER	5160	5100	2
Germany	BREMEN	3960	3800	2
Germany	DILLINGEN	4790	2760	2
Germany	DUISBURG	11600	11560	4
Germany	EISENHÜTTENSTADT	2340	2400	2
Germany	SALZGITTER	4800	5200	3
Germany	VÖLKLINGEN BOF only	(3240)		0
Hungary	DUNAUIJVAROS	1310	1650	2
Italy	TARANTO	9590	11500	4
Netherlands	IJMUIDEN (Velsen-Noord)	6310	7500	2
Poland	DABROWA GORNICZA	4500	5000	2
Poland	KRAKOW	1310	2600	1
Romania	GALATI	3250	3200	2
Slovakia	KOSICE	2850	4500	2
Spain	AVILES BOF only	(4200)		0
Spain	GIJON	4480	1200	2
Sweden	LULEA	2200	2200	1
Sweden	ÖXELÖSUND	1800	1700	2
United Kingdom	PORT TALBOT	4770	4900	2
United Kingdom	SCUNTHORPE	3590	3200	3
Total		99340	104290	57
Mean		3973.6	4345.42	2.11

Tonnes per BF per year	1742.80702
Kg per BF per year	1742807.02
Kg per BF per Second	0.05522622
*Mean BF Energy Requirement (GJ/year)	23527.8947
*Mean BF Energy Requirement (J/sec)	745553.995
Heat of Combustion of Metallurgical Coke (J/kg)	29306204.4
**Modern Coke Rate (Kg/thm)	277.15
**Traditional Coke Rate (Kg/ thm)	565

* Liquid Iron Only
** thm = Tonnes of Liquid Iron

Table 10.3
Western European Hot Metal Capacity

It is immediately striking that a typical European furnace outputs a mere 55.2 grams of liquid metal in a second. The figure seems preposterously tiny. But back-multiplication proves the figures check out in their own terms.

Correspondingly the Mean European Blast Furnace Energy Requirement in GJ per Year is 23527.89474 or 745553.9945 J/s which is 0.7455539945 Megawatts.

So according to this an average West European Blast Furnace has a power of three-quarters of a megawatt!

Equally, I suppose you agree that less than sixty grams of hot iron per second is a derisory dribble.

Suppose that the sump (hearth) of a a furnace is a cylinder of five meters radius and it is allowed to accumulate a depth of one meter of molten iron before the molten iron pond is tapped at its base.

Then, given that liquid iron at its $1811°K$ melting point has a mass density of 6980 Kg/m^3, the liquid iron accumulated is 548.207918051419 tonnes. (The trailing digits are not significant from the industrial viewpoint but are included so that students can follow the working and check their own exercises).

Simple division then confirms that the tap is due at 114.891057971526 days.

This is at odds with the Georgian blast furnace tapped twice every twenty-four hours; the maxim "once every two to three hours"; or the modern model of tapping the iron seven times in a 24-hour period.

Seven times a day is in SI terms 0.000081018518519 Hertz whereas 114.89-day frequency implies 0.000000100739555 Hertz, an 804-fold difference.

These perplexing rate anomalies may be factually accurate but are outrageous enough to warrant extended enquiries.

The Bhattacharya-Muthusamy Model of
Heat Energy Balance[10.4]

Table 10.4 is the summative account of a blast furnace static sources-sinks model for an iron-smelting blast furnace.

This model leads forward, in terms of the European fifty-seven furnace averages, to Table 10.5.

Sources	ΔH (Kcals/THM)	ΔH (J/THM)	%
Energy from Fuel	3533420	14783829280	85.918902
Energy from Combustion Air	528378	2210733552	12.848079
Energy from Slag Production	50708	212162272	1.233019
Total Heat Input	**4112506**	**17206725104**	**100.000000**

HEAT OUTPUTS

Sources	ΔH (Kcals/THM)	ΔH (J/THM)	%
Sensible Heat of Hot Metal	696730	2915118320	16.941738
Sensible Heat in Volatile Matter	3091	12932744	0.075161
Heat needed to vaporize Moisture	18929	79198936	0.460279
Latent Heat in BF Gas	1480740	6195416160	36.005783
Sensible Heat in BF Gas	88809	371576856	2.159486
Total Heat in Dust (Dust Catcher + GCP)	39084	163527456	0.950369
Heat Carried away by Water	170143	711878312	4.137210
Sensible Heat in Slag	153299	641403016	3.727630
Summation of BF Reactions	359261	1503148024	8.735817
Calculated Heat Output	**3010086**	**12594199824**	**73.193474**
Other heat Losses	**1102420**	**4612525280**	**26.806526**
Total Heat Output	**4112506**	**17206725104**	**100.000000**

Table 10.4
The Bhattacharya-Muthusamy Model Heat Balance Account

In Table 10.5 the totals and averages are mine, and I have rubricated the Total Heat Output of the average furnace as 0.950262417 MW, according to the statistics of the published Bhattacharya-Muthusamy Model Heat Balance Account.

Nation	Location	Hot Metal Capacity (tonnes *10³/year)	Finished Steel Capacity (tonnes *10³/year)	Number of Furnaces
Austria	DONAWITZ (Leoben)	1370	1570	2
Austria	LINZ	430	6000	3
Belgium	GHENT	4430	5000	2
Chechia	OSTRAVA – BF only	3200		3
Chechia	TRINEC	2100	2400	2
Finland	RAAHE	2400	2600	2
France	DUNKERQUE	6800	6750	3
France	FOS-SUR-MER	5160	5100	2
Germany	BREMEN	3960	3800	2
Germany	DILLINGEN	4790	2760	2
Germany	DUISBURG	11600	11560	4
Germany	EISENHÜTTENSTADT	2340	2400	2
Germany	SALZGITTER	4800	5200	3
Germany	VÖLKLINGEN BOF only	(3240)		0
Hungary	DUNAUIJVAROS	1310	1650	2
Italy	TARANTO	9590	11500	4
Netherlands	IJMUIDEN (Velsen-Noord)	6310	7500	2
Poland	DABROWA GORNICZA	4500	5000	2
Poland	KRAKOW	1310	2600	1
Romania	GALATI	3250	3200	2
Slovakia	KOSICE	2850	4500	2
Spain	AVILES BOF only	(4200)		0
Spain	GIJON	4480	1200	2
Sweden	LULEA	2200	2200	1
Sweden	ÖXELÖSUND	1800	1700	2
United Kingdom	PORT TALBOT	4770	4900	2
United Kingdom	SCUNTHORPE	3590	3200	3
Total		99340	104290	57
Mean		3973.6	4345.42	2.11

Tonnes per BF per year	1742.80702
Kg per BF per year	1742807.02
Kg per BF per Second	0.05522622
Total Heat Input (J/year)	2.9988E+13
Total Heat Input (J/s = W)	950262.417
Total Heat Input (MW)	0.95026242
Heat of Combustion of Metallurgical Coke (J/kg)	29306204.4
**Modern Coke Rate (Kg/thm)	277.15
**Traditional Coke Rate (Kg/ thm)	565

* Liquid Iron Only
** thm = Tonnes of Liquid Iron

Table 10.5
European Metal Averages according to the Static Model

The He Et Al Blast Furnace Model[10.5]

Table 10.6 is my presentation of the relevant parts of the He, Guan, Zhu and Lee report on steel process energies.

INPUTS				
Sources	Mass (Kg)	Energy (MJ)	Energy (J)	Implied ΔHc (J/kg)
Coke	273.6	7779.8	7.78E+09	2.84E+07
Injected Coal	194.8	5130.5	5.13E+09	2.63E+07
Electricity		376.2	3.76E+08	
Hot Blast		1795.5	1.80E+09	
Total	468.4	15082	1.51E+10	5.48E+07
Mean	234.2	3770.5	3.77E+09	2.74E+07

OUTPUTS				
Sinks	Mass (Kg)	Energy (MJ)	Energy (J)	Implied ΔHc (J/kg)
Hot Iron	950	9195.7	9.20E+09	9.68E+06

| IMPLIED EFFICIENCY (%) | | | 60.97 | |

Table 10.6
The He Et Al Blast Furnace Energy Account

In terms of the European furnaces this leads forward to Table 10.7:-

Nation	Location	Hot Metal Capacity (tonnes *10³/year)	Finished Steel Capacity (tonnes *10³/year)	Number of Furnaces
Austria	DONAWITZ (Leoben)	1370	1570	2
Austria	LINZ	430	6000	3
Belgium	GHENT	4430	5000	2
Chechia	OSTRAVA – BF only	3200		3
Chechia	TRINEC	2100	2400	2
Finland	RAAHE	2400	2600	2
France	DUNKERQUE	6800	6750	3
France	FOS-SUR-MER	5160	5100	2
Germany	BREMEN	3960	3800	2
Germany	DILLINGEN	4790	2760	2
Germany	DUISBURG	11600	11560	4
Germany	EISENHÜTTENSTADT	2340	2400	2
Germany	SALZGITTER	4800	5200	3
Germany	VÖLKLINGEN BOF only	(3240)		0
Hungary	DUNAUIJVAROS	1310	1650	2
Italy	TARANTO	9590	11500	4
Netherlands	IJMUIDEN (Velsen-Noord)	6310	7500	2
Poland	DABROWA GORNICZA	4500	5000	2
Poland	KRAKOW	1310	2600	1
Romania	GALATI	3250	3200	2
Slovakia	KOSICE	2850	4500	2
Spain	AVILES BOF only	(4200)		0
Spain	GIJON	4480	1200	2
Sweden	LULEA	2200	2200	1
Sweden	ÖXELÖSUND	1800	1700	2
United Kingdom	PORT TALBOT	4770	4900	2
United Kingdom	SCUNTHORPE	3590	3200	3
Total		99340	104290	57
Mean		3973.6	4345.42	2.11

Tonnes per BF per year	1742.80702
Kg per BF per year	1742807.02
Kg per BF per Second	0.05522622
He Et Al Total Energy (J)	1.51E+10
He Et Al Product Iron (Kg)	950
He Et Al Energy of Liquid Iron (J/Kg)	1.59E+07
He Et Al Furnace Power (W)	876759.87
He Et Al Furnace Power (MW)	0.87675987
Heat of Combustion of Metallurgical Coke (J/kg)	29306204.4
**Modern Coke Rate (Kg/thm)	277.15
**Traditional Coke Rate (Kg/ thm)	565

* Liquid Iron Only
** thm = Tonnes of Liquid Iron

"Assessment on the energy flow and carbon emissions of integrated steelmaking plants"
Huachun He, Hongjun Guan, Xiang Zhu, Haiyu Lee
9 January 2017
Table 3: Page 32
pp 8

Table 10.7
European Metal Averages according to the He Et Al Model

In these terms, the estimate of rated furnace power is 0.876759871 MW.

The Fruehan Et Al Model[10.6]

As approved by the United States Government, the Fruehan, Fortini, Paxton and Brindle assessment of blast furnace energies leads forward, again in terms of the well-documented and almost fiducial European furnaces, to Table 10.9

I rubricated the implied Furnace Power of 0.478977022 MW.

<u>Furnace Rateable Powers according to Examined Reports</u>

Coke Rates have reduced drastically in the last three hundred years. Gone are the flaming, smoking, searing stacks of doom our great-great-grandfathers admired, or at least respected.

Now we have highly-evolved, highly technological blast furnaces, cooly gleaming but not glowing in the moonlight.

It is these to whom new-fangled, non-coal-based iron furnaces must compare, and compete.

When I speak of "Traditional" Coke Rates of furnaces I mean surviving European iron furnaces still in operation (of a kind) and built circa 1950-1975AD. By "Modern" Coke Rates and furnaces I mean 2024AD state-of-the-art East Asian megafurnaces with a yield of maybe 5000 or 10000 tonnes per annum of hot metal.

Table 10.8 is a comparison of Modern and Traditional furnaces in these terms and if the Energy Rate of the older furnaces is double that of the new it betokens not twice the efficiency but twice the waste:-

	Coke Rate (kg/thm)	Coke Feed (kg/year)	Energy Rate (J/year)	Energy Rate (J/s = W)	Energy Rate (MW)
Modern	277.15	483018.96	1.416E+13	448559.22	0.448559
Traditional	565	984685.96	2.886E+13	914436.08	0.914436

thm = tonnes of liquid iron

Table 10.8
Modern and Traditional Furnaces Compared

Finally, I offer you a summary of apparent Thermal Powers in Megawatts for sundry iron smelting blast furnaces and competent models thereof whose mean power is computed to be 0.7925 MW for a population standard deviation of 0.2397 MW. There is insufficient data to justify the calculation of limits of error for even the least sophisticated stochastic models.

This summary constitutes Table 10.10

Nation	Location	Hot Metal Capacity (tonnes $*10^3$/year)	Finished Steel Capacity (tonnes $*10^3$/year)	Number of Furnaces
Austria	DONAWITZ (Leoben)	1370	1570	2
Austria	LINZ	430	6000	3
Belgium	GHENT	4430	5000	2
Chechia	OSTRAVA – BF only	3200		3
Chechia	TRINEC	2100	2400	2
Finland	RAAHE	2400	2600	2
France	DUNKERQUE	6800	6750	3
France	FOS-SUR-MER	5160	5100	2
Germany	BREMEN	3960	3800	2
Germany	DILLINGEN	4790	2760	2
Germany	DUISBURG	11600	11560	4
Germany	EISENHÜTTENSTADT	2340	2400	2
Germany	SALZGITTER	4800	5200	3
Germany	VÖLKLINGEN BOF only	(3240)		0
Hungary	DUNAUIJVAROS	1310	1650	2
Italy	TARANTO	9590	11500	4
Netherlands	IJMUIDEN (Velsen-Noord)	6310	7500	2
Poland	DABROWA GORNICZA	4500	5000	2
Poland	KRAKOW	1310	2600	1
Romania	GALATI	3250	3200	2
Slovakia	KOSICE	2850	4500	2
Spain	AVILES BOF only	(4200)		0
Spain	GIJON	4480	1200	2
Sweden	LULEA	2200	2200	1
Sweden	ÖXELÖSUND	1800	1700	2
United Kingdom	PORT TALBOT	4770	4900	2
United Kingdom	SCUNTHORPE	3590	3200	3
Total		99340	104290	57
Mean		3973.6	4345.42	2.11

Tonnes per BF per year	1742.80702
Kg per BF per year	1742807.02
Kg per BF per Second	0.05522622
Specific Super Heat Required (MJ/tonne)	8673
Specific Super Heat Required (J/Kg)	8673000
Super Heat Required (J)	478977.022
Fruehan Et Al Furnace Power (MW)	0.47897702
Heat of Combustion of Metallurgical Coke (J/kg)	29306204.4
**Modern Coke Rate (Kg/thm)	277.15
**Traditional Coke Rate (Kg/ thm)	565

* Liquid Iron Only

** thm = Tonnes of Liquid Iron

Fruehan-Fortini-Paxto-Brindle Model

Page 3

Liquid Iron Super Heat to 1873°K

Requires 8673 MJ/tonne

(Ore Reduction + Sensible Heat + Latent Heat of Fusion)

Table 10.9
European Metal Averages according to the Fruehan Model

Model or Account	Thermal Power (MW)
European Modern Blast Furnaces	0.4486
European Traditional Blast Furnaces	0.9144
Bhattacharya-Muthusamy Static Model	0.9503
Fruehan Model	0.4790
He Et Al Model	0.8768
Mean	**0.7338**
Population Standard Deviation	**0.2219**

Table 10.10
Blast Furnace Power Summary

The phrase "Hydrogen Direct Reduction Furnace" (H_2DRF) seems to me somewhat of a misnomer, because the reduction of iron ore with hydrogen is not direct. Hydrogen does not occur free and stable in Nature, so the process comprises two distinct and sequential parts:-

(a)   The Electrolysis of Water to produce Hydrogen and Oxygen

(b)   The Thermal Reduction of Iron Ore by Reaction with Hydrogen

The total process of Conformation I, Reduction with Mechano-Electric Heat Recovery, is represented schematically in Figure 11.1, whilst Conformation II, Reduction with Heat Exchange Heat Recovery is presented in Figure 11.2

Both schemes are naive.

Firstly, we must not think of the Electrolyser as a simple glass H-tube like that demonstrated by schoolmasters in a school laboratory. It is purely illustrative. Industrial water electrolysers are assemblies of concentric metal cylinders separated by appropriately engineered permeable membranes that prevent physical gas mixing whilst preserving a moist Ion Bridge that enables electrical flows with minimal impedances.

Secondly, the interposition of moving parts (literal or figurative) vitiates any system, introducing delay, waste and inefficiency. Therefore, any scheme of recovery involving traps, pumps or engines, or any permutation of the three, is inherently sub-optimal.

Thirdly, any well-designed and well-managed furnace system will exhaust cool gases and minimise the Total Sensible Heat contained in the system. This renders Heat Recovery measures redundant. Neither of my schemes present valve work. A viable scheme would include adequate valves to manage the flow of input and output gasses, control steam, and manage safety routings. It would also include, in addition to rectification, controls to change and gradualise electrical inputs. A modern system would include Computer Control to Optimise Production and impose timely Safety Controls.

Hydrogen is a fugacious gas in the common rather than the thermodynamic sense of that word: It is more penetrating than

WD40 or indeed any other cool gas and will insinuate itself through the narrowest hairline seal or fracture. The safe and economical management of bulk hydrogen is a specialty of its own and it is not the type of chemical that may safely be stored near residential property.

From a chemical point of view we can draw a distinction between monatomic "nascent" hydrogen and diatomic common hydrogen. Monatomic hydrogen (H) is very highly reactive. It is sometimes appropriate to consider its action at or near electrodes, certainly from the viewpoint of possible electrode degradations, but it is diatomic hydrogen (H_2) that we can expect to be of practical interest in storage and smelting. Exhaust steam from a H_2DRF may be condensed and feed back to the electrolyser but inevitably some hydrogen and almost all oxygen is voided the process and accordingly copious amounts of clean water must be available to the electrolyser.

Iron Ore and Hydrogen are separately and when in contact thermodynamically-stable at ambient temperatures and pressures, so heat must be added to the reagents, presumably by burning some of the hydrogen in some of the oxygen. The resulting steam (H_2O) is not passive and its management must be undertaken.

Sophisticated H_2DRF experiments continue, especially in Sweden.

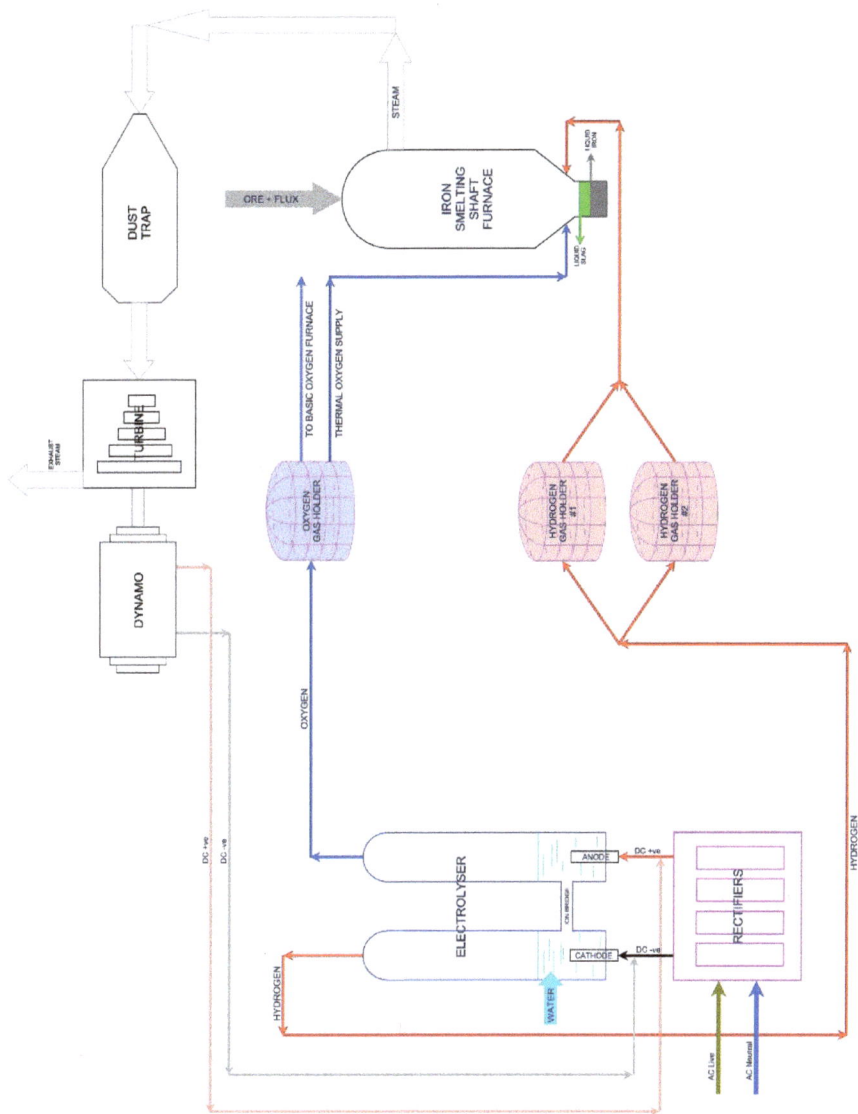

Figure 11.1
Conformation I
A H₂DRF System with Turbo-Electric Heat Recovery

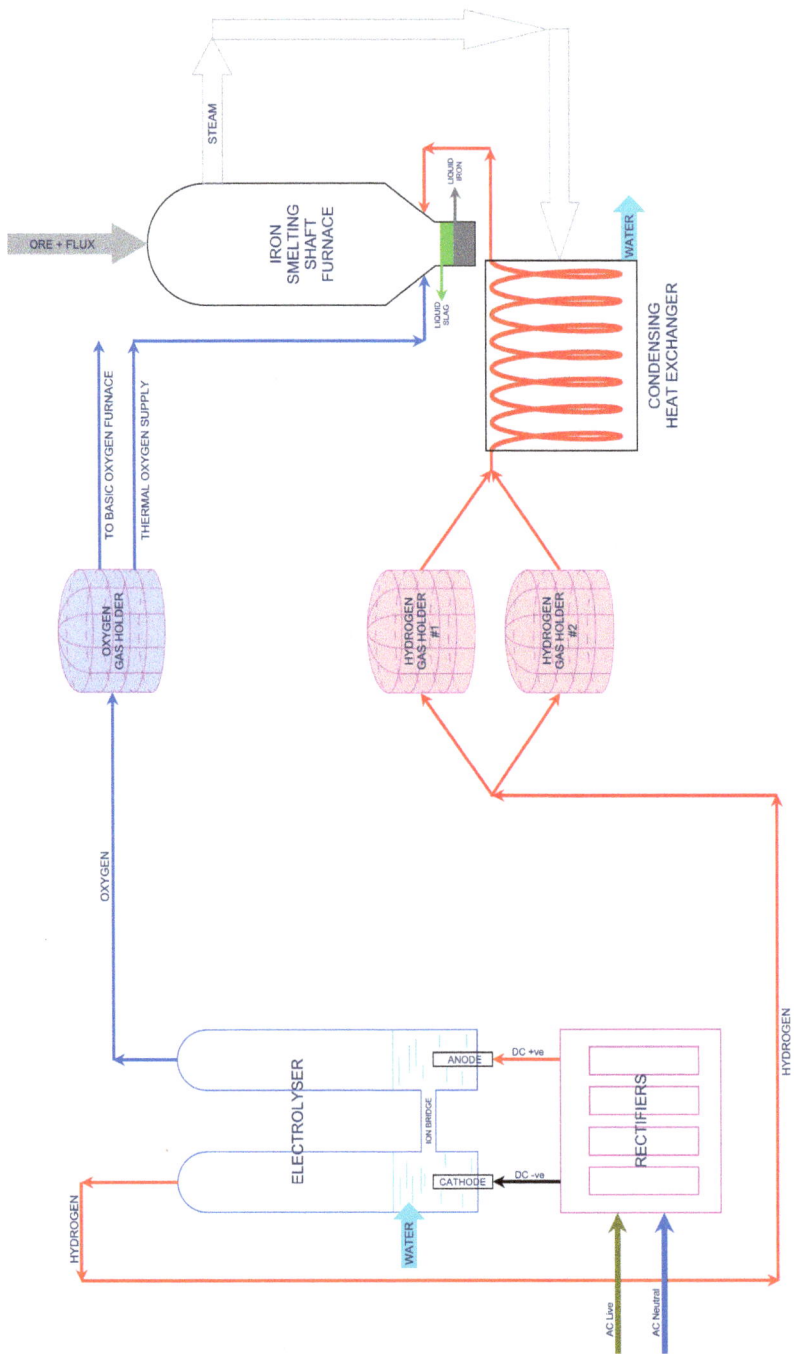

Figure 11.2
Conformation II
A H₂DRF System with Heat Exchange Heat Recovery

The Use of Hydrogen for Iron Smelting

Aside from the issues already adumbrated special problems attend the application of hydrogen to our purpose.

Steam, the inevitable by-product, tends to penetrate the surfaces of ore particles, inhibiting the reduction reactions. It may be possible to "activate" the ore particles by dusting them with quicklime (CaO) or some other strong desiccant, or incorporating similar agents into pelletised compositions.

The ore tends to sinter at intermediate temperatures above the main locus of smelting. The sinter is highly vesicular, being a blend of unreduced ore, iron and trapped gases, unfit for further reduction without secondary processing. This sinter problem reminds us of the Classical and medieval bloomery technology that yielded "sponge iron" requiring extended and expensive re-working.

He, Li, Zhang and Conejo[11.1] draw our attention to the step-wise reduction of iron ore with hydrogen reminiscent of the similar CO-mediated reductions in the traditional carbon furnace. These multi-stage sub-reactions, which effect by a variety of mechanisms at a variety of temperature ranges, were further resolved by Daniel Spreitzer and Johannes Schenk[11.2]

Several intermediate reactions, and the summative Hematite-Iron pathway are endothermic, so a proportion of the energy input must avail from some secondary source, presumably the Hydrogen-Oxygen gas reaction. The direct Hematite-Iron pathway is weakly exothermic.

Joswiak and his colleagues[11.3] are especially emphatic that the conventional Hydrogen-Iron pathways are all highly provisional, and that some display unstable species interchange even at low temperatures.

(a)   Hematite to Magnetite

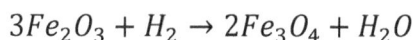

$$3Fe_2O_3 + H_2 \rightarrow 2Fe_3O_4 + H_2O$$

Pathway 11.1
The Reduction of Hematite to Magnetite

This reaction takes place below 570°C. But also between 25 and 650 Celsius at phase boundaries and between 250 and 800 Celsius by random nucleation.

(b)   Magnetite to Iron Direct Reduction

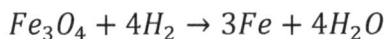

$$Fe_3O_4 + 4H_2 \rightarrow 3Fe + 4H_2O$$

Pathway 11.2
Magnetite to Iron Direct Reduction

This reaction takes place above 570°C, and according to Daniel Spreitzer and Johannes Schenk between 220 and 430 Celsius.

(c)   Magnetite to Wüstite

$$2Fe_3O_4 + 2H_2 \rightarrow 6FeO + 2H_2O$$

Pathway 11.3
The Reduction of Magnetite to Wüstite

This takes place between 800-900 Celsius at phase boundaries.

(d)   Wüstite to Iron

$$6FeO + 6H_2 \rightarrow 6Fe + 6H_2O$$

Pathway 11.4
The Reduction of Wüstite to Iron

For Carbon Monoxide this takes place above 800 Celsius by nucleation and growth. I have no information for the hydrogen reaction. There is also an unstable reaction involving the decomposition of FeO to magnetite and iron above 150°C.

As with the carbon-based furnaces the actual reaction cascades are highly complex. Except where there are special reasons to do otherwise I shall abridge these steps to the reduction of iron with the summative pathway for Hematite to Iron:-

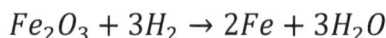

$$Fe_2O_3 + 3H_2 \rightarrow 2Fe + 3H_2O$$

Pathway 11.5
The Direct Reaction for the Reduction of Hematite to Iron

This direct treatment is valid in the terms of Hess's Law of thermodynamic equivalences and the Conservation of Matter.

Rates of Reaction and Extents of Reaction

Unfortunately, we cannot identify Extent of Reaction, a largely theoretical concept, with furnace Yield because so many adventitious factors impose in the latter.

Parameters and Dimensions

Table 11.1 defines some of the parameters of the empirico-theoretical constructs we are about to explore.

The Arrhenius Equation

The Arrhenius Equation relates the Rate of Reaction to Temperature according to:-

$$r = A.e^{-\frac{E_a}{RT}} \equiv A.\exp\left(-\frac{E_a}{RT}\right)$$
Equation 11.1

By taking logarithms:-

$$\ln(r) = \ln(A) - \frac{E_a}{R}.\frac{1}{T}$$
Equation 11.2

You can see that Equation 11.2 has the form of a First Order Polynomial Algebraic Equation:-

$$y = \alpha + \beta.x$$
Equation 11.3

and as such an equation it is presumably solvable using Linear Regression or some such method to give α = ln(A) and β = -E_a/R.

Parameter	Symbol	Units	Dimensions	Value
Percentage Specific Defect	$PSD(x,y)$		$M^0L^0T^0\Theta^0$	
Linear Regression Intercept	α		$M^0L^0T^0\Theta^0$	
Linear Regression Grade	β	seconds	T^{+1}	
Correlation Coefficient	R^2		$M^0L^0T^0\Theta^0$	
Temperature (°K)	T	°K	Θ^{+1}	
Universal Gas Constant	R	J/(Kg.mol)	$ML^2T^{-2}\Theta^{-1}mol^{-1}$	8.3144626181531240
Activation Energy	E_a	Joules	ML^2T^{-2}	
Arrhenius Exponent	$E_a/(RT)$		$M^0L^0T^0\Theta^0$	
Pre-Exponential Coefficient	A		$T-1$	
Absolute Grade (E_a/R)	κ	mol.°K	$\Theta^1 mol^1$	
Exponential Numerator Composite	ρ	mol	mol^1	
Rate of Reaction (k_T) at Temperature T	r_T	seconds^{-1}	T^{-1}	
Stoichiometric Coefficient	λ			
Molar Mass (Formula Weight)	FW_x	grams/mol	$Mmol^{-1}$	
Ratio of Solid to Gas Stoichiometric Coefficients	b		$M^0L^0T^0\Theta^0$	
Molar Concentration of Gas in One Cubic Meter at STP	z_g	mols/m3	ML^{-3}	
Fraction of Solid Converted	ω		$M^0L^0T^0\Theta^0$	
Fraction Change Coefficient	S_a	grams/(m^3.°K)	$ML^{-3}T^{-1}$	
Relative Time	t	seconds	T^{+1}	
Fraction Factor	f_ω		$M^0L^0T^0\Theta^0$	
Fraction Factor Rate Coefficient	x		$M^0L^0T^0\Theta^0$	

Table 11.1
Notation and Dimensions

If this can be arranged, perhaps by an experiment that measures the fraction of product produced at different temperatures, then we immediately obtain:-

$$A = exp(\alpha)$$
Equation 11.4

and:-

$$E_a = -\beta.R$$
Equation 11.5

Since the Universal Gas Constant R is publicly-known to high precision we have an immediate knowledge of the Activation Energy E_a.

He Et All performed the experiments and I reproduce their Table 1 as Table 11.2:-

T, °C	Rate, %/min	ln(Rate), (min^{-1})	1/T (K^{-1})
600	1.93	0.66	0.00115
650	2.96	1.08	0.00108
700	3.77	1.33	0.00103
750	4.67	1.54	0.000979
800	5.41	1.69	0.000933
850	6.78	1.91	0.000898
900	8.22	2.11	0.000854
950	10.97	2.4	0.000819
1000	11.72	2.46	0.000787
1050	10.97	2.4	0.000757

Table 11.2
He Et Al Arrhenius Data as Given[11.1]

I have rendered their result to SI in Table 11.3

Figure 11.3 is the plot of the SI experimental Points (in blue) with my fitted EXCEL® linear regression line as a red solid straight line.

The regression α is 0.101354557 and the regression slope β is -77.58647208. (The minus sign merely means that the line is showing decreasing ln(rate) with increasing reciprocal temperature 1/T).

Mathematically:-

$$\ln(r) = \alpha + \frac{\beta}{T}$$

Equation 11.6

so that:-

$$r = A.e^{-\frac{E_a}{RT}} \equiv A.\exp\left(-\frac{E_a}{RT}\right) = exp\left(\alpha + \frac{\beta}{T}\right) \equiv exp(\alpha).exp\left(\frac{p}{T}\right)$$

Equation 11.7

In Equation 11.7, the local constant p is defined by:-

$$p = -\kappa = -\frac{E_a}{R} = \beta$$

Equation 11.8

Hence:-

$$\ln(r) = \ln(A) - \kappa.\frac{1}{T} = \alpha + \frac{\beta}{T}$$

Equation 11.9

Figure 11.4 shows the extended variation of Rate of Reaction (y-axis) plotted against the reciprocal temperature.

T_r °K	Rate, (fraction /second)	ln(Rate), (sec^{-1})	1/T (K^{-1})	ln(Rate), (sec^{-1})	Computed ln(Rate)	Fractional Specific Defect (FSD)
873.15	0.000321667	0.011	0.00115	0.011	0.01220399	-0.1094532
923.15	0.000493333	0.018	0.00108	0.018	0.01771627	0.01576255
973.15	0.000628333	0.022166667	0.00103	0.022166667	0.02167199	0.02231612
1023.15	0.000778333	0.025666667	0.000979	0.025666667	0.02572267	-0.0021818
1073.15	0.000901667	0.028166667	0.000933	0.028166667	0.02938999	-0.0434315
1123.15	0.00113	0.031833333	0.000898	0.031833333	0.03218912	-0.0111766
1173.15	0.00137	0.035166667	0.000854	0.035166667	0.03571884	-0.0157015
1223.15	0.001828333	0.04	0.000819	0.04	0.03853518	0.0366205
1273.15	0.001953333	0.041	0.000787	0.041	0.04111682	-0.0028494
1323.15	0.001828333	0.04	0.000757	0.04	0.04354294	-0.0885736
						-0.0198668 Mean FSD
						0.98405941 R^2

Table 11.3
He Et Al Arrhenius Data in SI Units

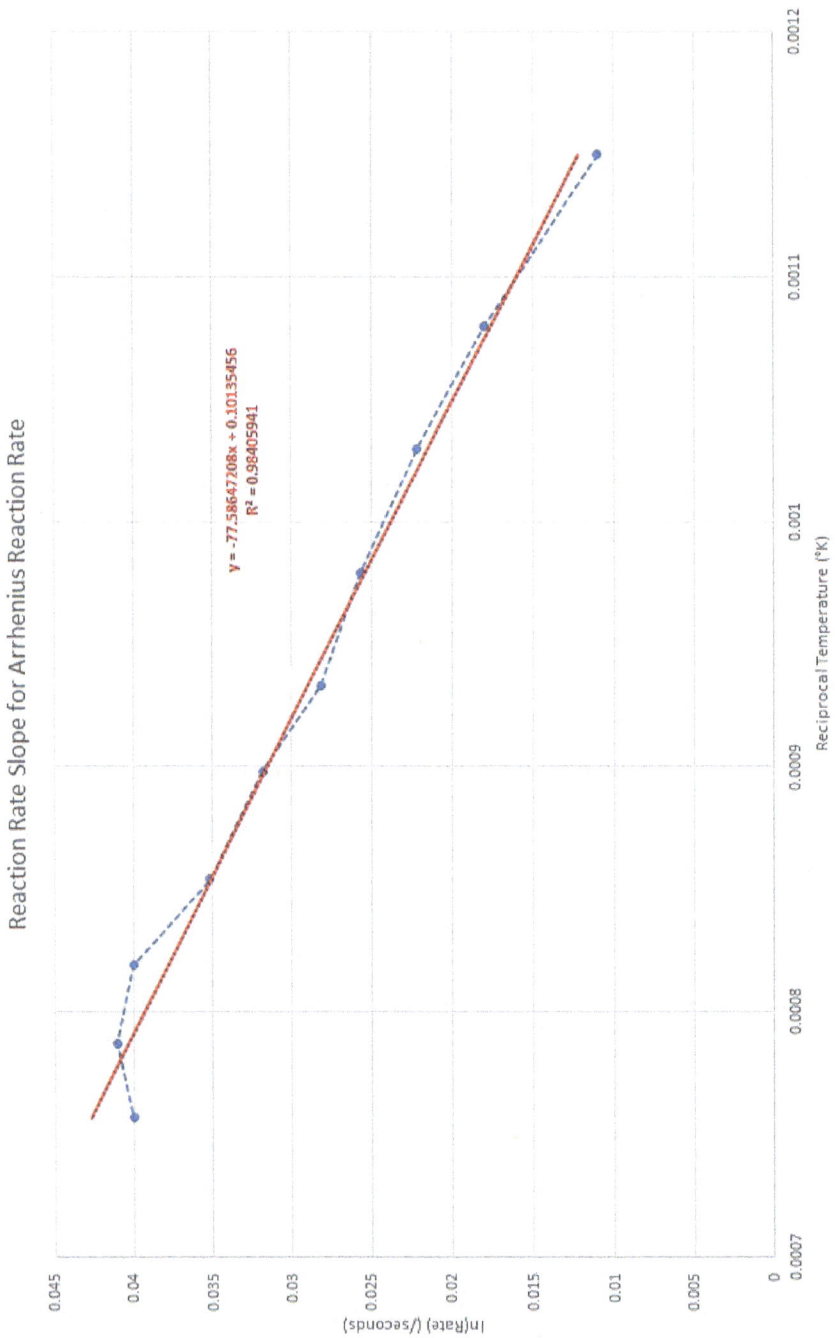

Figure 11.3
ln(reaction rate) Plotted Versus Reciprocal Temperature (°K)

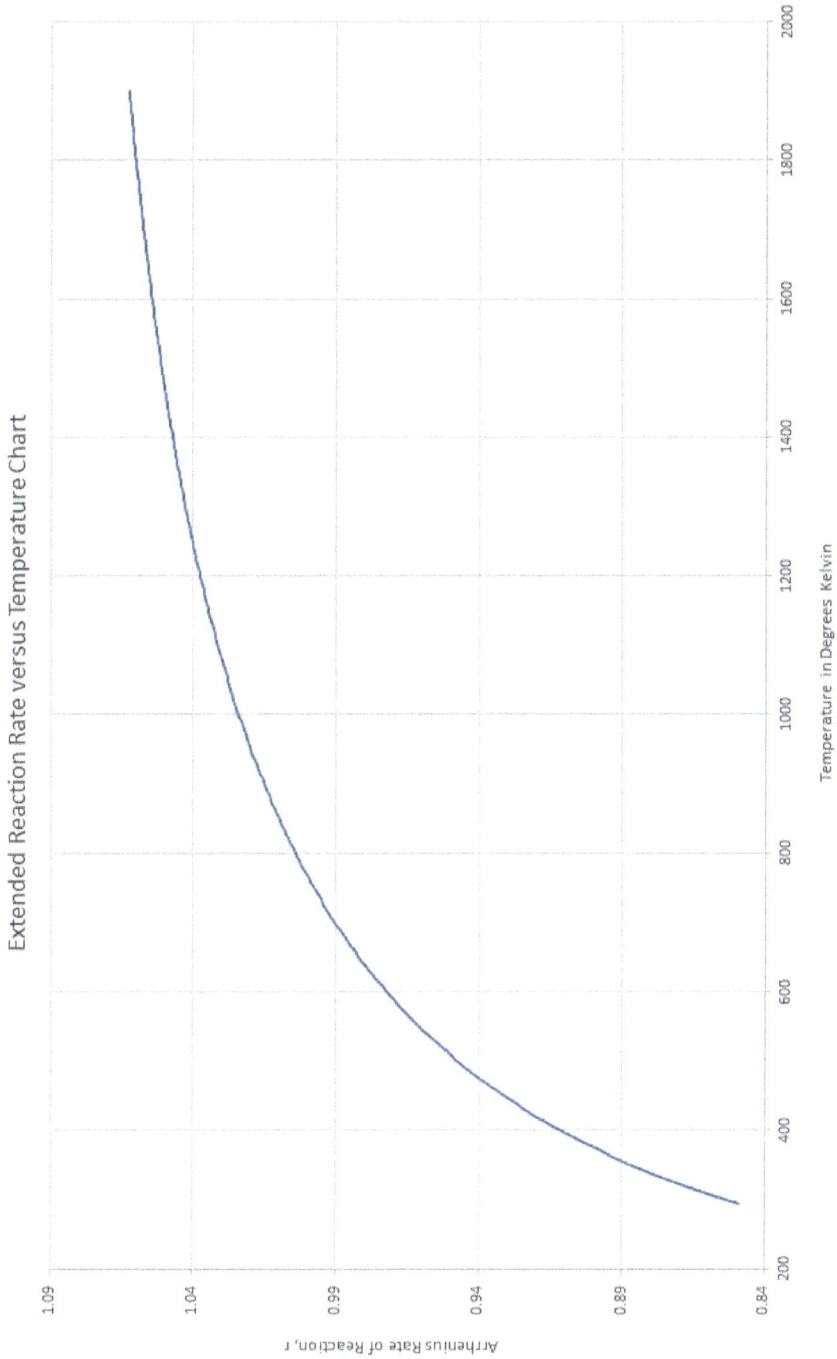

Figure 11.4
Extended Rate Chart

Loosely speaking, the variation of rate is Asymptotic: That is to say that the curve rises to a horizontal limit called the Asymptote. We cannot rightly see where the limit is but it is going to be around 1.07 mols/second when the Temperature reaches 2000°K.

What *is* clear is that there is a limit to the Rate of Reaction and that an indefinite increase in applied heat is not necessarily going to force an optimum.

Reaction Extent

We will assume for simplicity that Pathway 11.4 applies to this reaction:-

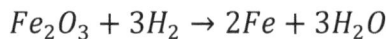

$$Fe_2O_3 + 3H_2 \rightarrow 2Fe + 3H_2O$$
Pathway 11.4
The Direct Reaction for the Reduction of Hematite to Iron

and that therefore:-

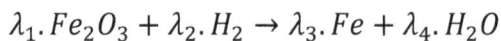

$$\lambda_1.Fe_2O_3 + \lambda_2.H_2 \rightarrow \lambda_3.Fe + \lambda_4.H_2O$$
Pathway 11.5
The Lambda Stoichiometric Coefficients for the
Direct Reaction for the Reduction of Hematite to Iron

so that the respective Stoichiometric Coefficients are λ_1 = 1; λ_2 = 3; λ_3 = 2; λ_4 = 3.

Only the Reagents are of interest at this time because we want to define the Ratio of the Reagent Stoichiometric Coefficient, b, as:-

$$b = \frac{\lambda_1}{\lambda_2} = \frac{1}{3} = 0.333333333333'$$
Equation 11.10

The Molar Concentration of Gas, z_g, is given by:-

$$z_g = \frac{1000}{22.4} = 44.6428571428571$$
Equation 11.11

This is because there is one mole of *any* ideal gas in 22.4 liters of the gas at STP, and there are 1000 liters in a cubic meter.

ω is the Fraction of Solid (in this case Hematite)
Converted which by arbitrary example is 0.05, i.e. 5%
So:-

$$\frac{d\omega}{dt} = b.r_T.z_g.(1-\omega) = S_a(1-\omega)$$

Equation 11.12

Clearly the range of ω is 0≤ω≤1 and the range of the Relative Reaction Duration t is also 0≤t≤1

The red dot-dash curve representing the degree of completion at the liquid iron temperature of nineteen hundred degrees Kelvin shows that the Pathway 11.4 reaction completes at entire conversion ω = 1 and unity relative reaction time t = 1.

The solid blue cold curve represents reaction behavior at 20°C (293.15°K) where ω = 1 is not attained until t = 1.177401858. In real life we expect nothing to happen if we mix hematite and hydrogen at room temperature.

Table 11.4 summarises the data and results of this review of the He Et Al report. All figures other than Formula Weights are SI.

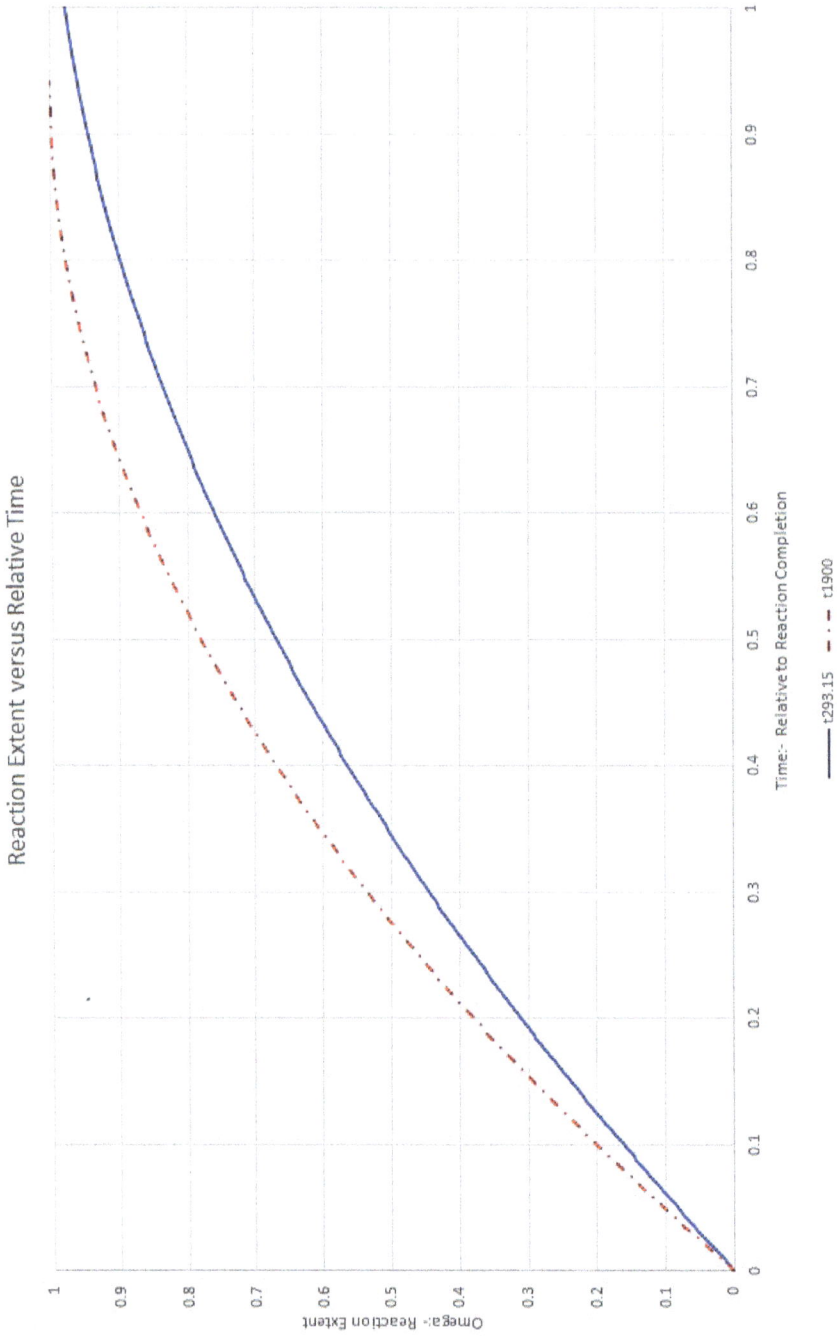

Reaction Extent versus Relative Time

Omega:- Reaction Extent

Time:- Relative to Reaction Completion

——— t293.15   — · · — t1900

Figure 11.5
Time versus Omega Chart

Intercept of Rate Chart Line, α	0.101354557
Slope of Rate Chart Line, β	-77.58647208
Regression Coefficient, R^2	0.984059407
Universal Gas Constant, R	8.314462618
Pre-Exponential Constant, A	1.106668949
Activation Energy, E_a	645.0898217
E_a/R	77.58647208
$p=-E_a/R$	-77.58647208
Minimum Temperature (°K)	293.15
Maximum Temperature (°K)	1900
Number of Intervals	256
Temperature Interval (°K)	6.276757813
Minimum ω	0
Maximum ω	1
Number of Intervals	256
ω Interval	0.00390625
Specimen Temperature (°K)	293.15
Specimen Reaction Rate, r_T	0.84932769
Specimen ω	0.05
Fraction Factor f_ω	0.025320566
Relative Time of Reaction	0.029812481
Stoichiometric Coefficient of Solid Reagent (λ_1)	1
Stoichiometric Coefficient of Gas Reagent (λ_2)	3
Ratio of Stoichiometric Coefficients (b)	0.333333333
Molar Gas Concentration (mol/m3) (z_g)	44.64285714
Fraction Change Coefficient Sa	12.63880491
$d\omega/dt$	12.00686466

Table 11.4
He Et Al Arrhenius-Related Data and Outcomes in SI

Pathway Stoichiometry

Formula Weights (or Molar Masses) are conventionally measured in *grammes*/mole. We must bear this in mind if we are working in any Kilogram-Meter-Second (KMS) system including SI.

Table 11.5 lists the chemical formula weights *in grammes/mole* relevant at this time:-

Chemical	Symbol	Formula Weight (grams/mol)
Monatomic Hydrogen	H	1.008
Diatomic Hydrogen	H_2	2.016
Monatomic Oxygen	O	15.999
Diatomic Oxygen	O_2	31.998
Water	H_2O	18.01528
Hematite	Fe_2O_3	159.687
Magnetite	Fe_3O_4	231.533
Wüstite	FeO	71.844
Iron	Fe	55.845

Table 11.5
Formula Weights for Iron Smelting with Hydrogen

In reviewing pathway stoichiometry we will attempt to marry the water electrolysis and hematite reduction equations in terms of equivalent hydrogen mass according to $6H_2$ (that is 12.09528 gram moles of hydrogen).

Therefore the appropriate water electrolysis pathway is:-

$$6H_2O \rightarrow 6H_2 + 3O_2$$
Pathway 11.6

and the corresponding hematite reduction pathway is:-

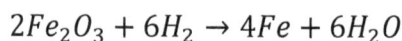

$$2Fe_2O_3 + 6H_2 \rightarrow 4Fe + 6H_2O$$
Pathway 11.7

In the case 11.6 you can see that the respective reactant Stoichiometric Coefficients λ_i are:- $\lambda_1 = 6$, $\lambda_2 = 0$, $\lambda_3 = 6$ and $\lambda_4 = 3$. Hence:-

$$\lambda_1.FW_{H2O} + \lambda_2.0 = \lambda_3.FW_{H2} + \lambda_4.FW_{O2} = 108.09168$$
Equation 11.13

Therefore the moles of each species in a kilogram of that species are:-

$$\sigma_{H2} = \frac{1000}{FW_{H2}} = 496.061273488501$$
Equation 11.14a

$$\sigma_{O2} = \frac{1000}{FW_{O2}} = 31.251171918947$$
Equation 11.14b

$$\sigma_{H2O} = \frac{1000}{FW_{H2O}} = 55.508435061792$$
Equation 11.14c

Hydrogen Volume in 1 Kg of Water at STP

To establish how many moles of H_2 are required we may use:-

$$mols_{H2} = \sigma_{H2}.\frac{\lambda_3}{\lambda_1} = 496.061273488501$$
Equation 11.15

and the grams of water needed are confirmed by:-

$$m_{H2O} = \sigma_{H2}.\frac{\lambda_3}{\lambda_1}.FW_{H2O} = 1000$$
Equation 11.16

There are $v_{mol} = 22.414$ liters of *any* ideal gas for a mole of that gas.

Therefore:-

$$v_{H2} = v_{mol}.\sigma_{H2}.\frac{\lambda_3}{\lambda_1} = 11118.7173839713 \; liters$$
Equation 11.17

Accordingly the Volume of Hydrogen needed at Standard Temperature and Pressure (STP) for one kilogram of water is 11.1187 cubic meters.
Furthermore:-

$$6\left(\frac{\sigma_{H2O}.FW_{H2O}}{\lambda_1}\right) = 2\left(\frac{\sigma_{H2}.FW_{H2}}{\lambda_3} + \frac{\sigma_{O2}.FW_{O2}}{\lambda_4}\right) = 1000 \; grams$$
Equation 11.18

Hematite Reduction Stoichiometry

According to Pathway 11.7 the respective reactant Stoichiometric Coefficients λ_i are:- $\lambda_1 = 2$, $\lambda_2 = 6$, $\lambda_3 = 4$ and $\lambda_4 = 6$.
Therefore we may write the appropriate stochiometric equation as:-

$$\lambda_1.FW_{Fe2O3} + \lambda_2.FW_{H2} = \lambda_3.FW_{Fe} + \lambda_4.FW_{H2O} = 331.47128$$
Equation 11.19

The mols of each species in one kilogram of that species as per the respective stoichiometric coefficients λ_x are:-

$$\sigma_{Fe2O3} = \frac{1000}{FW_{Fe2O3}} = 6.26221131205851$$
Equation 11.20a

$$\sigma_{H2} = \frac{1000}{FW_{H2}} = 496.061273488501$$
Equation 11.20b

$$\sigma_{Fe} = \frac{1000}{FW_{Fe}} = 17.90670606142$$
Equation 11.20c

$$\sigma_{H2O} = \frac{1000}{FW_{H2O}} = 55.508435061792$$

Equation 11.20d

In terms of the moles of iron in one kilogram of iron as per λ_3:-

$$\sigma_{Fe}.\lambda_1.FW_{Fe2O3} + \sigma_{Fe}.\lambda_2.FW_{H2} = \sigma_{Fe}.\lambda_3.FW_{Fe} + \sigma_{Fe}.\lambda_4.FW_{H2O}$$
$$= 5935.55877876265$$

Equation 11.21

or:-

$$(\lambda_1.FW_{Fe2O3} + \lambda_2.FW_{H2}) = (\lambda_3.FW_{Fe} + \lambda_4.FW_{H2O})$$

Equation 11.22

From Equation 11.21 it follows that the Absolute Ideal Mass Yield of Fe in grams per gram of H_2 is:-

$$Y_A = \frac{\sigma_{Fe}.\lambda_3.FW_{Fe}}{\sigma_{Fe}.\lambda_2.FW_{H2}} = \frac{\lambda_3.FW_{Fe}}{\lambda_2.FW_{H2}} = 18.4683612119769$$

Equation 11.23

and the Relative Ideal Mass Yield of Species when FW_{Fe} is unity is given by:-

$$Y_R = \frac{\lambda_1.FW_{Fe2O3} + \lambda_1.FW_{H2}}{\lambda_3} = \frac{\lambda_3.FW_{Fe} + \lambda_4.FW_{H2O}}{\lambda_3} = 82.86792$$

Equation 11.24

Requirements in Kilograms for One Kilogram of Iron

The Species Kilogram Ratios, $ReKg_i$ are given by:-

$$ReKg_1 = \frac{\lambda_1}{\lambda_3}.\frac{FW_{Fe2O3}}{FW_{Fe}} = 1.42974303876802$$

Equation 11.25a

$$ReKg_2 = \frac{\lambda_2}{\lambda_3}.\frac{FW_{H2}}{FW_{Fe}} = 0.054146655922643$$

Equation 11.25b

$$ReKg_3 = \frac{\lambda_3}{\lambda_3} \cdot \frac{FW_{Fe}}{FW_{Fe}} = 1$$
Equation 11.25c

$$ReKg_4 = \frac{\lambda_4}{\lambda_3} \cdot \frac{FW_{H2O}}{FW_{Fe}} = 0.483891485361268$$
Equation 11.25d

from which:-

$$ReKg_1 + ReKg_2 = ReKg_3 + ReKg_4$$
Equation 11.26

Accordingly:-

$$\frac{1}{\lambda_3}\left(\lambda_1 \frac{FW_{Fe2O3}}{FW_{Fe}} + \lambda_2 \frac{FW_{H2}}{FW_{Fe}}\right) = \frac{1}{\lambda_3}\left(\lambda_3 \frac{FW_{Fe}}{FW_{Fe}} + \lambda_4 \frac{FW_{H2O}}{FW_{Fe}}\right)$$
$$= 1.48388969469066$$
Equation 11.27

and:-

$$\frac{1}{\lambda_3 . FW_{Fe}}(\lambda_1 FW_{Fe2O3} + \lambda_2 FW_{H2})$$
$$= \frac{1}{\lambda_3 . FW_{Fe}}(\lambda_{13} FW_{Fe} + \lambda_4 FW_{H2O})$$
$$= 1.48388969469066$$
Equation 11.28

Hydrogen Production by the Electrolysis of Water

By Faraday's Second Law the Ideal Mass Yield (in Kilograms) of Hydrogen is:-

$$M_{H2} = \frac{Q.FW_{H2Kg}}{\lambda_{H2}.F} = 0.499970238169286$$
Equation 11.29

M_{H2} being the Mass of Liberated Hydrogen in Kilograms; Q the liberative Electric Charge in Coulombs; FW_{H2Kg}

being the Formula Weight of H_2 in Kilograms; $\lambda_{H2} = 6$ as the Stoichiometric Coefficient of Hydrogen and F being the Faraday Constant in Coulombs per Mol.

The value of F is 96485.3321233100184 C/mol and of FW$_{H2KG}$ 0.00201588 kilograms.

The Mass is one half kilogram due to $\lambda_3 \equiv \lambda_{H2}$ being double its LCM value of three.

Transposition of Equation 11.29 for the Charge, Q_{H2} yields:-

$$Q_{H2} = \frac{M_{H2}\lambda_{H2}F}{FW_{H2Kg}} = 143579363.3\ Coulombs$$
Equation 11.30

Now one Kilowatt-Hour (KWh) of energy as charge is:-

$$KWh = 1000 \times 60 \times 60 = 3600000\ Coulombs$$
Equation 11.31

Therefore the energy to liberate one kilogram of hydrogen is given in KWh as:-

$$E_{KWh} = \frac{Q_{H2}}{Kwh} = 39.8831564722222\ KWh$$
Equation 11.32

Many of the estimates of E_{KWh} in the online literature center about 39.4 KWh to liberate one kilogram of hydrogen, though this is a theoretical Faraday's Law figure for 100% efficient systems. More realistic empirical estimates hover between 50 and 55 KWh. The PSD(39.4, E_{KWh}) = -1.22628546249296

<u>The Mass of Hydrogen Required to liberate</u>
<u>One Kilogram of Iron</u>

Pathway 11.8 is:-

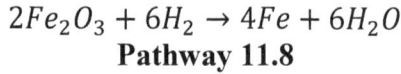

$$2Fe_2O_3 + 6H_2 \rightarrow 4Fe + 6H_2O$$
Pathway 11.8

and in this formulation you can see that the respective reactant Stoichiometric Coefficients λ_i are:- $\lambda_1 = 2$, $\lambda_2 = 6$, $\lambda_3 = 4$ and $\lambda_4 = 6$.
Accordingly:-

$$\lambda_1.FW_{Fe2O3} + \lambda_2.FW_{H2} = \lambda_3.FW_{Fe} + \lambda_4.FW_{H2O} = 331.47128$$
Equation 11.33

and the Number of Kilograms of Hydrogen, Kg_{H2}, ideally required to produce one kilogram of iron is:-

$$Kg_{H2} = \frac{\lambda_3.FW_{Fe}}{\lambda_2.FW_{H2}} = 18.4683612119769 \; kilograms$$
Equation 11.34

<u>The Reaction Thermodynamics of Hydrogen</u>

Pathway 11.9 may be specified as:-

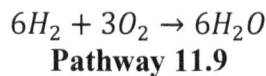

$$6H_2 + 3O_2 \rightarrow 6H_2O$$
Pathway 11.9

Clearly, $\lambda_1 = 6$; $\lambda_2 = 3$; $\lambda_3 = 6$ and $\lambda_4 = 0$ and the appropriate Heats of Formation Δ_fH_x (KJ/mol) may be selected from Table 11.6.

Substance	Formula	Formula Weight (g/mol)	Heat of Formation $\Delta_f H_x$ (KJ/mol)	Standard Entropy ΔH (KJ/mol)	Specific Heat C_p at T°K (J(mol.°K))	State of Substance at T°K
Hydrogen	H_2	2.01588	0.000	29.089	31.33934500	Gas
Oxygen	O_2	31.99880	0.000	32.378	35.90654323	Gas
Water	H_2O	18.01528	-241.826	37.738	44.62273084	Gas
Hematite	Fe_2O_3	159.68800	-825.500	139.568	142.04970432	Solid
Magnetite	Fe_3O_4	231.53300	-1120.890	202.448	200.83200013	Solid
Wüstite	FeO	71.84400	-249.530	54.927	68.19920000	Solid
Iron	Fe	55.84500	0.000	37.203	39.00284868	Solid

Table 11.6
Selected Thermodynamic Data
for Iron Smelting with Hydrogen

Accordingly, the Pathway Heat of Reaction $\Delta_r H_{H2}$ in KJ/mol is given by:-

$$\Delta_r H_{H2} = \frac{(\lambda_3 . \Delta_f H_{H2O} + \lambda_4 . 0) - (\lambda_1 . \Delta_f H_{H2} + \lambda_2 . \Delta_f H_{O2})}{\lambda_3}$$
$$= -241.826$$
Equation 13.35

If $\Delta_r H_{H2}$ is +ve water is divided and if -ve water is formed: This is in agreement with Pathway 11.9
Allow that:-

$$\tau_{H2} = \frac{1000}{\lambda_1 . FW_{H2}} = 82.6768789147502$$
Equation 11.36a

$$\tau_{O2} = \frac{1000}{\lambda_2 . FW_{O2}} = 20.8341146126313$$
Equation 11.36b

$$\tau_{H2O} = \frac{1000}{\lambda_3 . FW_{H2O}} = 9.251405843632$$
Equation 11.36c

where τ_x is the mols of species x in one kilogram of x as per λ_x, then the Heat of Formation for One Kilogram of Water, $\Delta_f H_{H2OKg}$, in KJ/Kg is:-

$$\Delta_f H_{H2Okg} = \sigma_{H2O} . \Delta_r H_{H2} = \tau_{H2O} . \lambda_3 . \Delta_f H_{H2O}$$
$$= -13423.3828172529$$
Equation 11.37

The Reaction Thermodynamics of Hematite

Pathway 11.10 may be specified as:-

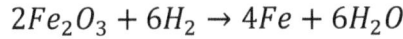

$$2Fe_2O_3 + 6H_2 \rightarrow 4Fe + 6H_2O$$
Pathway 11.10

Clearly, $\lambda_1 = 2$; $\lambda_2 = 6$; $\lambda_3 = 4$ and $\lambda_4 = 6$ and the appropriate Heats of Formation $\Delta_f H_x$ (KJ/mol) may be selected from Table 11.6

Accordingly, the Pathway Heat of Reaction $\Delta_r H_{H2}$ in KJ/mol is given by:-

$$\Delta_r H_{Fe} = \frac{(\lambda_3 . \Delta_f H_{Fe} + \lambda_4 . \Delta_f H_{H2O}) - (\lambda_1 . \Delta_f H_{Fe2O3} + \lambda_2 . \Delta_f H_{H2})}{\lambda_3}$$
$$= +50.011$$
Equation 11.38

If the Heat of Reaction is +ve then iron is formed: If it is -ve iron is consumed. The reaction of Equation 11.38 is feebly exothermic.

Allow that:-

$$\tau_{Fe2O3} = \frac{1000}{\lambda_1 . FW_{Fe2O3}} = 3.13110565602926$$
Equation 11.39a

$$\tau_{H2} = \frac{1000}{\lambda_2 . FW_{H2}} = 82.6768789147502$$
Equation 11.39b

$$\tau_{Fe} = \frac{1000}{\lambda_3 . FW_{Fe}} = 4.476676515355$$
Equation 11.39c

$$\tau_{H2O} = \frac{1000}{\lambda_3 . FW_{H2O}} = 9.251405843632$$
Equation 11.39d

where τ_x is the mols of species x in one kilogram of x as per λ_x, then the Heat of Formation for One Kilogram of Iron, $\Delta_f H_{FeKg}$, in KJ/Kg is:-

$$\Delta_r H_{Fe} = \left(\lambda_3 . \Delta_f H_{Fe} + \lambda_4 . \Delta_f H_{H2O}\right) - \left(\lambda_1 . \Delta_f H_{Fe2O3} + \lambda_2 . \Delta_f H_{H2}\right)$$
$$= +200.044$$

Equation 11.40

PART II
REPORTED RESEARCHES

Rosner Et Al on the Iron-Hydrogen Heat of Reaction[11.4]

Rosner and his collaborators noted the Heat of Reaction of the pathway:-

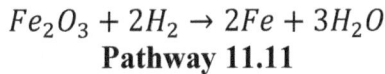

$$Fe_2O_3 + 2H_2 \rightarrow 2Fe + 3H_2O$$
Pathway 11.11

as +99 KJ/mol.

This pathway is essentially half the weight of Pathway 11.10

Accordingly, we should compare twice +99, i.e. 198 with the Δ_rH_{Fe} figure of Equation 11.41, which is +200.044

PSD(Δ_rH_{Fe}, 2×99) is 1.02177520945397%, indicating a good agreement between these independent estimates of Iron-Hydrogen Heat of Reaction.

Vogl, Åhman and Nilsson on Power Consumption per Tonne[11.5]

Vogl, Åhman and Nilsson interestingly remarked that the power consumption of a traditional Hot Blast Carbon Furnace was 3.68 MWh per tonne of liquid iron, whilst the corresponding figure for a Hydrogen Direct Reduction Furnace (H_2DRF) was 3.48 MWh of electricity per *steel* ton. It was not clear to me whether by "ton" they meant American Short Tons or Metric tonnes. I discounted the British Long Ton as a possibility. I presumed that they intended the smelt phase only without considering Basic Oxygen refining or other steel post-processing. I assumed tonnes.

Vogl and his colleagues adopted Pathway 11.11 as germane and quoted the Heat of Reaction as +99.5 KJ/mol

On this basis the doubled pathway weight of Equation 11.40 would indicate a ΔH_r of 2×99.5 = 199 kJ/mol

PSD(Δ_rH_{Fe}, 199) is 0.521885185259291%, showing a discrepancy of about one in two hundred.

Vogl, Åhman and Nilsson indicate for the H_2DRF an electric power consumption of 3.48×1000000×60×60 = 12528000000 Joules for 1000 kilograms of ferrous product (I presume pig iron in liquid state) that implies:-

$$JoulesPerKg = \frac{E_{LHS}}{M_{RHS}} = 12528000$$
Equation 11.41

where E_{LHS} is the Energy per Tonne in Joules, and M_{RHS} is the Mass of Ferrous Product, that is 1000 kilograms.

If we assume the consensus that it takes 39.4 KWh to form one kilogram of *hydrogen* that is 141840000 *coulombs*.

Accordingly, the Estimated Kilowatt hours of power for one kilogram of H_2, E_{KWh} is given by:-

$$E_{KWh} = \frac{Q_{H2}}{Kwh} = 39.8831564722222 \; KWh$$
Equation 11.42

whereupon:-

$$PSD(39.4, EKWh) = -1.22628546249296$$
Equation 11.43

The standard Heat of Reaction ΔH_r for the pathway:-

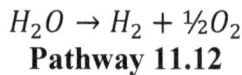

$$H_2O \rightarrow H_2 + \tfrac{1}{2}O_2$$
Pathway 11.12

is +242 KJ/mol

Moving on by way of running checks:-

$$\text{Stoichiometric Balance} = (\lambda_1.FW_{H2} + \lambda_2.FW_{O2}) - (\lambda_3.FW_{H2O} + 0) = 0$$
Equation 11.44

and the Vogl Et Al estimate of the Pathway 11.12 Heat of Reaction, $\Delta_r H_{Vogl}$, is:-

$$\Delta_r H_{Vogl} = (\lambda_3.\Delta_f H_{H2} + \lambda_4.\Delta_f H_{O2}) - (\lambda_1.\Delta_f H_{H2O} + 0) = 241.826$$
Equation 11.45

and:-

$$PSD(242, \Delta_r H_{Vogl}) = 0.071900826446284$$
Equation 11.46

The Vogl Group and Gross Energy Estimates

Vogl, Åhman and Nilsson provide discussions of large-scale estimates of energy consumptions per tonne of iron melt.

At this juncture it is useful to reproduce estimates of European carbon blast furnace consumption and production in terms of the Mean Blast Furnace Energy Requirement, $E_{\mu BFyear}$, which is 23527.8947 GJ/year; and the Mean Tonnes of BF Iron per Year, $M_{\mu BFyear}$ which is 1742.80702 tonnes.

These data are presented in full on Table 11.7

It follows that the Energy, $E_{BFtonne}$ *for carbon blast furnaces in Europe* is:-

$$E_{BFtonne} = \frac{E_{\mu BFyear}}{M_{\mu BFyear}} = 13.4999483190055 \; GJ/tonne$$
Equation 11.47

This provides a reference point for the comparison of novel furnace efficiencies.

To convert one GigaJoule to one MegaWatt Hour we may employ the conversion factor K_{Vogl} as:-

$$K_{Vogl} = \frac{1000000000}{1000 \times 1000 \times 60 \times 60}$$
$$= 0.277777777777778 \; MWhours$$
Equation 11.48

This enables us to estimate the Energy Consumption of European Blast Furnaces per Tonne of Pig Iron, E_{MWhFe}, as:-

$$E_{MWhFE} = K_{Vogl}E_{BFtonne} = 3.7499856441682 \; MWh \; per \; Tonne$$
Equation 11.49

The ABB Discussion of Iron and Steel Making[11.6]

ABB note that, in practical terms, it takes 50-55 KWh to produce one kilogram of hydrogen, which differs from the lower theoretical figure of E_{KWh} which is 39.8831564722222 Kilowatt Hours.

A further ABB assertion is that 94 million tonnes of *steel* demands 296 TWhours of energy.

The number of Kilowatt Hours expended in the smelting of One Tonne of Iron, $E_{KWhTonneFe}$, is 2159.668

Therefore, the number of Kilowatt Hours for 94 Million tonnes of iron, $E_{KWh94MFe}$, is given by:-

$$E_{KWh94MFe} = 94 \times 1000000 \times E_{KWhTonneFe} = 203008792000$$
Equation 11.50

A TeraWatt is 10^{12} Watts.

Accordingly, the TeraWatt Hours for 94 million tonnes of iron should be:-

$$E_{TWh94MFe} = \frac{E_{KWh94MFe}}{10^9} = 203.008792$$
Equation 11.51

The large discrepancy between 203.008792 and 296 is almost certainly due to ABB costing *steel* output, which engages Basic Oxygen Furnace refinement of the pig iron, and possibly other downstream processing of the raw product.

Nation	Location	Hot Metal Capacity (tonnes $*10^3$/year)	Finished Steel Capacity (tonnes $*10^3$/year)	Number of Furnaces
Austria	DONAWITZ (Leoben)	1370	1570	2
Austria	LINZ	430	6000	3
Belgium	GHENT	4430	5000	2
Chechia	OSTRAVA – BF only	3200		3
Chechia	TRINEC	2100	2400	2
Finland	RAAHE	2400	2600	2
France	DUNKERQUE	6800	6750	3
France	FOS-SUR-MER	5160	5100	2
Germany	BREMEN	3960	3800	2
Germany	DILLINGEN	4790	2760	2
Germany	DUISBURG	11600	11560	4
Germany	EISENHÜTTENSTADT	2340	2400	2
Germany	SALZGITTER	4800	5200	3
Germany	VÖLKLINGEN BOF only	(3240)		0
Hungary	DUNAUIJVAROS	1310	1650	2
Italy	TARANTO	9590	11500	4
Netherlands	IJMUIDEN (Velsen-Noord)	6310	7500	2
Poland	DABROWA GORNICZA	4500	5000	2
Poland	KRAKOW	1310	2600	1
Romania	GALATI	3250	3200	2
Slovakia	KOSICE	2850	4500	2
Spain	AVILES BOF only	(4200)		0
Spain	GIJON	4480	1200	2
Sweden	LULEA	2200	2200	1
Sweden	ÖXELÖSUND	1800	1700	2
United Kingdom	PORT TALBOT	4770	4900	2
United Kingdom	SCUNTHORPE	3590	3200	3
Total		99340	104290	57
Mean		3973.6	4345.42	2.11

Tonnes per BF per year	1742.80702
Kg per BF per year	1742807.02
Kg per BF per Second	0.05522622
*Mean BF Energy Requirement (GJ/year)	23527.8947
*Mean BF Energy Requirement (J/sec)	745553.995
Heat of Combustion of Metallurgical Coke (J/kg)	29306204.4
**Modern Coke Rate (Kg/thm)	277.15
**Traditional Coke Rate (Kg/ thm)	565

* Liquid Iron Only
** thm = Tonnes of Liquid Iron

Table 11.7
Recent Consumption and Production Statistics
For the Surviving (2023AD) European Blast Furnaces

Thomas Koch Blank and Green Steel[11.7]

TK Blank's primary assertion is that one tonne of steel requires 2633 KWh, which is itself markedly different to the $E_{KWhTonneFe}$ ABB-derived estimate estimate of 2159.668

By TK Blank's assertion the Energy required in Joules per Kilogram of Iron, $E_{TKBJperKg}$, is:-

$$E_{TKBJperKg} = \frac{2633 \times 1000 \times 60 \times 60}{1000} = 9478800$$
Equation 11.52

Hence the equivalent in KWh per Kilogram of Iron is:-

$$E_{TKBKWhperKg} = \frac{2633 \times 1000 \times 60 \times 60}{1000 \times 1000 \times 60 \times 60} = 2.633$$
Equation 11.53

as per the TK Blank assertion.

Now according to our own derivations the Energy for One Kilogram of Iron, $E_{KWhperKgFe}$, is:-

$$E_{KWhperKgFe} = \frac{1}{1000} \times \frac{Q_{H2}}{KWh} \times 1000 \frac{\lambda_2.FW_{H2}}{\lambda_3.FW_{Fe}}$$
$$= \frac{1}{1000} \times \frac{Q_{H2}}{60.60} \times \frac{\lambda_2.FW_{H2}}{\lambda_3.FW_{Fe}}$$
$$= 2.15953955061035$$
Equation 11.54

so that:-

$$PSD\left(E_{KWhperKgFe}, E_{TKBKWhperKg}\right) = -21.9241388404225$$
Equation 11.55

I do not know the source of this large discrepancy.

Maria Gallucci on Clean Energy and Green Steel[11.8]

Maria Gallucci asserts, on behalf of Canary Media, that it takes 380 kilotons of H_2 to make 7 million tonnes of *steel* per annum.

I thus computed that the Asserted Joules per Kilogram of Iron, $E_{MGJperKgFe}$, was 186750000, or in terms of KWh per Kg, $E_{MGKWhperKgFe}$, 51.875

Using our Stoichiometric Hydrogen-Iron Ratio:-

$$\rho_{H2toFeKg} = 1000 \frac{\lambda_2 . FW_{H2}}{\lambda_3 . FW_{Fe}} = 54.146655922643$$

Equation 11.56

we determine that the mutual Percentage Specific Defect is:-

$$PSD\left(\rho_{H2toFeKg}, E_{MGKWhperKgFe}\right) = 4.19537621286981$$

Equation 11.57

Discussion of the Ijmuiden Works, Nederland[11.9]

Please note that "Common Futures" of Utrecht in 2024AD discuss the Ijmuiden Steel Works in terms of *steel* production using *carbon blast furnaces only*.

The assertion is that 380 kilotons of H_2 would be consumed to manufacture 7 million tonnes of *steel* per annum.

I calculated that this represented an Output, $\Omega_{KGFeperKgH2}$, of 18.4210526315789 kilograms of iron per kilogram of hydrogen.

The appropriate Absolute Yield, Y_A, equation is:-

$$Y_A = \frac{\lambda_3 . FW_{Fe}}{\lambda_2 . FW_{H2}} = 18.4683612119769$$

Equation 11.58

The resulting PSD is:-

$$PSD\left(Y_A, \Omega_{KGFeperKgH2}\right) = 4.19537621286981$$

Equation 11.59

The values and discrepancies of our calculations and recent Reported Researches into the Hydrogen Direct Reduction Furnace (H2DRF) prospectus are summarised in Table 11.7

On the whole, given the context, we are doing very well if the PSD's of home and foreign workers' results betray a ±5% discrepancy.

If the discrepancy between two measures is several thousand percent it does not mean that this or that researcher is idle or stupid. We all make different assumptions about what is admissible, in particular chemical reaction and or sensible heat energies, and different assumptions are made about process efficiencies, questions that are highly device-dependent.

Applied thermodynamics is a daunting science. Errors are normal (A note to statisticians: No pun intended, and I trust none taken). In my professional past students often mislaid whole orders of magnitude so that Watts could easily be mistaken for Megawatts or Terawatts, and I was sometimes guilty of such myself. It is another good argument for indicial computations.

The whole point of this book is to encourage deep critical evaluation of new technologies.

I am seventy-two. If you gave me a dollar for every error of my life I would be a millionaire.

Reference	Reference Assertion	JRW Value	Reference Value	Units	PSD (Value_JRW / Value_Ref)	Comments
11.4	$3H_2+Fe_2O_3=3H_2O+2Fe$: $\Delta H=+99.(5)KJ/mol$	200.044	198	KJ/mol	1.0218	ΔH doubled to 198 for my 6H₂ reaction: Sponge Iron leaves at 700°C: **Heat of Reaction Only**
11.5	H2DRF: 3.48 MWh electricity per *steel* ton	7713560	12528000	J/Kg	-62.4153	Has JRW omitted to consider *steel* processing such as BOF demands?
	Remark that Carbon Hot Blast BF is 3.68 MWh/tonne Fe	3.749986	3.68	MWh/tonne	1.8663	
	$H_2O=H_2+\frac{1}{2}O_2$ $\Delta H_2=+242$ KJ/mol	241.826	242	KJ/mol	-0.0720	Checked by JRW
	$Fe_2O_3+3H_2=2Fe+3H_2O$ $\Delta H_2=+99.5$ KJ/mol	200.044	199	KJ/mol	0.5219	ΔH doubled to 199 for my 6H₂ reaction: **Heat of Reaction Only**
	One tonne of *steel* needs 1504 Kg of iron ore pellets	1429.743	1504	Kg	-5.1937	
	51 Kg of H₂ is needed per tonne of *steel*	54.14666	51	Kg	5.8114	
11.6	50-55 KWh for 1 kg H₂	39.57133	52.5	KWh/Kg	-32.6718	**Implied Heat of Reaction Only**
	94 million tonnes of *steel* requires 296 TWh of Electricity	203.0088	296	TWh	-45.8065	ABB figure may include BOF refining and or other post-processing
11.7	One Tonne Fe needs 2,633 KWh	2.15954	2.633	KWh	-21.9241	Source of Discrepancy doubtful to JRW
11.8	160 kilotons of H₂ needs 8.3 Terawatt-Hours	54.14666	51.875	KWh/Kg	4.1954	
11.9	380 kilotons of H₂ for 7 million tonnes *steel* per annum	18.46836	18.4210526	Kg/Kg	0.2562	

Table 11.8
Critical Comparisons of Literature Estimates with the
Present H₂DRF Computations

Please consult Table 11.9 for the Molar Masses and Heats of Formation of germane species.

Substance	Formula	Formula Weight (g/mol)	Heat of Formation $\Delta_f H_x$ (KJ/mol)	Standard Entropy ΔH (KJ/mol)	Specific Heat C_p at T°K (J/(mol.°K))	State of Substance at T°K
Hydrogen	H_2	2.01588	0.000	29.089	31.33934500	Gas
Oxygen	O_2	31.99880	0.000	32.378	35.90654323	Gas
Water	H_2O	18.01528	-241.826	37.738	44.62273084	Gas
Hematite	Fe_2O_3	159.68800	-825.500	139.568	142.04970432	Solid
Magnetite	Fe_3O_4	231.53300	-1120.890	202.448	200.83200013	Solid
Wüstite	FeO	71.84400	-249.530	54.927	68.19920000	Solid
Iron	Fe	55.84500	0.000	37.203	39.00284868	Solid

Table 11.9
The Molar Masses and Heats of Formation
of Selected Chemical Species

According to Kovtun and his colleagues the following three stepwise reactions pertain to the reduction of Hematite to Iron:-

(a) Hematite to Magnetite
$$3Fe_2O_3 + H_2 \rightarrow 2Fe_3O_4 + H_2O$$
$\lambda_1 = 3$ $\lambda_2 = 1$ $\lambda_3 = 2$ $\lambda_4 = 1$
$$\Delta_r H_{Fe}a = \left(\lambda_3 . \Delta_f H_{Fe3O4} + \lambda_4 . \Delta_f H_{H2O}\right) - \left(\lambda_1 . \Delta_f H_{Fe2O3} + \lambda_2 . \Delta_f H_{H2}\right)$$
$\Delta_r H_{Fe}a$ = -7.106 KJ/mol

(b) Magnetite to Wüstite
$$Fe_3O_4 + H_2 \rightarrow 3FeO + H_2O$$
$\lambda_1 = 1$ $\lambda_2 = 1$ $\lambda_3 = 3$ $\lambda_4 = 1$
$$\Delta_r H_{Fe}b = \left(\lambda_3 . \Delta_f H_{FeO} + \lambda_4 . \Delta_f H_{H2O}\right) - \left(\lambda_1 . \Delta_f H_{Fe3O4} + \lambda_2 . \Delta_f H_{H2}\right)$$
$\Delta_r H_{Fe}b$ = +130.474 KJ/mol

(c) Wüstite to Iron
$$FeO + H_2 \rightarrow Fe + H_2O$$
$\lambda_1 = 1$ $\lambda_2 = 1$ $\lambda_3 = 1$ $\lambda_4 = 1$
$$\Delta_r H_{Fe}c = \left(\lambda_3 . \Delta_f H_{Fe} + \lambda_4 . \Delta_f H_{H2O}\right) - \left(\lambda_1 . \Delta_f H_{FeO} + \lambda_2 . \Delta_f H_{H2}\right)$$
$\Delta_r H_{Fe}c$ = +7.704 KJ/mol

I suppose that the relative proportions of the three pathways could be established by the solution of linear equations or something, given that the heat of the overall Hematite-Iron reaction is known.

Experimental H₂DRFs in 2024AD

It is currently thought that there are some twenty-seven organised explorations of H₂DRF feasibility world-wide, though many of these have not yet reached the experimental stage.

At least two projects are at or near the level of commercial production:-

(a) **The HYBRIT Initiative**[11.11]
(Hydrogen Breakthrough Ironmaking Technology)
A Sveco-Finnish project to phase Swedish and Finnish BF steelworks to H₂DRF facilities by the mid-2020s

(b) **The ArcelorMittal Bremen-Eisenhüttenstadt Project**[11.12]
This is a European Union effort to harmonise the production of ArcelorMittal Bremen and Eisenhüttenstadt furnaces and convert them from natural gas firing to hydrogen

HYBRIT is a collaborative venture between three state-owned Swedish enterprises that avowedly exploit the special conjunction of natural resources available in North-East Sweden and North-Western Finland: Vast deposits of high-quality iron ore, excess availability of cheap hydro-electric power, and navigable seaboards. The three firms involved are the mining company Luossavaara-Kiirunavaara Aktiebolag (LKAB); SSAB, Svenskt Stål AB; and the wonderfully-named Vattenfall AB, a hydro-electric supplier.

In 2018AD the construction of a H_2DRF pilot plant commenced at Luleå, a port on the Gulf of Bothnia in Arctic Northern Sweden. I recollect sixty years ago my geography master Mr Jack Turner telling me of Luleå, an iron port, not ice free, but the Eastern terminus of a strategic railway between the Baltic and Narvik, an ice-free port in Northern Norway. I thought then and think now Luleå: "what a euphonious name for a town!". The undertakers intend to store hydrogen in a lined underground cavern.

Before I forget, I should state that the basic thermodynamic equation referenced by Pei Et Al[11.11] is:-

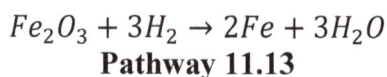

$$Fe_2O_3 + 3H_2 \rightarrow 2Fe + 3H_2O$$
Pathway 11.13

where the Heat of Reaction ΔH_{298} is -95.8 KJ/mol which betokens a weak *endothermic* reaction: But note that 298°K is roughly 25°C, a warm room temperature, not the approximate 600°C looked for when making sponge iron.

The Luleå pilot plant commenced production on 31 August 2020 and a plant to demonstrate industrial-scale H_2DRF iron smelting commenced work on 24 March 2021 at Gällivare, an iron mining town about 151 miles North of Luleå.[11.13]

The overall aim is to replace the 2.2M tonnes per annum BF production at Luleå; the 1.8M tonnes per annum at Oxelösund and the Finnish Rautaruukki Mill 2.4M tonnes per annum at Raahe entirely with hydrogen produced steel.

The existing pilot at Luleå has a output of one tonne per hour of sponge iron (Pei Et Al state "ton/h" but they almost certainly mean metric tonnes). For 24-7 operation this equates to 8766 tonnes per year or 0.008766M tonnes per annum.

In Figure 7 of the Reference 11.11 Pei Et Al reproduce the data of Reference 11.14 to compare the gross energetics of the SSAB Blast Furnace route with that of the HYBRIT "concept". I was unable to decide whether the latter was purely theoretical or whether it owed something to measurement.

BLAST FURNACE			HYBRIT		
Resource	Units	Amount	Resource	Units	Amount
CO_2	Kg	1600	CO_2	Kg	25
Oil	MJ	292	Carbon	MJ	151
Coal	MJ	19836	Bio	MJ	2016
Electricity	kWh	235	Electricity	kWh	3488

Table 11.10
Blast Furnace and Energetic Comparisons
According to Pei Et Al[11.11]

Table 11.11 is the Pei Et Al data conformitised to MJ by me, taking the Heat of Formation of Carbon Dioxide to be -393.51 KJ/mol whilst there are 22.72237 mols of CO_2 in a Kilogram of CO_2. Also, one Kilowatt-Hour is 3.6MJ.

BLAST FURNACE		HYBRIT			
Resource	Energy (MJ)	Resource	Energy (MJ)		
CO_2	14488.14	CO_2	226.3773		
Oil	292	Carbon	151		
Coal	19836	Bio	2016		
Electricity	846	Electricity	12556.8		
	35462.14		14950.18	Total Energy (MJ)	
	34616.14		2393.377	Total Carbonaceous Energy (MJ)	
	846		12556.8	Electrical Energy (MJ)	
	0.976144		0.16009	Ratio: Carbonaceous Energy to Total Energy	

Table 11.11
Blast Furnace and Energetic Comparisons
In Terms of Inherent Heat (MJ)

This Table 11.11 data is represented on the column chart of Figure 11.6

Judging by the orders of magnitude illustrated I have taken the liberty of presuming that energy quantities are given per *tonne of sponge iron* produced.

Clearly, much more heat is taken away as reactive heat of formation in carbon dioxide by a BF than is the case with the HYBRIT conformation of a H_2DRF. A BF consumes ten times the amount of carbonaceous material as a H_2DRF, whilst a H_2DRF consumes fifteen times as much electricity as a blast furnace.

I assess these proportions to approximately be general to BF versus H_2DRF relativities.

Furthermore, I have identified the Pei Et Al category "Bio" with wood fuel, and the category "Oil" with kerosene. On the grounds that these workers show that oil supplemented coke in Swedish furnaces prior to the mid-1970s and that coke supplements PCI (pulverised coal injectate) today; I have identified "Coal" with PCI and or coke.

Table 11.11 makes manifest that for this Sveco-Finnish pilot project HYBRIT consumes 42.158% as much total energy as the compared carbon-based blast furnace. Also, 6.914% as much carbonaceous energy. Carbon consumed in graphite electrodes, and in the EAF post-process are added to HYBRIT carbon consumption by Pei Et Al.

Pei Et Al stress the importance of controlled-composition pelleted ore as an essential feedstock of H_2 furnaces, in their case including olivine as a self-flux with reduced-phosphorous ore. Historically, phosphorous was a serious inhibitor of iron smelting development which made the massively-bedded ores of Central Sweden and elsewhere in the Mesozoic facies of Europe, especially Germany, Lorraine and Britain, worthless.

In 1877, British cousins Percy Carlyle Gilchrist and Sidney Gilchrist Thomas invented the Gilchrist-Thomas Process for the de-phosphatisation of pig iron melt. This process opened up the vast phosphatic iron ores of Europe as an economic resource.

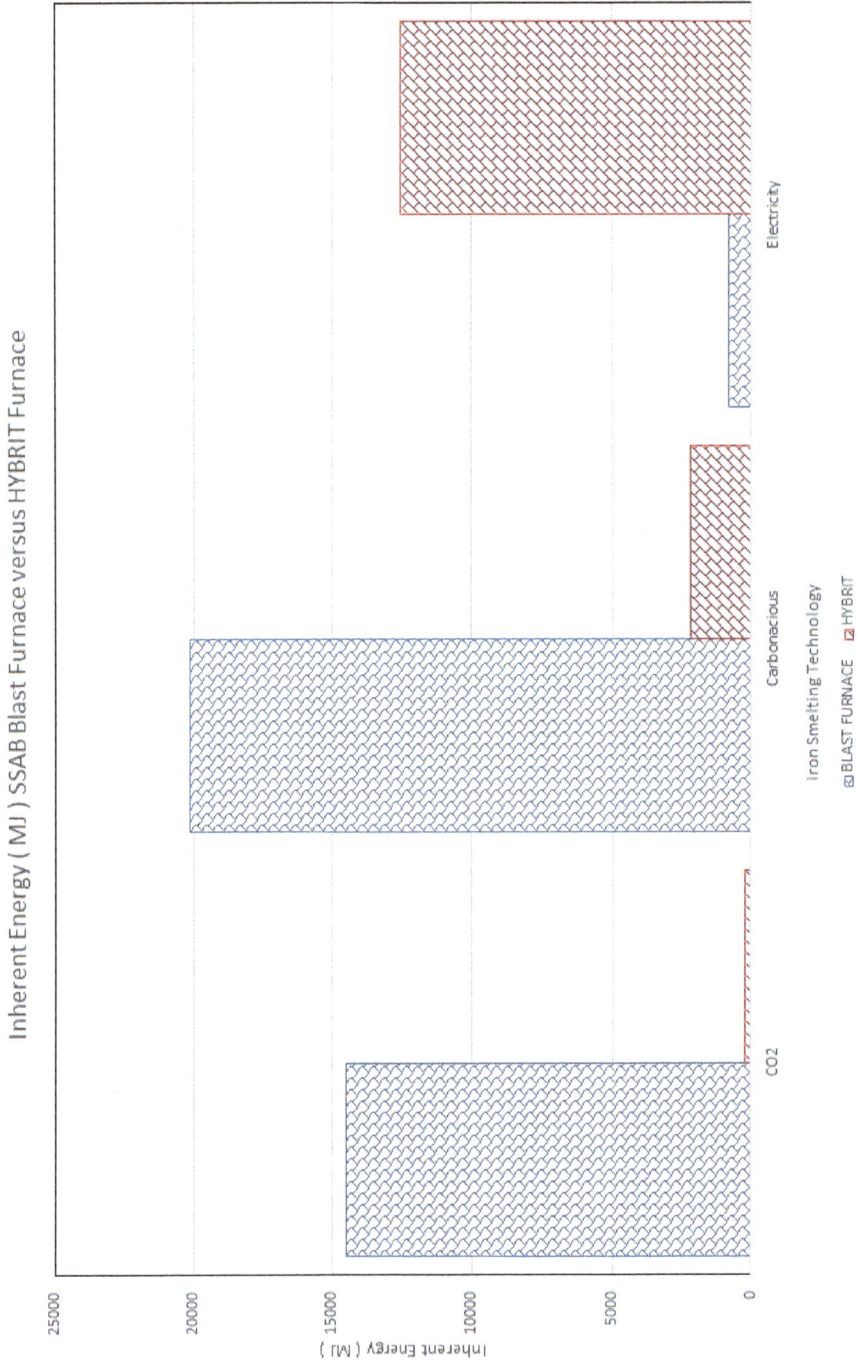

Figure 11.6
Blast Furnace and Energetic Comparisons
In Terms of Inherent Heat (MJ)

Pei ET Al list eight desiderata of the HYBRIT process:-

(1)   Non fossil fuels are used in pellet production
(2)   Hydrogen is produced with electrolysis
       using fossil-free electricity
(3)   Storage of hydrogen in a specially designed unit
       is used as a buffer to the grid
(4)   A shaft furnace is used for iron ore reduction
(5)   Tailor-made pellets are used as iron ore feed
(6)   The reduction gas/gas mixture is preheated
       before injection into the shaft
(7)   The product can either be DRI or HBI
       free of carbon or carburized
(8)   The DRI/HBI is melted together with
       recycled scrap in an electric arc furnace

I cite these statements verbatim in their English-language terms in the Pei Et Al paper[11.11]

With regard to (1) I am not sure whether wood or other bio-material is included in the pellet composition. Point (2) seems to me to be key to the viability of the H_2DRF. Certain countries blessed with a low population density and with high rainfall, at least locally, such as Sweden, Brazil, Canada and New Zealand commend themselves to H_2DRF iron smelting. In the case of lowland or population-concentrated nations like Britain, nuclear power is the only practicable power source other than coal. (3) HYBRIT employs sealed subterranean caverns as stores of hydrogen, but various other strategies for storage have suggested themselves, including liquefaction and metal hydride formation. If palladium were as cheap as gold to the Incas then maybe palladium shavings would be an excellent sequestrant for hydrogen!

In regard to (5), pelletisation is a great net economiser by permitting the optimal incorporation of fluxes and of pre-conditioned iron ore, optimised for comminution, texture and composition, and as noted, pre-conditioned for the elimination of phosphorus and other pig iron vitiants, traditionally removed by complex and expensive post-processing.

Point (6) seems to me to be an excellent idea, analogous to the traditional BF hot blast, an accelerant of reactions and an energy economiser.

HBI is Hot Briquetted Iron and DRI is Direct Reduced Iron. Both are unfused "sponge" iron. As per (8) both can be fused in an Electric Arc Furnace (EAF) to produce liquid iron for further refining.

CHAPTER TWELVE
THE DIRECT ELECTROLYSIS FURNACE

PART ONE
INTRODUCTORY REMARKS

Reductive Electrolysis of metallic salts or oxides to metallic elements and gases can take place in two environments:-

(a)   Aqueous Solution
This is the reduction of a dissolved salt or salts in liquid water at or near ambient temperatures, around 300-400°K. The water may be pH-adjusted by dissolving minor amounts of acid or alkali.

Majid Et Al[12.1] contributed an experimental study of Fe_2O_3 dust reduction in an electrolyte of aqueous sodium hydroxide at 110°C and draw attention to the low ohmic resistance and potential current economies of the system. They asserted the optimality of aqueous NaOH in comparison to potassium and lithium hydroxides on the one hand and acidic solutions on the other.

In another 2023AD laboratory study Fayaz Et Al[12.2] explored the influence of cell conformations ("form factors") on reduction efficiency. Cell design, including electrode and electrolyte morphology and composition are crucial to the viability of oxide electrolysis.

(b)   Molten Oxide or Salt
The appropriate oxide or salt is melted in a furnace with or without an added flux or catalyst. In the case of iron smelting the requisite temperature, which may be induced electrically, is around 1900°K.

The electrolytic reduction of Ferric Oxide as Hematite, Fe_2O_3, is a type of Molten Salt Electrolysis designated Molten Oxide Electrolysis (MOE).

Hematite breaks down spontaneously to Magnetite (Fe_3O_4) and Oxygen (O_2) before the melting point of Iron (1810°K) is attained. Accordingly, it is thought efficient to dissolve minor amounts of hematite in a hot alkaline artificial magma which I loosely describe as "basic slag".

Allanore, Ortiz and Sadoway[12.3] defined the thermodynamic and electrolytic principles bearing upon hematite reduction in a molten solvent whilst Weincke Et Al[12.4] described experiments with varying concentrations of iron oxide (0% to 10%) in a 66% SiO_2, 20% Al_2O_3, 14% MgO melt (when Fe is 0%) at a temperature of 1823°K.

In a lecture delivered at Charleston[12.5, 12.6], Carolina, Khetpal, Ducret and Sadoway described their experiments with an (artificial) diopsidic pyroxene melt of 25.59% CaO; 26.15% MgO; and 49.26% SiO_2 (Mixture S1). Superadded was FeO at 0%, 5%, 10%, 15% and 20%. The liquidus of this material is less than 1700°K.

In all instances reductive energy is supplied by a large electric current at a small thermodynamically-calculated voltage of maybe four or nine Volts. Fine electroplating applications, not relevant here, may require low currents.

In every instance of electrolysis the electropositive deposit (usually a metallic element and or hydrogen if available) forms at the cathode from which electrons issue, that is to say the cathode is conventionally "negative". Meanwhile, the relatively electronegative product, often a gas, forms at the anode which is conventionally "positive", and a sink for the current electrons.

From mechanical points of view, metallic solids might adhere to a graphite cathode, or in high temperature melts pond as molten liquid in a cathodic sump. Meanwhile, the anode product percolates away as a gas. In addition metal refining may throw down low-temperature "anode slimes" or high-temperature sublimates. These secondary anodic products are often recovered for being highly valuable concentrates of rare metals, including gold. Gold is often present in high-quality hematite iron ore and artisanal miners harvest it in nugget form from lateritic desert regoliths.

It is not impossible that future MOE smelting may be economically-enhanced by the isolation of precious by-products, just as the Ancients separated silver from lead by oxidation of the molten crude lead.

Iridium (Ir)[12.7] is not attacked by synthetic alkaline magma, whilst Tungsten (W) is highly refractory. Therefore, for the purpose of outline calculations, I choose to assume that the anode is of iridium-plated tungsten, whilst the cathode is a pool of liquid iron, the primary product of the process. Meanwhile, the electrolyte is a Weincke "slag" of composition Fe#10: 10% Fe_3O_4; 12.6% MgO; 18% Al_2O_3; 59.4% SiO_2[12.8]

In all electrolysis it is important that the liquid substrate is in a semi-ionised state that chemists prefer to describe as "polar", so that while the actual molecule of the liquid compound is covalent it has a distinct charge separation across the molecule so that electricity can be conducted. For example, water is a "polar covalent" compound. Perfectly pure water is a dielectric (an insulator); the addition of a little sodium chloride (common salt) or a trace of alkali makes it a superb conductor, like seawater. Hydrocarbons like gasoline are doggedly covalent and cannot be electrolysed. On the other hand many organic oxides, including ethanol (common alcohol) are polar covalent.

Figure 12.1 is a schematic of a basic high-temperature Metal Oxide Electrolysis furnace for the reduction of Hematite Iron Ore, Fe_2O_3, to liquid elemental Iron, Fe, and gaseous elemental Oxygen, O_2.

The problems of Metal Oxide Electrolysis, like most things in life, are simple to state but hard to solve.

The exact composition and temperature of the electrolyte affects the conductivity of the melt and the quantity and quality of any resolved products, whether at anode or cathode. So does the Area and the Separation Distance of the two electrodes. There is a distinction between electronic lysis and mass transfer which affects the rate of electrolysis. The chemistry of the furnace lining greatly affects the melt behavior: The melt may cause expensive spoilation and lining damage. Furnace technologists try to achieve conditions in which appropriate electrolyte "creeps" between the melt and the lining in order to safeguard the latter.

As illustrated, furnace designers try to pond liquid products to protect the cathode, but the anode is especially vulnerable to thermal and chemical attack: Experiments continue with platinum and iridium anodes which are prohibitively expensive for commercial iron smelting. On the other hand, graphite electrodes themselves erode and participate in oxygen reactions, hardly desirable if you wish to minimise carbon dioxide evolution.

As with BF and H_2DRF technologies simple analysis and design of the actual chemical reaction is frustrated by the fact that the hot iron switches between divalent and trivalent atomic states and in the extreme a high Wüstite melt can with an almost captious irony become such a splendid conductor of electricity that it entirely circumvents the chemical processes desired by removing the resistance necessary for reduction.

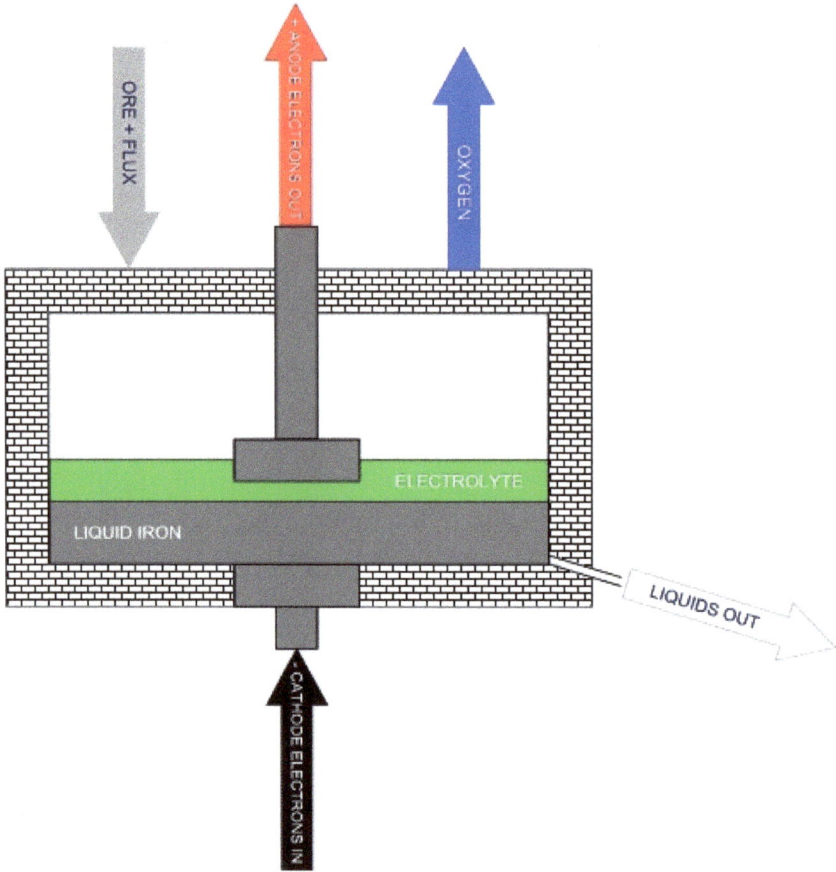

Figure 12.1
The Basic Conformation of a
Metal Oxide Electrolysis (MOE) Furnace

   I have only mentioned a fraction of the many technical difficulties which beset practical MOE.

   As in life, a simple, obvious and economical solution to a plain problem like smelting iron: Upon further exploration becomes fraught and complex, invoking unsuspected, unanticipated vagaries.

PART TWO
DATA

Fundamental Constants

Selected Fundamental Constants of physics are tabulated below:-

Constant	Units	Symbol	VALUE	
			Mantissa	Denary Exponent
Rest Mass of the Electron	Kg	m_e	9.109383713900000	-31
Charge on the Electron	Coulombs	e	1.602176634000000	-19
Avogadro's Number	Count/mol	N_A	6.022140760000000	23
Faraday Constant	Coulombs.mol^{-1}	$F = e.NA$	9.648533212331000	4
Boltzmann Constant	J.K^{-1}	k_B	1.380649000000000	-23
Universal Gas Constant	J.K^{-1}.mol^{-1}	R	8.314462618153240	0

Table 12.1
Fundamental Constants

Furnace Geometrical Factors

Geometrical Factors may be called Geometrical Coefficients or "Form Factors", or be noted in other terms in literature.

As remarked above, the geometry and size of the furnace cell; the composition of the electrolyte; and the geometry and materials of both the anode and the cathode are critical to both aqueous and MOE reduction technologies.

The primary dimensions of the simple Figure 12.1 furnace are given in meters in the annotations of Figure 12.2, and explicated in Table 12.2

In our simple furnace the furnace itself and all pools and electrodes are cylindrical.

Element x	Nominal Gross Depth (Depth)	Nominal Electrode Contact Diameter	Nominal Depth in Contact d_x	Area in Contact A_x	Pouillet Ratio f_x
Electrolyte	0.86	2.58	0.43	5.22792433	0.08225062
Anode	0.86	2.58	0.43	5.22792433	0.08225062
Cathode	1.26	10.32	0.63	83.64678936	0.00753167

Table 12.2
The Geometrical Factors of the Basic Furnace
and its Electrodes
with Resistivity Factors f_x

ORE + FLUX

ANODE ELECTRONS OUT

OXYGEN

0.43 m   0.86 m   ELECTROLYTE   0.86 m

1.26 m   LIQUID IRON   10.32 m

2.58 m

0.86 m

LIQUIDS OUT

CATHODE ELECTRONS IN

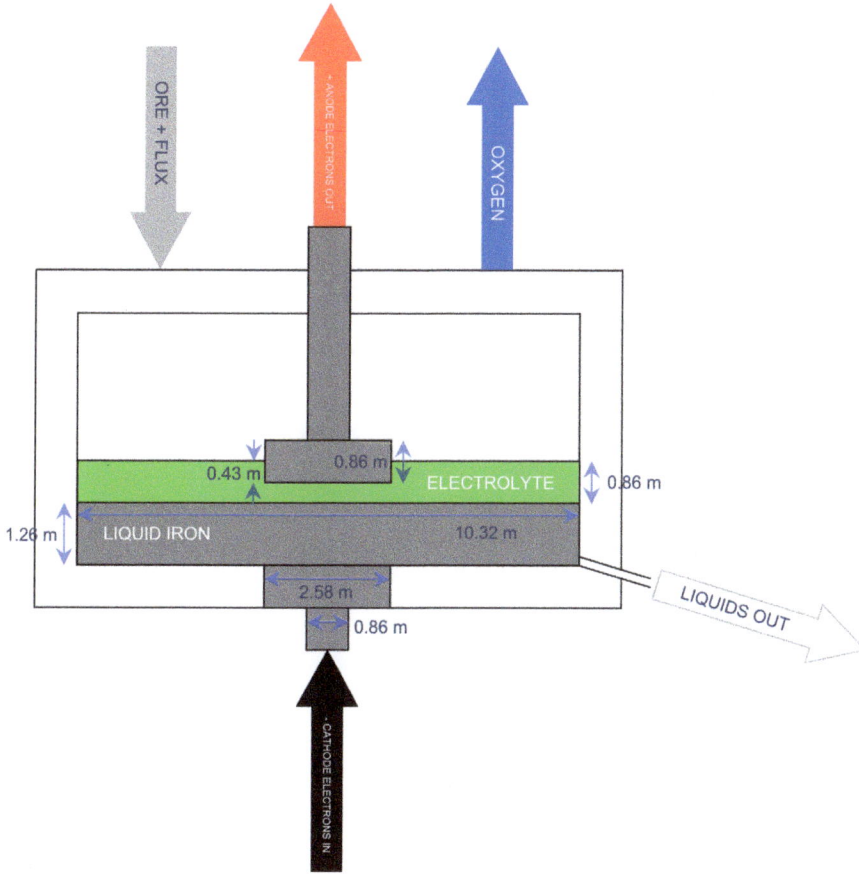

Figure 12.2
The Geometrical Factor Dimensions of
The Basic Conformation of a
Metal Oxide Electrolysis (MOE) Furnace

Conductivities and Resistivities[12.9,12.10]

Conductivity σ is the facility with which a given substance at a given temperature can conduct electricity. In the SI system it is measured in Siemens per Meter and thus has the dimensions $M^{-1}L^{-3}T^3I^2$.

Resistivity ρ is the reciprocal of Conductivity and is measured in Ohm.Meters (Ω.m) and accordingly has dimensions $ML^3T^{-3}I^{-2}$.

$$\rho = \frac{1}{\sigma}$$

Equation 12.1

According to Pouillet's Law the actual body Resistance, Ω, of the object passing a current is:-

$$\Omega = \rho . \frac{\ell}{A}$$

Equation 12.2

where ℓ is the Length of the Conductor (exposed to electric current); and A is the Characteristic Area of the Conductor.

In terms of the electrodes and electrolyte of our specimen furnace of Figure 12.1 we may write:-

$$\Omega_x = \rho_x . \frac{d_x}{A_x}$$

Equation 12.3

where $d \equiv \ell$ and x is some element of the electrolyser: cathode, anode or electrolyte.

Since all these elements are cylindrical it follows that all A are circles and hence:-

$$A_x = \pi \left(\frac{d_x}{2}\right)^2$$

Equation 12.4

Allow that f_x is the quotient of d_x by A_x, then:-

$$f_x = \frac{d_x}{A_x}$$

Equation 12.5

and:-

$$\Omega_x = \rho_x f_x = \rho_x . \frac{d_x}{A_x} = \frac{\rho_x}{\pi} . \frac{d_x}{\left(\frac{d_x}{2}\right)^2} = \frac{\rho_x}{\pi} . \frac{4}{d_x}$$

Equation 12.6

The f-factors for each of the furnace elements x are declared in Table 12.2 above.

Table 12.3 shows the conductivities and resistivities for the selected basic slag electrolyte (containing 10% FeO) and the eight candidate electrode materials identified from the literature.

Electrolyte Conductivity[12.5,12.8,12.9]

Electrolyte conductivity, and its reciprocal, are heavily dependent upon electrolyte composition and temperature.

Please consult Jan Wiencke, Hervé Lavelaine, Pierre-Jean Panteix, Carine Petitjean, Christophe Rapin[12.8] Table V for data correlating FeO% concentrations in a basic slag electrolytic excipient with Conductivity σ at the melt temperature 1823°K.

I plotted these data using EXCEL® and then used the intrinsic tool to fit an exponential regression to both the conductivity and resistivity curves.

The given data are reproduced in Table 12.4 and the respective plots are Figures 12.3 and 12.4

The solvent is defined as S1 (a diopsidic pyroxene) of molar proportions 24.59% CaO; 26.15% MgO and 49.26% SiO_2. The liquidus of this substance is less than 1700°K.

Weight Percent FeO_x	σ_{MO} (S cm^{-1})	σ_{MO} (S m^{-1})	ρ_{MO} (Ω cm)	ρ_{MO} (Ω m)
0	0.03	300	33.33333	0.003333
1	0.032	320	31.25	0.003125
2	0.035	350	28.57143	0.002857
3.5	0.039	390	25.64103	0.002564
5	0.043	430	23.25581	0.002326
7.5	0.052	520	19.23077	0.001923
8.75	0.057	570	17.54386	0.001754
10	0.063	630	15.87302	0.001587

Table 12.4
Conductivities and Resistivities
of the
Chosen 10%FeO Electrolyte at 1823°K

Material	Symbol	Melting Point (°K)	Gage Temperature (°K)	Conductivity (S m⁻¹)	Resistivity (Ω m)	Notes
MOE Electrolyte		1700	1823	626.81581684	0.0015953650	MIT S1 = 24.59% CaO;26.15% MgO; 49.26% SiO2: Then 10% FeO superadded
Graphite	C	3915	2300	117647.05882353	0.0000085000	Nuclear Graphite Grade PCEA: Sublimes: Reference 12.10
Liquid Iron	Fe	1811	1823	719424.46043166	0.0000013900	
Molybdenum	Mo	2894	1823	1872659.17602996	0.0000005340	
Nickel	Ni	1728	1823	719424.46043166	0.0000013900	Unsuitable Electrode due to Low Melting Point
Chromium	Cr	1857	1823	7900000.00000000	0.0000001266	
Tantalum	Ta	3290	1823	7633587.78625954	0.0000001310	
Tungsten	W	3695	1823	18939393.93939390	0.0000000528	
Iridium	Ir	2719	1823	21231422.50530790	0.0000000471	

Table 12.3
Conductivities and Resistivities

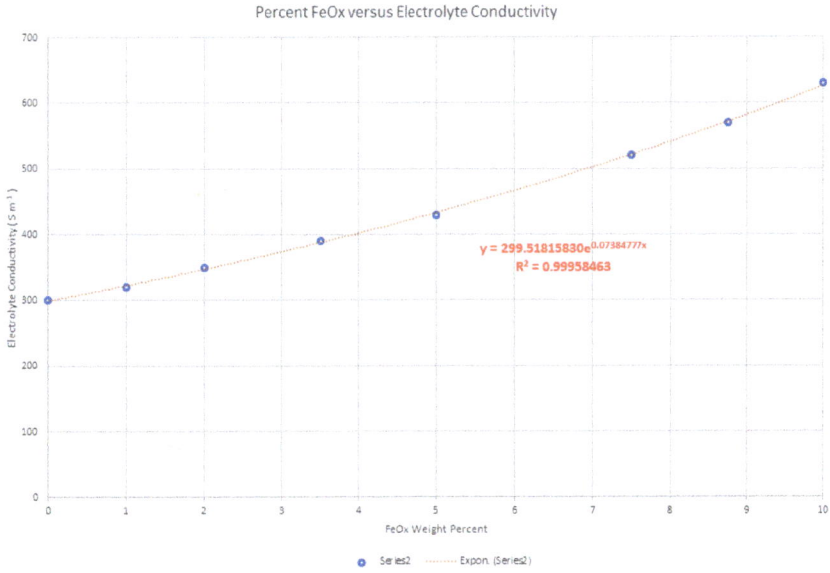

Figure 12.3
Electrolyte Conductivity versus FeO%
at 1823°K

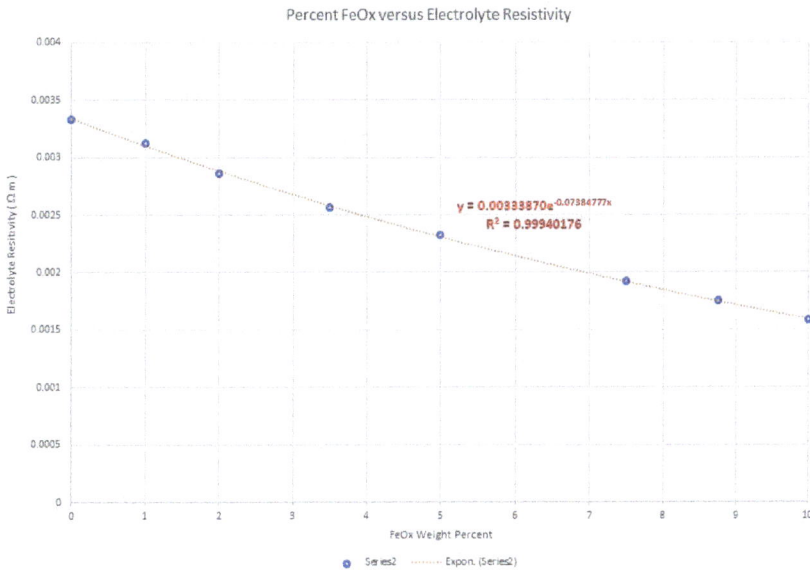

Figure 12.4
Electrolyte Resistivity versus FeO%
at 1823°K

As per the plot, the empirical equation for Electrolyte Conductivity at 1823°K is:-

$$\sigma_{electrolyte} = 299.51815830. e^{0.07384777\ FeOx}$$
Equation 12.7

and the Electrolyte Resistivity is:-

$$\rho_{electrolyte} = 0.00333870. e^{-0.07384777\ FeOx}$$
Equation 12.8

When FeO_x = 10 (%): $\sigma_{electrolyte}$ = 626.815816839618 Siemens per Meter, and $\rho_{electrolyte}$ = 0.001595364975061 Ωm

Nuclear Graphite PCEA[12.10]

Graphite does not melt at viable pressures but it does sublime at 3915°K. Graphite is carbon and carbon reacts chemically with oxidic ferruginous melts, leading to power loss, expense and undesirable emissions.

The resistivity of graphite, $\rho_{graphite}$ at 2300°K is 0.0000085 giving a corresponding conductivity $\sigma_{graphite}$ of 117647.058823529 S/m.

Liquid Iron

The melting point of iron is 1811°K.

At 1823°K, the resistivity of liquid iron, $\rho_{liquidFe}$ is 0.00000139 Ωm giving a conductivity $\sigma_{liquidFe}$ of 719424.460431655

PART THREE
MINIMUM VOLTAGE AND CELL OVERVOLTAGE

We shall assume that the ambient Standard Temperature T_0 is 298.15°K, and that the Melt Temperature, T_{melt} is 1900°K.

Our chosen Pathway is:-

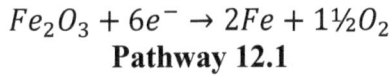

$$Fe_2O_3 + 6e^- \rightarrow 2Fe + 1½O_2$$
Pathway 12.1

<u>The Shomate Equations</u>[12.12]

Define the Shomate Temperatures t_0 and t_{melt} as:-

$$t_0 = \frac{T_0}{1000} = 0.29815$$
Equation 12.9a

and:-

$$t_{melt} = \frac{T_{melt}}{1000} = 1.9$$
Equation 12.9b

The Shomate Function for Enthalpy, H, may then be defined as:-

$$Enthalpy(t,x) = x_1.t + \frac{x_2.t^2}{2} + \frac{x_3.t^3}{3} + \frac{x_4.t^4}{4} - \frac{x_5}{t} + x_6 - x_8$$
Equation 12.10

and similarly for the Shomate Function for Entropy:-

$$Entropy(t,x) = x_1.\ln(t) + x_2.t + \frac{x_3.t^2}{2} + \frac{x_4.t^3}{3} - \frac{x_5}{2t^2} + x_7$$
Equation 12.11

Utilising these data and functions the pathway Enthalpy Change, ΔH, may be defined as:-

$$\Delta H = Enthalpy(t_{melt}, d) - Enthalpy(t_0, c)$$
$$= 230.04052943665 \, Joules$$

Equation 12.12

whilst Entropy Change ΔS is:-

$$\Delta S = Entropy(t_{melt}, d) - Entropy(t_0, c)$$
$$= 257.833493381125 \, Joules$$

Equation 12.13

c and d are Arrays of Shomate Coefficients, respectively for 298.15°K and 1900°K.

These Coefficient Arrays, together with other inputs and outcomes may be consulted in Table 12.5

It follows that the Gibbs Free Energy ΔG for Temperature T = 1900°K is given by:-

$$\Delta G = \Delta H + T.\Delta S = -489653.571090107 \, Joules$$

Equation 12.14

Knowing that Valency z is 6; and Faraday's Constant is 96485.33212331, the Minimum Voltage, U_0, is given by:-

$$U_0 = \frac{-\Delta G}{z.F} = 0.845816941488268$$

Equation 12.15

Literature values vary from 1.0 to 1.3, but everything depends upon the furnace design and the electrolyte.

Overvoltages

In this part we shall use Allanore[12.11] and the online NIST Chemistry WebBook[12.12] to establish the thermodynamic Minimum Electrolytic Dissociation Voltage, U_0, that is the lower theoretical limit of the voltage that can be applied across an electrolyte for dissociation, and then we shall move forward to ascertain the several Overvoltages that must be added to U_0 in a practical electrolysis. The sum of U_0 and these several Overvoltages is the Cell Voltage U_{cell}.

This sum may be explicated as:-

$$U_{cell} = U_0 + \eta_{anode} + \eta_{cathode} + U_{ohmic}$$
Equation 12.16

where η_{anode} is the Potential Difference (Voltage) due to Anode Resistance; $\eta_{cathode}$ is the Potential Difference due to Cathode Resistance and U_{ohmic} is the Potential Difference due to the Resistance of the Electrolyte.

U_{ohmic} may further be be resolved to:-

$$U_{ohmic} = I.R_{ohmic}$$
Equation 12.17

where I is the applied Current, the sole determinant of Metal Yield; and R_{ohmic} is the effective Electrolyte Resistance. A proper determination of R_{ohmic} would engage the solution of a set of Laplace Equations descriptive of the force field between the electrodes, given the presence and composition of the electrolyte.

In my simple-minded way I shall just assume that the anode and cathode are of equal area, and the electrolytic field is neatly confined to the cylindrical space between.

As you have observed, our cathode is much bigger in area than our anode.

The Anode is iridium-plated tungsten. Given that the Resistivity of Tungsten, ρ_W, is 0.0000000528 Ω.meters, d_{anode} is 0.43 meters and A_{anode} is 5.22792433483878 square meters we find that Pouillet's Quotient $f_{anode} = d_{anode}/A_{anode} = 0.082250616584959$.

Hence, by Pouillet's Law:-

$$\eta_{anode} = I.\rho_W.f_{anode} = 0.000108570813892 \; Volts$$
Equation 12.18

Correspondingly, for the cathode, which is liquid iron we have the Resistivity of Liquid Iron $\rho_{liquidFe}$ is 0.00000139 Ω.meters, $d_{cathode}$ is 0.63 and $A_{cathode}$ is 83.6467893574204 square meters. Hence $f_{cathode}$ =0.007531669832634 and:-

$$\eta_{cathode} = I.\rho_{WliquidFe}.f_{cathode} = 0.000261725526684 \; Volts$$
Equation 12.19

				Shomate Coefficients for the Range 298°K to 950°K		Shomate Coefficients for the Range 1050°K to 2500°K	
T_0	298.15000						
t_0	0.29815						
T	1900.00000						
t_{melt}	1.90000						
H_{low}	-0.025804595						
H_{high}	230.040529437						
ΔH	230.066334032						
S_{low}	87.33209902						
S_{high}	345.1655924						
ΔS	257.8334934						
ΔG	-489653.5711						
z	6						
F	96485.33212						
U_0	0.845816941						
			c_1	93.43834	d_1	110.9362	
			c_2	108.3577	d_2	32.04714	
			c_3	-50.86447	d_3	-9.192333	
			c_4	25.58683	d_4	0.901506	
			c_5	-1.61133	d_5	5.433677	
			c_6	-863.2094	d_6	-843.1471	
			c_7	161.0719	d_7	228.3548	
			c_8	-825.5032	d_8	-825.5032	

Table 12.5
Shomate Computations for the
Determination of U_0

With regard to the electrolyte we know from Equation 12.8 that $\sigma_{electrolyte}$ is 626.815816839618 Ω.meters. Also, $d_{electrolite}$ = 0.43 (height in contact with anode) and $A_{anode} \equiv A_{electrolyte} \equiv A_{anode}$ = 5.22792433483878 square meters.

As aforementioned, I am modelling the active zone as a simple cylinder with Depth, $d_{electrolyte}$, equal to the InterElectrode Gap g and with $A_{electrolyte} \equiv A_{anode}$.

Hence R_{ohmic} is by Pouillet's Law:-

$$R_{ohmic} = \rho_{electrolyte} \cdot \frac{d_{electrolyte}}{A_{anode}} = 0.000131219920197 \, Ohms$$
Equation 12.20

from which:-

$$U_{ohmic} = I.R_{ohmic} = 3.28049800493064$$
Equation 12.21

Hence:-

$$U_{cell} = U_0 + \eta_{anode} + \eta_{cathode} + U_{ohmic} = 4.12668524275948$$
Equation 12.16

The excess voltages sum $\eta_{anode} + \eta_{cathode} + U_{ohmic}$ represents process Heat Loss dure to resistance of the system (i.e. I^2R heat losses) which vitiate process efficiency. The Faradaic Efficiency of the electrolytic furnace, ϕ (or in some literature ξ) is the dividend of U_0 by U_{cell}:-

$$\phi = \frac{U_0}{U_{cell}} = 0.204962795011397$$
Equation 12.22

This efficiency of merely some 20½% is very disappointing when compared with our aluminum smelter efficiency of about 95% but we must remember that my MOE smelter analysis is very crude, as is my very basic "back-of-a-fag-packet" electrolytic cell design.

A competent student could do a lot better.

Current Density

The Current Density j is a measure of the concentration of current at the anode, interesting from the point of view of quantifying the rate of anode erosion at differing currents, temperatures and electrolyte chemistries.

$$j = \frac{I}{A_{anode}} = 4782.01259214877 \; Amperes \; per \; Square \; Meter$$
Equation 12.23

Computation of U_{ohmic} via Current Density

As a check on our algebra, let us define U_{ohmic2}, the Overvoltage due to Electrolyte, in these terms:-

$$U_{ohmic2} = \frac{j \cdot d_{electrolyte}}{\sigma_{electrolte}} = 3.28049382191979$$
Equation 12.24

The PSD(U_{ohmic},U_{ohmic2}) is 0.000127511458412

U_{ohmic} must equal or exceed 1.25 Volts for autothermal heating (Allamore, Ortiz and Sadoway[12.3]). The Autothermal Heat (as Power) Output in the Electrolyte Active Region is:-

$$P_{auto} = I^2 \cdot R_{ohmic} = 82012.4501232659 \; Watts$$
Equation 12.25

Specific Energy Consumption

The Formula Weight (in grams/mol) of Iron is 55.845. Accordingly, the Tonne Formula Weight of Iron, $FW_{FeTonne}$ is 0.000055845

Allamore, Ortiz and Sadoway[12.3] give the Specific Energy in Joules per Tonne of Iron produced as:-

$$E_{specAllanore} = \frac{z \cdot U_{ohmic} \cdot F}{\phi \cdot FW_{FeTonne}}$$
Equation 12.26

I compute $E_{specAllanore}$ as $1.65917805312727 \times 10^8$ Joules per Tonne.

In a BF context Fruehan Et Al cite 5% C liquid iron as Absolute Minimum Energy, $E_{specFruehan}$, as 9.8×10^6 BTU/ton which equals 1.13974×10^{10} J/tonne.

The PSD($E_{specAllanore}$,$E_{specFruehan}$) is 32.9702257269181 or about 33%.

Faraday Energy

$E_{faraday}$ is the Indicated Faraday Energy applied defined by:-

$$E_{faraday} = I.U_{cell}.t$$
Equation 12.27

We shall work on the basis that energy is applied for one Year so that t = $365.25 \times 24 \times 60 \times 60$ = 31557600 seconds. We have already computed the U_{cell} to be 4.12668524275948 volts.

Therefore, as Joules per Annum, $E_{faraday}$ is computed to be $3.25570705542266 \times 10^{12}$.

To render this as KWh per Annum we may write:-

$$E_{faradayKWh} = \frac{I.U_{cell}.t}{3600000}$$
$$= 9.0436307095074 \times 10^5 \, KWh \, per \, Year$$
Equation 12.28

PART FOUR
IRON YIELD

Thinking about Faraday's Laws of Electrolysis

Faraday's Second Law gives an accurate prediction of the metallic cathode product mass, if we have prior knowledge of the Balanced Chemical Pathway, and of the Faradaic Efficiency ϕ.

First compute the Molar Mass of Metal, M_{Fe} = $\lambda_3 \times AW_{Fe}$, where AW_{Fe} is the Atomic Weight of Iron represented in grams.

We know that λ_3 is 2 and AW_{Fe} is 55.845

Hence:-

$$M_{Fe} = \lambda_3 \times AW_{Fe} = 111.69$$
Equation 12.29

This figure is in <u>grams per mole</u>.

To render Faraday's Second Law for a product yield for a Standard one Coulomb of passed charge Q = 1, and one second of current duration t = 1:-

$$m_{Fe} = \phi . \frac{Q . M_{Fe}}{z . F} = 0.000039543652337 \ grams$$
Equation 12.30

So for this pathway the Amount of Iron precipitated at the cathode is roughly 40 micrograms for one second and one ampere applied.

To clarify the structure of Faraday's Second Law let us resolve it into three simple divisions:-

$$Div_1 = \frac{\phi}{z} = 0.034160465835233$$
No Units (dimensionless)
Equation 12.31a

$$Div_2 = \frac{Q}{F} = 0.000010364269656$$
mols
Equation 12.31b

$$Div_3 = M_{Fe} = 111.69$$
grams per mol
Equation 12.31c

Confirm that:-

$$m_{Fe} = \frac{\phi}{z} . \frac{Q}{F} . M_{Fe} = \prod_{i=1}^{3} Div_i = 0.000039543652337$$
Equation 12.32

The constant of proportionality Z is the First Law Electrochemical Equivalent (ECE) such that:-

$$Z = \frac{\phi}{z} \cdot \frac{Q}{F} = 0.000000354048279$$
Equation 12.33

Z can be in any units independently of M_{Fe}.

But in this example Cancellation of units confirms that m_{Fe} is in grams.

So m_{Fe} is in grams when M_{Fe} is likewise in grams/mol: But m_{Fe} can be in any mass unit compatible with the units of M_{Fe}.

Alternatively:-

$$m_{Fe} = Z . M_{Fe} = 0.000039543652337$$
Equation 12.34

Iron Yield by Faraday's Second Law (SI units)

As a reminder, our chosen Pathway is:-

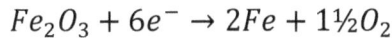

$$Fe_2O_3 + 6e^- \rightarrow 2Fe + 1\tfrac{1}{2}O_2$$
Pathway 12.1

As for the Boston Metal® pilot furnace currently operational let us assume that the Applied MOE Furnace Current I = 25000 Amperes; Valency (λ_2) = 6; z ≡ λ_2; and that *initially* Duration of Current t = 1 second.

Faraday's Constant is 96485.33212331 coulombs per mole.

Faradaic Efficiency is 0.204962795011397: **Not** for the Boston Metals® Furnace; for my design.

Then we may state Charge Q to be:-

$$Q = I.t = 25000 \times 1 = 25000 \; Coulombs$$
Equation 12.35

The Number of Moles of Electrons transacted, n_e, is given by:-

$$n_e = \frac{Q}{z.F} = 0.043184456901092$$

Equation 12.36

And:-

$$m_{Fe} = \phi.n_e.mols_{Fe}.AW_{Fe} = \phi.\frac{Q.mols_{Fe}.AW_{Fe}}{z.F} = \phi.\frac{Q.M_{Fe}}{z.F}$$
$$= 0.988591308433552$$

Equation 12.37

So *for this pathway* the Amount of Iron precipitated at the cathode is roughly 1 gram for one second of 25000 Amperes applied.

The European mean iron production per Blast Furnace (2023AD) is 1742.80702 Tonnes per Annum.
One Year comprises $t = 365.25 \times 24 \times 60 \times 60 = 31557600$ seconds.
Again allow that the Current Applied is 25000 amperes. Then by Faraday's Second Law:-

$$m_{Fe} = \phi.\frac{I.T.M_{Fe}}{z.F} = 31197569.0750226 \; grams \; per \; annum$$

Equation 12.38

To render this as Tonnes of Iron per Annum we divide by one million:-

$$m_{FeTonnes} = 0.000001 \times m_{Fe} = 31.1975690750226$$

Equation 12.39

Therefore, the mass of iron liberated is about 31.198 tonnes per annum.

Faradaic Energy Demand

To calculate the Faradaic Energy Demand of this furnace, $E_{faraday}$, it is interesting and useful first to calculate the rated Power in Watts using:-

$$P = I.U_{cell} = 103167.131068987 \ Watts$$
Equation 12.40

      revealing that this 25000A furnace design has a power of about 103.167 KiloWatts: It is only a feeble furnace.

      We may move forward to compute $E_{faraday}$ in Joules per Annum as:-

$$E_{faraday} = I.U_{cell}.t = P.t = 3255707055422.66$$
Equation 12.41

      or to express the Faradaic Energy in KWh per Annum:-

$$E_{faradayKWh} = \frac{I.U_{cell}.t}{3600000} = 904363.07095074$$
Equation 12.42

      To calculate the Number of Furnaces, $nf_{elec,}$ to match a typical European blast furnace of output 1742.80702 tonnes of liquid iron per year we use:-

$$nf_{elec} = \frac{m_{EuroBF}}{m_{FeTonnes}} = \frac{1742.80702}{31.1975690750226} = 55.8635519264008$$
Equation 12.43

      nf_{elec} may only be an integer so we may round up to 56 using ceil(nf_{elec}) or whatever functional syntax is convenient using your software. We may compare this figure of 56 parallel cells ("pots") with the 60 cells available in the Lochaber Aluminum Smelter at Fort William in Scotland, our specimen smelter for electrolysis examples.

      We may adjust our Faradaic Energy with this integer value of furnace numbers in these terms:-

$$E_{KWh} = E_{faradayKWh} \times nf_{elec} = 904363.07095074 \times 56 = 50644331.9732414$$
Equation 12.44

      So the suite of 56 iron smelting electrolytic cells demand about 50.6 million KWh during the course of a year's operation.

This equates to E_{MWh} = 50644.3 MWh per year or E_{GWh} = 50.6 GWh per year.

MegaWatt Hours per Tonne

The Faradaic Megawatt Hours of electricity demand per tonne of iron, $E_{MWHPerTonne}$, is given by:-

$$E_{MWhPerTonne} = \frac{E_{MWh}}{m_{EuroBF}} = 29.0590589732886$$
Equation 12.45

According to Boston Metals[13.2] this figure should be nearer 4 MWh per tonne, *but remember*, their design is much more efficient than mine and their Faradaic Efficiency ϕ will approach unity much more closely.

Ore Ratio

The Ore Ratio, Ore_{rat}, is given by:-

$$Ore_{rat} = \frac{\lambda_{Fe2O3} \cdot FW_{Fe2O3}}{\lambda_{Fe} \cdot AW_{Fe}} = 1.42973542841794$$
Equation 12.46

Iron Yield by Simplified Route Two

By Pathway 12.1:-

$$\lambda_{Fe} \equiv \lambda_3 = 2$$
Equation 12.47

whilst:-

$$\lambda_e \equiv \lambda_2 = 6$$
Equation 12.48

and the Standard Time t is unity (i.e. t is one second). We know that the (gram) Atomic Weight of Iron is 55.845 and so the Kilogram Atomic Weight of Iron, FW_{Fe} must be 0.055845

Therefore the Mass of Reduced Iron in Kilograms, m_{Fe}, is by Faraday's Second Law:-

$$m_{Fe} = \frac{Q.\lambda_{Fe}.FW_{FeKg}}{\lambda_e.F} = \frac{I.t.\lambda_{Fe}.FW_{FeKg}}{\lambda_e.F} = 0.004823271991283$$

Equation 12.49

Meanwhile, the Energy to Reduce m_{Fe} (in Joules) is given by:-

$$E = P.t = I.U_{cell}.t = 103167.131068987$$

Equation 12.50

where P is Power and U_{cell} is summative Cell Overpotential.

Accordingly the Energy to Reduce One Kilogram of Iron, E_{JperKg} is:-

$$E_{jperKg} = \frac{E}{m_{Fe}} = 21389449.1655123$$

Equation 12.51

To express this in KWh per Kilogram we write:-

$$E_{KWhperKg} = \frac{1}{3600000} \times \frac{E}{m_{Fe}} = 5.94151365708674$$

Equation 12.52

Note that KWh per Kilogram is numerically equivalent to MegaWatt Hours per Tonne.

So by Route Two, about 5.94 MegaWatt hours of electric power is expended to obtain one tonne of iron metal by MOE, *if there is 100% Faradaic Efficiency.* (That never obtains in real life)

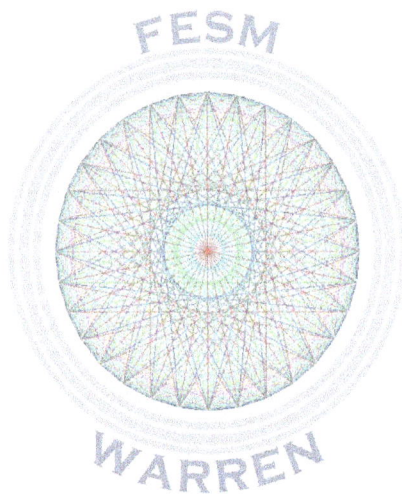

CHAPTER THIRTEEN
WAYS AHEAD

PART ONE
THE ENERGY EFFICIENCIES OF
IRON SMELTING METHODS

Table 13.1 presents the yield efficiencies of the three salient iron smelting methods established or in advanced piloting in 2024AD: The Coal-based Blast Furnace (BF); the Hydrogen-based Direct Reduction Furnace (H_2DRF); and the Metal Oxide Electrolysis Furnace (MOE).

The efficiency, or rather Efficacy, is represented as MegaWatt-hours of total thermal or electrical energy required per tonne (metric ton) of hot iron produced. We know that the H_2DRF method does **not** output iron in a liquid state.

You can see at once that the three methods consume about 3.74 MWh per tonne and the PSD percentage deviation is at worst 13% between them.

This is about what we would expect for a well-designed, well-managed furnace of any sort which must obey the basic laws of science: Stoichiometric chemistry; the Laws of the Conservation of Mass and of Energy (as exemplified by Hess's Law); and the laws of elementary chemical thermodynamics.

Further study emphasises the fundamental inadequacy of my JRW MOE design which is about 51% more energy-expensive than the Boston Metal® MOE furnace, largely due to my high U_{cell}, which in turn reflects the naive quality of my electrodes and electrolyte design. You recall that the Faradaic Efficiency, ϕ, of my design is around 20%.

So there is little objectively to choose between these several fuels and formats. Subjective considerations are therefore salient: Dollar Costs, Hydrocarbon Feedstocks Conservation; Environmental preferences; and above all considerations of National Security.

All of these subjective constraints implicate humane privations which should be addressed through rational policy trade-offs.

The Yield of Iron and MegaWatt-years

Unfortunately, it is impractical to operate a commercial iron furnace for a second, or even for a whole hour.

It is much more convenient for us to work in MegaWatt-years, MWy, noting that:-

$$MW_y = 24 \times 365.25 \times MW_h = 8766 MW_h$$

Equation 13.1

where MW_h is in MegaWatt-hours.
MW_y and MW_h are both dimensionally Energy $[ML^2T^{-2}]$.
Numerically speaking, we can resolve Iron Yield, Y_{Fe}, into:-

$$Y_{Fe} = \xi . MW_y = \xi . s_c . k_m$$

Equation 13.2

where ξ is the Overall System Efficiency (dimensionless $0 \leq \xi < 1$); s_c is the Scale Constant (i.e. 8766); and K_m is the value of the Amount of Energy needed to Reduce One Tonne of Iron.

If the Requirement of the furnace is Rq tonnes of Iron then the Number of Years, y_{Rq}, to achieve this output is given by:-

$$y_{Rq} = \frac{Rq}{\xi . MW_y} = \frac{Rq}{\xi . s_c . k_m}$$

Equation 13.3

Table 13.1 displays the various data and results for three species of iron reduction furnace.

Table 13.1 uses estimated Overall Efficiencies from the literature except that the efficiency of 20% I calculated for my MOE furnace "design".

In every case the *single* furnace requirement was 70,000 tonnes of Iron Product for reasons that I shall later clarify.

The Overall Efficiency of the H2DRF furnace will be somewhat less than 53%, which is the faradaic efficiency of water electrolysis considered in isolation with none of the attendant gas management or furnace heat wastages taken into account.

The Cost of Electricity

The cost of public electricity in the UK makes heavy industry uneconomic, except in circumstances where extensive sites can generate their own power (as applies, for example, to blast furnace integrated steel plants); or where a monopoly provider such as a railway can pass on costs to the customer.

Also, around 0% to 24.5% of electricity on the National Grid is provided by wind turbines, and this unsteady supply has to be re-enforced by electricity purchased from Norway and France.

In August 2024, the cost of electricity in the UK was £79.3 per MWhour which equates to $103.34 per MWhour. This is roughly the same as for other major West European states, except Italy.

For the end consumer, however, the cost of electricity in the UK is 0.40 € per KWh (£341.13581 per MWh), and 0.25 € per KWh in the EU. On 21 August 2024, 0.40 Euros was $0.44, and 0.25 Euros was $0.28

The cost of electricity only marginally affects the BF, but is key to the viability of Hydrogen Direct Reduction furnaces and to Metal Oxide Electrolysis.

Method Code	Source	Furnace Technology	Value K_m (MWh/tonne)	Assumed Efficiency ξ (fraction)	Annual Iron Yield Y (tonnes Fe)	Annual Iron Requirement Rq (tonnes)	Years to Smelt Rq tonnes	Notes and References
a	Boston Metal	MOE	4.00	0.45	15778.80	70000.000	4.4363	Ch13: Ref 13.2
b	European BF	BF	3.75	0.12	3944.68	70000.000	17.7454	Ch11: Eqn 11.46
c	Vogl	H_2DRF	3.48	0.53	16168.01	70000.000	4.3295	Ch11: Table 11.8 Ref 11.5 (Vogl Et Al)
d		Mean	3.74	0.37	12031.81	70000.000	5.8179	
e	JRW	MOE	6.23	0.20	10917.07	70000.000	6.4120	Ch12: Route 1: $\phi = 1$: Eqn 12.38
f	JRW	MOE	5.94	0.20	10416.66	70000.000	6.7200	Ch12: Route 2: $\phi = 1$: Eqn 12.52
g		Mean	6.08	0.20	10666.87	70000.000	6.5624	

PSD(a,b)   6.25 h
PSD(a,c)   13.00 i
PSD(a,g)   -52.11 j

All Figures are for One Furnace Cell only
One MWyear = 24*365.25 = 8766 MWhours

ξ H_2DRF: Figure for Water Electrolysis Only (Bhasker Et Al)

Table 13.1
The Energetic Iron Yield of
Three Iron Smelting Methods

PART TWO
THE PROBLEM OF CHOICE

As you know I am no big fan of doctrine: I consider that it stifles Enquiry rather than inspiring action.

But when I was a young student geologist one of the first things I was taught was the Doctrine of Uniformity: "The Present is the Key to the Past".

If the Present is the Key to the Past then I suppose that Predicament is the Key to the Future.

An old joke concerns a benighting wayfarer who asks a peasant for directions to the nearest hostelry. The peasant hesitates. The peasant re-lights his pipe and ponders, and the burgeoning smoke grows above the glow in the still and silent crepuscule. The peasant commences to explain. "If I were you, I would not start from here", he counsels.

Well that is fine and dandy, but if we are to start, we have to start somewhere.

Rate and Status

Whether we like it or not, we all have a stage or station, individually and collectively.

Whole nations have a qualitative status, and like men they sometimes change their fortune, for good or evil, sometimes unexpectedly, since there is competition as well as comity amongst all.

Some students will question the relevancy of this to iron smelting. I would counsel them to bear with me, because no-one ever smelted iron for no reason. Ever since the Hallstatt Cultures, men have sought iron assiduously whether to till the land; to defend they and theirs; to bridge gulphs, to build boats, or to pipe water. Or to furnish the table with tools other than dirty fingers. Nothing else is like iron.

When we look at men we see that they are of differing quality but fall into classes that have internal affinities greater than similarities with other groups. So it is with the associations themselves, whether local ironworks or whole empires.

If your class frames the words and works you choose, then quite assuredly the Predicament of nations will select the modes of iron working most suiting to them. For sure, men and nations will find reasons or rationalisations why they choose this or that, often educing economic or financial arguments, but equally sure it is that ultimately choice is socio-political or even anthropological at base.

There are at least three Rates or types of countries in the modern world briefly and for certain superficially described in Table 13.2

Predicament affects choice and choice is just as subjective as any antecedent, but it is inadequate to classify countries on the basis of Nominal Gross National Product, or any other arbitrary criterion selected for mathematical convenience or who knows why.

So Predicament furnishes the starting point, but nations and empires can prosper and decay just as men and women do, except that nationhood like literature offers itself some kind of immortality.

Russia, for example, demoted itself from Rate One power status to Rate Two on 8 December 1991 when the late Boris Yeltsin ratified the Belavezh Accords to terminate the Soviet Union. Similarly the United States promoted itself from Rate Two to Rate One on 8 June 1915 when William Jennings Bryan tendered his resignation to Woodrow Wilson ending American neutrality. But these are matters for others than I.

In fact organisations and individuals are in constant flux though I am not competent to discuss the ascents and vicissitudes of your country so I shall further illustrate with my nation.

Britain, or specifically England, promoted from a Fourth Rate sovereignty to a Third Rate some time towards the end of the fifteenth century, possibly on 3 November 1492 when Henry the Seventh, Tudor, ratified the Peace of Étaples with Charles the Eighth, Valois, which may have set England on the track of economic ascendancy in Europe.

Britain then promptly rose from a Rate Three kingdom to Rate Two when on 3 November 1534 the Act of Supremacy made England independent of The Holy See at Rome. England had to wait another 179 years until it graduated to being a First Rate sovereign power independent of Continental evolutions. This was achieved with the signing of the Treaty of Utrecht on 14 March 1713.

Change takes time. The specious precision of these transitions may be slightly comedic and the state transitions should be viewed as plus or minus ten years say. The Peace of Étaples was preceded by a seven-year campaign of English economic reconstruction following the Battle of Bosworth Field which ended one of the national civil wars and some say concluded the Middle Ages. And the Peace of Étaples was postceded by continuing reform and recovery in England.

Rate

1 Major populous global industrial powers
("Superpowers")
Major and diverse industrial and commercial activities
many of a transglobal character
Independent civil and military nuclear and space establishments
Several megacities in each
Adequate reserves of coking coal permit
continued use of Blast Furnace technology

2 Middle-ranking stable democracies
with significant industrial or extractive outputs
Actual or potential nuclear powers with
a developed independent civil nuclear sector
Difficulties with fossil fuel supply encourage
research into non-traditional iron smelting

3 Lower-ranking agrarian or extractive economies
Politically labile (except Canada and Australia)
Though mainly agrarian food processing is locally important
Low capital formation
Argentina, Britain and South Africa are former nuclear powers
Each state characterised by a single enormous megacity
which may or may not be the capital
These countries accommodate significant
ethno-cultural minorities and income differentials are stark
Dearths of fossil fuels and political problems with their supply
and of fickle electric power provision encourage
stored hydrogen applications including H_2 DRF iron smelting

4 Highly heterogenous non-"G22" territories
Many of these states or statelets, including residual
colonies, are very small
Few employ iron furnaces of any type and may rely upon
expensive oil fuel for general power supply.
On the other hand, larger Rate 4 polities may have
abundant hydropower rendering MOE furnaces feasible

Rate 3 and 4 polities are highly-dependent upon Rate 1
for military and economic sponsorships

Rate 1, 2 and 3 countries have active Blast Furnace ironworks in 2024AD

Table 13.2
Characteristics of Differently Rated Countries

On 18 July 1947 Britain reverted to being a Second Rate power and stepped down to Third Rate status on 1 January 1973 when Britain ceded sovereignty to an entity which is now the European Union. Britain dismantled its independent civil nuclear and military nuclear delivery capabilities; the majority of its steel and aluminum production; and made over its public electricity supply to foreign control led by Germany and France.

These apparently political or anthropological phenomena are mirrored by respective technological innovations. It is difficult to imagine an Act of Supremacy without printing, and the expansion of British trade facilitated by the 1651-1696 Navigation Acts and by the Utrecht protocols was complemented by the development of the Steam Engine between 1698 and 1712 and Darby's smelting of iron ore with coke invented on 10 January 1709 when an old charcoal furnace was blown-in with roasted coal at Coalbrookdale.

I do not mean to blame or disparage either France or Germany, or indeed Russia. All three are very excellent countries, worthy friends and rivals. If my remarks at any point cause offence I sincerely apologise.

If, as Francis Fukuyama argued, history ended in 1991, then it is remarkable that in the last thirty-three years Clio has picked up her skirts and raced. In that time there has happened an accelerating succession, not only of abominations and atrocities, but of striking discoveries and remarkable inventions.

Conveniently for us some twenty of the current wealthiest nations, all iron smelters, have appointed themselves the "G20" intergovernmental forum. They include the European Union (additionally to France, Germany and Italy) and also the African Union (in addition to South Africa), but technically exclude the Netherlands, Spain and Poland.

This book is already long and we lack space further to elaborate these fascinating aspects of history and election, because selection theory and decision theory have become massive intellectual adventures in their own right, comparable in scope to Computer Science to which they are intimately related.

One of the facets of decision arithmetic is examined in the essay "A Boolean Calculus of Moiety for Two Arguments" in my 2022 book "Mathematical Explorations"[13.1] amongst whole libraries of other works.

Table 13.3 presents a digest of statistics for these twenty-two national entities (other than the European and African Unions). I should like you also to examine, if you will, Table 13.4, which offers a quantitative breakdown of the sources of electrical power for the twenty leading iron-smelting countries.

We are still living in the age of the Blast Furnace and so I suppose we can look to quantify an "obvious" correlation between coal-fired electricity generation and crude steel output. This is illustrated by the plot Figure 13.1 which shows Crude Steel Output in 2023AD versus Coal-fired Electricity Output in the same year.

The plotted linear regression equation is:-

$$Steel\ Output = 2.39888614 + 0.18280263 \times Coal\ Generation$$
Equation 13.4

Note that the Coefficient of Determination is 0.96678262 which implies that 96⅔% of the variance in y with respect to x is "accounted for" or "explained" by Crude Steel Production.

This is an example of the distortion inherent in uncritical application of linear fitments. If we exclude China we obtain Figure 13.2

Now the equation becomes:-

$$Steel\ Output = 11.04996812 + 0.09129369 \times Coal\ Generation$$
Equation 13.5

with an R^2 Coefficient of Determination of 0.65158509 so that 65% of variation is "explained" by y.

This is the kind of selective trick that journalists and politicians perform all the time, whether through ignorance or intent to deceive.

For correlations of this degree of dispersion it is usually more useful, and much clearer for the long-suffering reader, if we use log-log regressions where we simply take the (natural) logarithm of both the x and y parameters, plot that, and fit a *linear* regression to the resulting data scatter.

For example, if x = Log_n(Coal-sourced Electricity) and y = Log_n(Crude Steel Production) we arrive at the equation:-

$$log_n(Steel\ Output)$$
$$= 1.38500247$$
$$+ 0.43724406 \times log_n(Coal\ Generation)$$
Equation 13.6

Nation	Rate	Crude Steel (Tonnes×10⁶)	Trade bil. USD [2022][64]	Nom. GDP mil. USD [2024][65]	PPP GDP mil. USD [2024][65]	Nom. GDP per capita USD [2024][65]	PPP GDP per capita USD [2024][65]	HDI (2022)[66]	Population (2022)[67]	Area km²	First Rate	Second Rate	Third Rate
Argentina	3	4.9	170.10	604,260	1,244,646	12,812	26,390	0.849	46,300,000	2,780,400			4.9
Australia	3	5.8	721.40	1,790,348	1,791,358	66,589	66,627	0.946	26,141,369	7,692,024			5.8
Brazil	2	36.2	626.40	2,331,391	4,273,668	11,352	20,809	0.760	217,240,060	8,515,767		36.2	
Canada	3	13	1179.10	2,242,182	2,472,227	54,866	60,495	0.935	38,743,000	9,984,670			13
China	1	1032.8	6309.60	18,532,633	35,291,015	13,136	25,015	0.788	1,411,750,000	9,596,960	1032.8		
France	2	13.9	1435.80	3,130,014	3,987,911	47,359	60,339	0.910	68,305,148	640,679		13.9	
Germany	2	40.1	3226.90	4,591,100	5,686,531	54,291	67,245	0.950	84,316,622	357,114		40.1	
India	2	118.2	1176.80	3,937,011	14,594,460	2,731	10,123	0.644	1,406,632,000	3,287,263		118.2	
Indonesia	3	14.3	529.40	1,475,690	4,720,542	5,271	16,861	0.713	279,088,893	1,904,569			14.3
Italy	3	24.4	1346.40	2,328,028	3,347,103	39,580	56,905	0.906	61,095,551	301,336			24.4
Japan	2	96.3	1644.20	4,110,452	6,720,962	33,138	54,184	0.920	125,592,404	377,930		96.3	
Mexico	3	18.5	1204.50	2,017,025	3,434,224	15,249	25,963	0.781	131,541,424	1,964,375			18.5
Netherlands	3	6.6	1444.00	1,143,000	1,329,000	63,750	74,158	0.946	18,146,200	41,865			6.6
Poland	3	8.5	381.52	844,623	1,801,000	23,014	49,060	0.881	36,620,970	312,696			8.5
Russia	2	75.6	772.30	2,056,844	5,472,880	14,391	38,292	0.821	145,807,429	17,098,242		75.6	
Saudi Arabia	3	8.7	598.80	1,106,015	2,354,392	33,040	70,333	0.875	36,168,000	2,149,690			8.7
Spain	3	14.2	817.62	1,647,000	2,516,000	34,045	52,012	0.911	48,797,875	505,990			14.2
South Africa	3	5	259.10	373,233	1,025,930	5,975	16,424	0.717	61,060,000	1,221,037			5
South Korea	3	70.4	1415.00	1,760,947	3,057,995	34,165	59,330	0.929	51,844,834	100,210			70.4
Turkey	3	40.4	617.90	1,113,561	3,831,533	12,765	43,921	0.855	85,551,932	783,562			40.4
United Kingdom	3	7.2	1353.30	3,495,261	4,029,438	51,075	58,880	0.940	68,492,933	242,495			7.2
United States	1	85.8	5441.00	28,781,083	28,781,083	85,373	85,373	0.927	337,341,954	9,833,517	85.8		
Count											2	6	14
Total											1118.6	380.3	241.9
Mean											559.3	63.38333	17.27857
Ratio												8.824086	3.66832

Table 13.3
Statistical Summaries for the Twenty-Two "G20PLUS" Nations

Location	Rate	Area (sq kms)	Population	Population Density (capita/km²)	GDP Nominal (USD×10⁹)	Crude Steel (Tonnes×10⁶) (2021)	Total (TWh)	Coal	Gas	Hydro	Nuclear	Wind	Solar	Oil*	Bio.	Geo.	Nuclear	Fossil	Elemental	Fraction Elemental
China	1	9596960	1409670000	147	18532633	1032.8	8849.00	5398.00	291.00	1303.00	418.00	763.00	428.00	72.00	177.00	0.00	418.00	5761.00	2671.00	0.3018
United States	1	9525067	335893238	35	28781083	85.8	4287.00	832.00	1687.00	249.00	772.00	434.00	205.00	39.00	52.00	18.00	772.00	2558.00	958.00	0.2234
Brazil	2	8510346	203080756	24	2331391	36.2	677.00	16.00	42.00	427.00	15.00	82.00	30.00	12.00	53.00	0.00	15.00	70.00	592.00	0.8744
France	2	543940	68468000	126	3130014	13.9	469.00	4.00	46.00	46.00	295.00	39.00	61.00	22.00	10.00	0.60	295.00	60.00	115.60	0.2456
Germany	2	357581	84673158	237	4591100	40.1	567.00	180.00	80.00	18.00	35.00	125.00	61.00	22.00	48.00	0.20	35.00	282.00	252.20	0.4431
India	2	3287263	1404910000	427	3937011	118.2	1858.00	1380.00	47.00	175.00	46.00	70.00	95.00	46.00	41.00	0.00	46.00	1431.00	381.00	0.2051
Japan	2	377915	123960000	328	4110452	96.3	1034.00	348.00	361.00	75.00	52.00	8.00	102.00	46.00	41.00	0.00	52.00	755.00	226.00	0.2188
Russia	2	17098246	146150789	9	2056844	75.6	1167.00	192.00	534.00	198.00	224.00	4.00	2.00	12.00	0.80	0.00	224.00	738.00	204.80	0.1755
Argentina	3	2780400	47067441	17	604260	4.9	151.00	2.00	80.00	24.00	7.00	14.00	3.00	17.00	2.00	0.00	7.00	99.00	43.00	0.2886
Canada	3	9984670	41012563	4	2242182	13.0	660.00	35.00	84.00	398.00	87.00	38.00	6.00	3.00	9.00	0.00	87.00	122.00	451.00	0.6833
Indonesia	3	1904569	281603800	148	1475690	14.3	334.00	205.00	57.00	27.00	0.00	0.40	0.40	6.00	21.00	17.00	0.00	268.00	65.80	0.1971
Italy	3	302068	58971479	195	2328028	24.4	280.00	23.00	141.00	28.00	0.00	20.00	28.00	16.00	18.00	6.00	0.00	180.00	100.00	0.3571
Mexico	3	1964375	129713690	66	2017025	18.5	341.00	22.00	192.00	36.00	11.00	20.00	19.00	34.00	7.00	6.00	11.00	248.00	82.00	0.2405
Netherlands	3	41865	17977676	429	1142513	6.6	120.00	15.00	48.00	0.05	4.00	21.00	17.00	5.00	10.00	0.00	4.00	68.00	48.05	0.4002
Saudi Arabia	3	2149690	32175224	15	1106015	8.7	402.00	0.00	269.00	0.00	0.00	0.01	0.80	131.00	0.00	0.00	0.00	400.00	0.81	0.0020
South Africa	3	1219090	63015904	52	373233	5.0	239.00	202.00	0.00	3.00	10.00	10.00	10.00	4.00	0.40	0.00	10.00	206.00	23.40	0.0977
South Korea	3	100432	51285153	511	1760947	70.4	620.00	211.00	175.00	4.00	176.00	3.00	27.00	7.00	17.00	0.00	176.00	393.00	51.00	0.0823
Spain	3	505370	48692804	96	1647114	14.2	286.00	8.00	86.00	18.00	59.00	62.00	36.00	12.00	7.00	0.02	59.00	106.00	123.02	0.4271
Turkey	3	783562	85372377	109	1113561	40.4	321.00	114.00	72.00	67.00	0.00	35.00	15.00	0.70	8.00	10.00	0.00	186.70	135.00	0.4196
United Kingdom	3	244376	67596281	277	3495261	7.2	326.00	6.00	125.00	5.00	48.00	80.00	14.00	13.00	35.00	0.00	48.00	144.00	134.00	0.4110
Australia	3	7741220	27364621	4	1790348	5.8	274.00	131.00	46.00	17.00	0.00	32.00	39.00	5.00	3.00	0.00	0.00	182.00	91.00	0.3333
Poland	3	312685	37582000	120	844623	8.5	179.00	125.00	11.00	2.00	0.00	20.00	8.00	5.00	8.00	0.00	0.00	141.00	38.00	0.2123
Totals		79331690	4766236954		89411328	1740.8	23441	9449	4474	3120.05	2259	1880.41	1166.2	475.7	568.2	51.82	2259	14398.7	6786.68	
Means		3605985.9	216647134.3	153.41523	4064151	79.127273	1065.5	429.5	203.364	141.82	102.682	85.4732	53.0091	21.6227	25.8273	2.35545	102.682	654.486	308.485455	0.31091459

Table 13.4
Electrical Power Sources and Quantities for the G20+ Iron-Smelting
Countries in the Year 2023AD

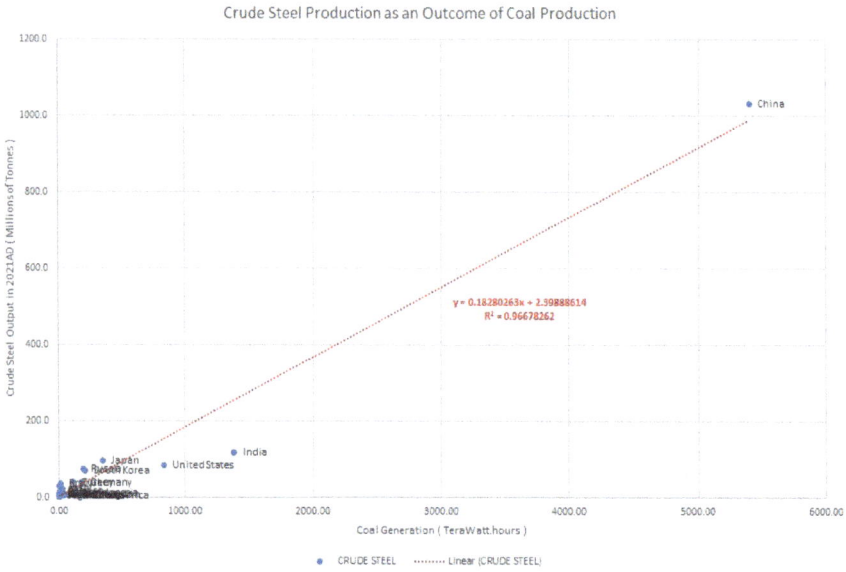

Figure 13.1
Coal Versus Crude Steel Production

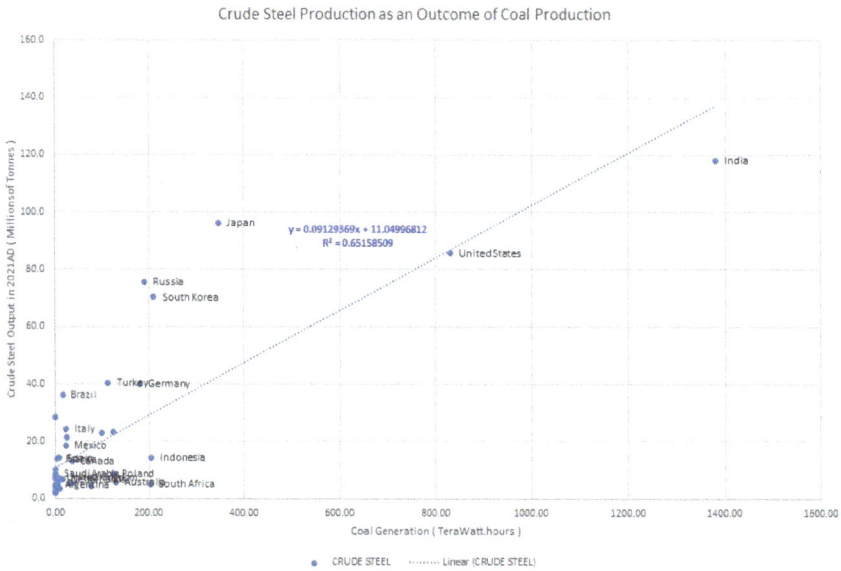

Figure 13.2
Coal Versus Crude Steel Production
Excluding China

Figure 13.3 is the log-log plot of these variables with the attaching linear regression line and its equation:-

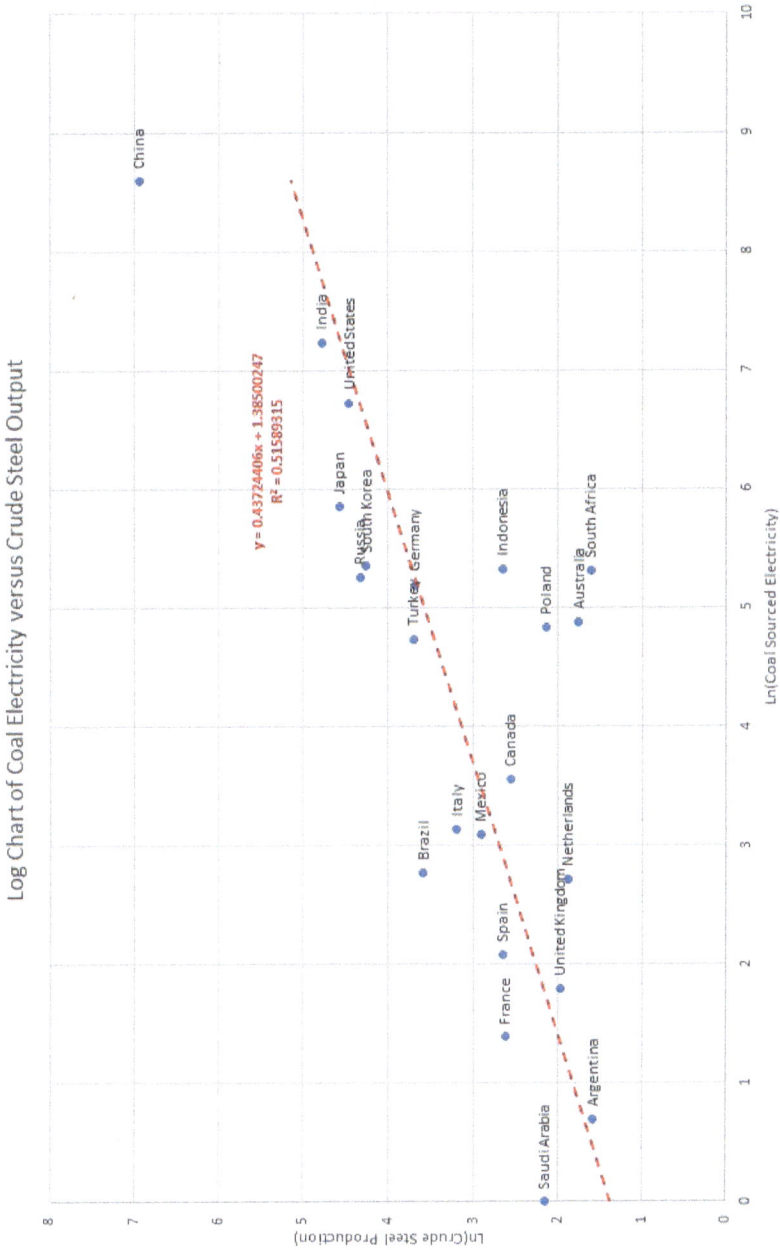

Figure 13.3
Coal Versus Crude Steel Production
as a Log-Log Correlation

This log-log relationship can be expressed by a fitted linear regression (shown on the plot):-

$$\ln(S) = a + b.\ln(C) = 1.38500247 + 0.43724406.\ln(C)$$
Equation 13.7

where S denotes Crude Steel Production and C is Coal Sourced Electricity. The R^2 is 0.51589315, virtually half the R^2 for the original arithmetic data.

(I assess steel production as equivalent to iron since crude steel is about 99% iron).

The numeric value of Equation 13.7 is 2.06501960629117

By exponentiating both sides of Equation 13.7 we obtain:-

$$e^{\ln(S)} = e^{[a+b.\ln(C)]} = e^{2.06501960629117} = 7.88545250061706$$
Equation 13.8

Equation 13.8 is equivalent to:-

$$S = e^a.C^b$$
Equation 13.9

where e^a is itself the constant k = 3.9948357720488
Hence:-

$$S = k.C^b = 7.88545250061706$$
Equation 13.10

Figure 13.2 is clear, and clearly suggestive, but is another meaningless illustration because the coal-based production of iron and coal-based production of electricity both arise from the availability of coal, but C and S are *not* cause and effect.

Sophistication is not proof.

An empirical equation like 13.10 is a handy look-up numerical tool but not a scientific law, or *of itself* a policy guide.

Rate, Steel Production and Electric Power

At this stage, I identify iron with crude steel. Mild Steel as decanted from the Basic Oxygen Furnace is less than one percent carbon and in well-managed circumstances does not yet include contaminants or additives.

Table 13.4 includes various quantitative sources of electric power supply in the several countries.

I have groped these sources into three classes:-

Nuclear	Nuclear
Fossil	Coal, Gas and Oil
Elemental	Hydro, Wind, Solar,
	Biomass and Geothermal

To express the proportion of Elemental Electricity to the electric output of a nation overall I further define:-

$$\frac{Fraction}{Elemental} = \frac{Elemental\ Electricity}{Nuclear + Fossil + Elemental} = \frac{Elemental}{Total}$$

Equation 13.11

These Elemental or Renewable sources of electric power arise from environmental forces like gravitating water, the wind or volcanic hot springs. They are essentially both local and fickle but where a country is blessed with such they are cheap and clean once the initial capital works are complete.

Nuclear and Fossil supplies are expensive and dirty, both in capital costs and ongoing logistical and procurement expenses, not to mention de-commissioning costs, but they offer steady and controllable "base load" input essential to iron smelting, or the thermal treatment of anything else, for example glass or cement.

Blast furnaces use coal or coal-derived fuels but very little electricity, so as we have seen we cannot say that because a country produces a lot of coal it also produces a lot of coal-fired electricity because the two factors do not cogently correlate.

First Rate powers with plentiful coking coal make lots of iron using BFs, but have low Fractions Elemental of electric power, despite their strong base-load sustentations involving Nuclear and Fossil fuels.

Second Rate powers offer a more mixed picture. For example, France which has low Elemental input (because it is essentially Nuclear) would be ideally suited to MOE though it is not currently using ferrous Metal Oxide Reduction on a commercial basis: Its legacy Blast Furnaces supply its internally-sourced iron. On the other hand, Brazil has abundant hydroelectric resources (for example the Falls of Iguazu Itaipu dam complex: 14GW shared with Paraguay) which make it a prime candidate for transition from BF to MOE smelting.

Third Rate powers usually have plentiful Elemental power in proportion to their population density. Whilst hydropower is usually quasi-steady in environments of plentiful rainfall with little or moderate freezing, wind and solar resources are very fickle and H_2DRF processes are better suited to such economies because hydrogen can be stored at bulk whereas electrical battery systems are both feeble and extremely expensive at scale. Spain has high wind and solar energy supplies whilst Britain has high gas and wind generation, but in neither case is it realistic to think that that they might return to coal for thermal supply.

Small Modular Reactors (SMRs) may be just the ticket for remote or isolated communities almost anywhere: But one thinks especially of oceanic islands, outback Australia, or the parts of New Zealand ill-favored for Hydropower or Geothermal power.

Further Data Regressions

Figure 13.4 and Figure 13.5 are both log-log plots of raw data.

Figure 13.4 relates x = $\log_n$(Population) of G22 states to y1 = $\log_n$ (Crude Steel Output) and y2 = $\log_n$(Total Electricity Generated).

Figure 13.5 relates x = $\log_n$(Nominal Gross National Product) which is in millions of dollars but could be in any arbitrary unit to y1 = $\log_n$ (Crude Steel Output) and y2 = $\log_n$(Total Electricity Generated).

It is readily seen that Steel and Electricity outputs correlate positively and separately both with national populations and national GDPs.

Perhaps it is more striking that the nations group by Rate.

Whilst they jostle a little amongst themselves like pollen cells on a microscope slide they tend, like kinetic particles, to retain a mean position. Like for the genera of pollen cells that average position, transferable between parameters, reflects the kind of countries they are.

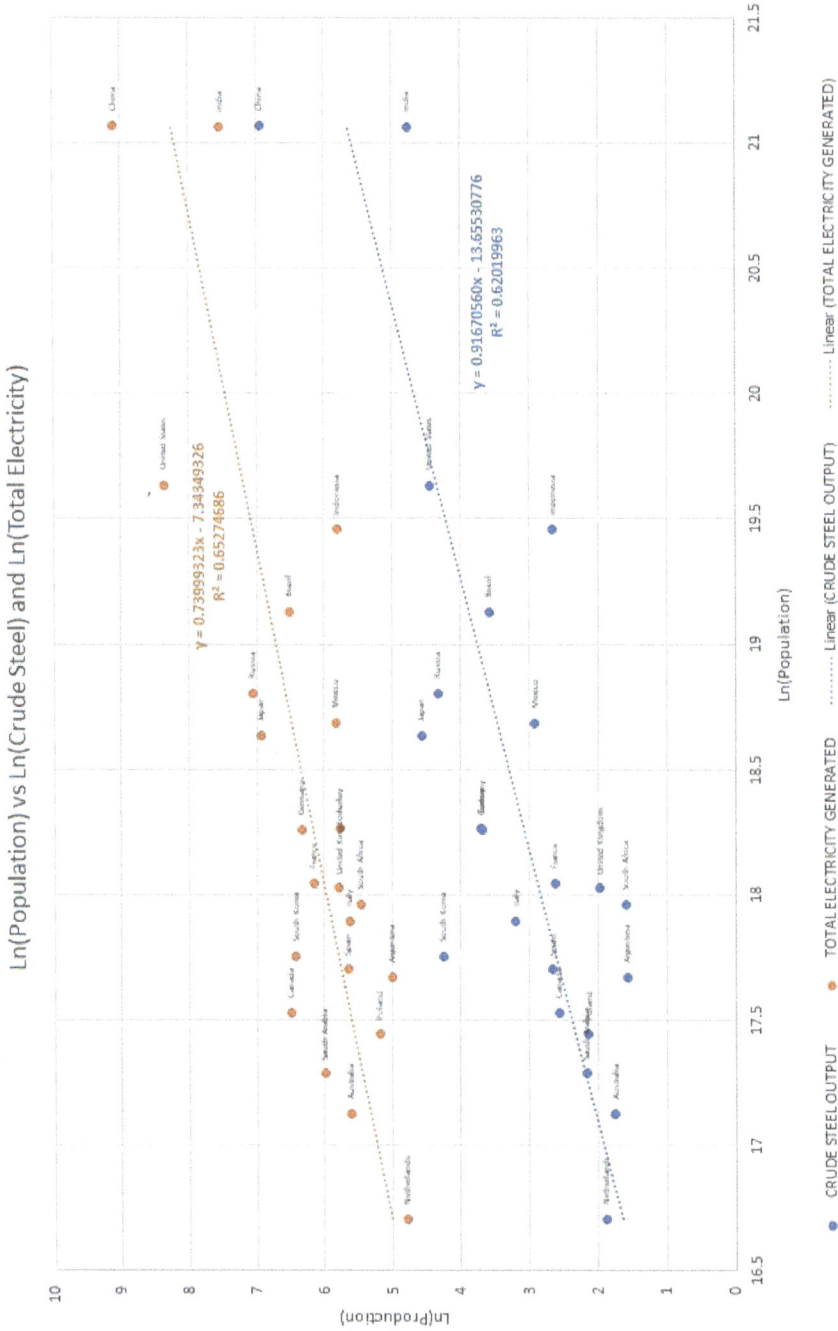

Figure 13.4
Population Versus Crude Steel Production
and Total Electricity Generation as a Log-Log Correlation

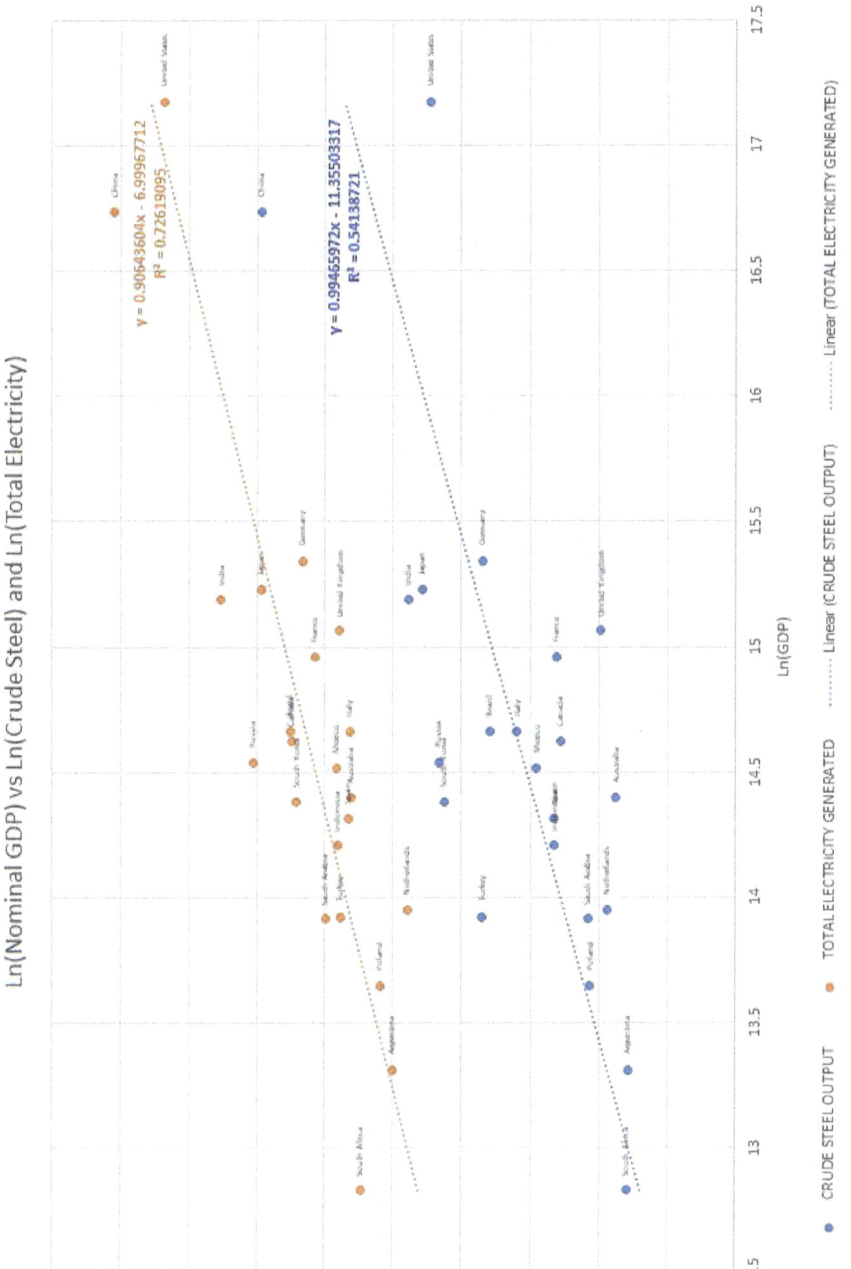

Figure 13.5
Nominal GDP Versus Crude Steel Production
and Total Electricity Generation
as a Log-Log Correlation

PART THREE
CHOOSING NEW STEEL FACILITIES
FOR BRITAIN

PLEASE NOTE: Most Dollar Costs and Legal Considerations
are Intentionally Unaddressed

The British Predicament

It is naughty of me to keep on writing about "Britain" when what I mean is the United Kingdom (UK) of Great Britain and Northern Ireland.

The people of Northern Ireland rightly object to this lazy use of "Britain" to denote a part of their country.

As I attempt to explain elsewhere Great Britain is a particular island in the North Atlantic ocean: The Isle of Wight is not part of Great Britain, and neither is Skomer Island. Portland may be, but the pedantic geographer would say that Foulness is not part of Great Britain. The word "Great" was affixed in an ancient attempt to discriminate this tract of land from the lesser land mass of Brittany in North-West France.

The political entity, the UK, comprises about 803 marine islands at all states of the tide, 188 of which are inhabited. There are about six thousand washed reefs.

The UK is a Rate Three polity.

The Nominal GDP in 2023 was 3.495 trillion dollars and Government revenues 1.408 trillion dollars of which 114.4 billion dollars was interest allocated to the servicing of foreign loans (7.3% of central government spending). The population is 68,682,962 of whom 33.09 million are gainfully employed.

One UK Pound (symbolically GBP or £) is approximately 1.28 Dollars (USD or $).

For most of the Nineteenth-Century there were exactly five US dollars in a Pound Sterling. During the Gold Crisis of September 1931 the Pound effectively devalued from $4.86 to $3.40. In response to the Great Depression, His Majestie's Government (HMG) decreed that all public servants (including sailors and soldiers) would receive a 6% pay cut, whether or not they could lawfully withdraw labor. This resulted in the Invergordon Naval Mutiny of 15-16 September 1931 and there was a "run" on the Pound which "knocked" Britain "off" the Gold Standard permanently. This means that you could no longer demand the equivalent value of gold metal for your pound fiduciary banknotes (treasury bills) if you took them to the Bank of England in Threadneedle Street for redemption.

On the 18 September 1949 the pound was devalued from \$4.03 to \$2.80 and on 18 November 1967 from \$2.80 to \$2.40. These respective devaluations roughly correlate with the UK transitions to Rate Two and Rate Three.

(My Late Father habitually referred to the British half-crown coin as "half a dollar". It was one eighth of one pound: Two and a half shillings. The half-crown was abolished on 1 January 1970).

59.2% of remaining British industry is foreign-owned (by turnover) including the entirety of public electricity provision (except Drax) and the entireties of iron and aluminum smelting.

Ethically, I need to declare that I am both a shareholder of the engine maker Rolls-Royce Holdings (2011) plc; and of the UK National Grid, an electricity distribution company.

In 1913, the UK brought to grass some 287 million (long) tons of coal which had been exploited since Roman times, and it is claimed that as of 2024AD four-hundred-years of supply remains in the ground. Today, however, the UK produces no coal, except approximately one hundred thousand tonnes a year raised by artisanal miners in the Swansea Valley, The Forest of Dean, Ayrshire and East Cumberland (this mine extends into Northumberland where production is logged). In 2023 the UK produced 93,243 tonnes of coal.

It is not realistic to think that Britain will ever again produce bulk coal in our lifetimes.

Correspondingly, the nine hundred UK blast furnaces of the Victorian Era have reduced to three, as the second of the Port Talbot furnaces was blown-out on 5 July 2024 whilst I was writing Chapter Eleven of this book. I understand that virgin steel production will cease at Port Talbot in September 2024, and Wales will produce no iron for the first time since the Iron Age. There are a number of antique blast furnaces in the British landscape, some of which are Scheduled Monuments, and a pair of eighteenth-century examples were recently unearthed (literally) at the site of the old Butterley Company at Ripley in Derbyshire. I do not think iron is currently mined in the UK, except for artists' ochre produced at the Florence Mine in Copeland, and possibly at the Clearwell Caves in the Forest of Dean. In 2023AD UK Iron Production was 5.6 million tonnes, all of the ore being imported by sea.

Britain's experience with civil nuclear power is mixed, and commercial technologies have been distorted by the need to produce fissile isotopic feedstocks for our plutonium refinery. (Plutonium is used to make the detonators of atomic weapons).

The first generation of ostensibly civil nuclear power stations to contribute power to the public Grid commenced with Calder Hall on 27 August 1956, a MAGNOX complex succeeded by ten similar MAGNOX sites yielding about 2.09 GW in total. The next distinct generation of UK nuclear power stations was of the larger AGR type (Advanced Gas-cooled Reactor) of which several reactors spread across six sites were commissioned to give around 8.85 GW. No foreigners opted to purchase either MAGNOX or AGR reactors, or to develop versions of their own. Other countries developed Pressurised Water Reactors (PWRs) of which the British installed but a single example, purchased from the Americans, and built at Sizewell in Suffolk. In 2009AD the British Government sold the UK operational civil nuclear installations (except Wylfa on Anglesey) to the French Government in the personality of Électricité de France (EDF).

Significant amounts of British power are generated from either natural gas combustion or off-shore wind farms. Total British electric power production in 2023AD was 326 TWh of which 125 TWh was gas, 48 TWh nuclear and 80 TWh wind. Hydropower is widespread but only 5 TWh all told.

Rolls-Royce Holdings plc is developing a Small Modular (Nuclear) Reactor (SMR) system which is intended to generate small amounts of dependable nuclear electricity at local sites. These are essentially adapted submarine motors. The current Astute class of British hunter-killer submarines, engined by Rolls-Royce have shaft powers of, I estimate, 200 MW plus or minus 20MW. The relevancy of this is that similar SMR systems may be suited to the generation of electricity for MOE iron smelters or indeed H2DRF smelters, and give excess power for BOF steel formation and even semi-fabrication mill works. Submarine motors or similar are too feeble for public Grid supplies in a country as populous as the UK. These Rolls-Royce motors are essentially Westinghouse-pattern PWRs coupled to steam turbines.

The UK is said to be the World's windiest inhabited territory and if you have ever visited you will believe it! A modern Grid-power generating wind turbine is rated at 7.9 MW and is typically deployed in farms of up to 165 machines yielding 1.3 GW. But the wind often drops to a flat calm even in Britain, so wind power is highly-unreliable and hydrogen is the only credible storage medium for smoothing aeolian power supplies.

A Modest Proposal

        In an illustrative fantasy scenario we will set ourselves the task of the Restitution of British Trade in the sense of awarding ourselves the same tonnage of merchant shipping that the UK registered on 30 June 1949 which was 18,093,159 according to Lloyd's. But instead of a rag-bag of tramp steamers, coastal colliers, ocean liners, oil tankers and refrigerator ships we shall for simplicity have a uniform fleet of modern container ships, standardised like the World War Two Liberty Ships, if you will.

        This figure of 18,093,159 Gross Registered Tons is *not* a gravimetric mass of tonnes weight, much less tonnes: That would be the Displacement Tonnage of the mass of water pushed aside by a floating object, according to the laws of Archimedean hydrostatics. Rather the Gross Registered Tonnage is a *volume*, but a highly-stylised surveyors' volume including this and excluding that calculated for fiscal purposes or the payment of harbor dues. Just for extra confusion, the Displacement Tonnage is not the actual mass of the steel in a floating ship. This actual mass of steel is the Lightweight Tonnage (LWT), but computing LWT *a posteriori* is a complex and uncertain business, outlined in Appendix D.

        Let the late Lord Teynham speak[13.3], not from the grave, but with words apposite to our time. At 1453 on Wednesday 9 December 1953 he:-

        ...rose to call attention to the depletion of the capital resources available for the progressive re-equipment of the Merchant Fleet, due to the inability of British ship owners to set aside adequate reserves to meet the cost of replacement; and to move for Papers. The noble Lord said: My Lords, in opening this debate to-day I think I should declare that I am a director of ship-owning companies, and therefore I am materially interested in their welfare...

        ...In the year 1951, the orders placed for new ships amounted to over 2¾ million tons gross;...

        So in this spirit we shall ask only that we create 2.75 million Gross Tons per annum, but naturally be content if we make more.[13.4 to 13.11]

The largest container ship (boxship) afloat today is the MSC Irena which contains about 70000 tonnes of iron (estimates vary from 64 to 69 LWT).

We will assume that each container ship to be built for Fleet Restoration will require 70000 of iron in the form of high-tensile steel.

Iron Requirement

Abramowski, Cepowski and Zvolenský[13.12] have provided accurate regression estimators for computation of the LWT Lightweight Tonnage iron requirement given GT (Gross Vessel Tonnage) and DWT (Deadweight Vessel Tonnage).

The relevant equation for LWT determination from knowledge of GT is:-

$$LWT = exp\left[\frac{1}{d}\left(2 + \frac{5}{6}\ln(GT - c)\right)\right] + f$$
Equation D.23

where d = 1.111111111; c = 2826.29 and f = 228.81
Please see Appendix D for details.

Before appreciable wartime losses, at the month of December 1939, the British Merchant Fleet had an iron content of about 1,664,242.94 metric tonnes. During December 2023 the same flag of shipping weighed 1,107,914.49 metric tonnes. Therefore to reinstate the tonnage an extra 556,329 tonnes of iron must be sourced. If we adopt Lord Teynham's modest 1953AD proposal of adding 2¾ million GT per annum it turns out that it would take 1.36205 years to furnish the metal, a little over one year and four months.

Note that we are assuming that the UK GRT figure was 17,891,134 which is a shade less than the Lloyd's figure of 18,093,159. I cannot explain the discrepancy, but I suspect the Lloyd's value of including foreign tonnage insured.

The proposal involves an increase of about 50.21% in the weight of British commercial shipping.

Naturally, as patriots we prefer that the iron is at least smelted if not mined in our respective countries to the benefit of our younger countrymen rather than imported.

<u>The State of Iron Smelting</u>

The current state of iron smelting in several countries is bleak, and such is certainly the case for Britain.

The de-industrialisation of Britain and Northern Ireland has, ironically, created a number of sites suited to the installation of small ironworks, especially near or at the coast.

Table 13.5 presents the salient desiderata of sites for modern iron smelting, which are naturally similar to but not the same as the ideal for traditional blast furnace integrated rolled steel production.

Table 13.6 gives estimates of the land footprints that are or would be occupied by British smelters or their prospective power sources.

It is visible that prospective MOE or H_2DRF plants would be much less demanding of land than BF integrated steel works, especially if gas and slag managements can be rationalised for off-site storage or disposal.

Britain and Ireland have very many sites that are, with due relaxation of stringency, suitable for small iron smelting installations, especially derelict or moribund twentieth-century industrial land.

I cite four examples of the most notorious tracts of actual or soon-to-be abandoned sites.

1	There should be adequate flat land adjacent to an adequate waterway
2	Notwithstanding (1) the land should not be susceptible to flooding
3	Rail services should exist on or near the site
4	Major roads should be near the site
5	Water coolant should be near and plentiful
6	Excess skilled labor should be nearby
7	The land should be of low elevation consistent with flood avoidance
8	The land should not be good-class agricultural land
9	The land should require the minimum of reconditioning
10	There should be no false ground (For example, subsidence-prone areas undermined by extractive industry)

Table 13.5
The Desiderata of Iron Smelting Sites

Astute Class Submarine (whole ship) Footprint:	0.0010961	km^2	L$_{oa}$×B
Theoretical Hundred-Furnace MOE Plant Footprint:	0.005233	km^2	(12.09*19)^2
Lochaber Aluminum Smelter Footprint:	0.09	km^2	
H$_2$DRF Footprint:	0.01	km^2	LKAB Luossavaara-Kiirunavaara AB
Port Talbot (Abbey Works) Footprint:	7	km^2	+3 km^2 of dumps
Scunthorpe Footprint:	10	km^2	

Table 13.6
The Ground Areas Occupied by Various
Installations Potentially Related to Iron Smelting

Name of Site:	Ratcliffe-upon-Soar	
Nearest Settlement:	Thrumpton	
Site Longitude:	52:52:06 N	
Site Latitude:	1:15:25 W	
Site Ordnance Survey NGR:	SK 50132 30324	
Elevation:	60	meters
Areal Extent of Site:	4	km^2
Distance to Navigation:	800	meters
Distance to Railway:	On Site	meters
Distance to Major Road:	500	meters
Owner or Tenant (19 Aug 2024):	Uniper SE, Düsseldorf	
Historical User:	CEGB	
Proposed Iron Smelting Method:	MOE	
Site Condition:	Coal Power Station to be Closed September 2024	
Comments:	Site on small plateau over looking the River Trent	
	flood plain immediately downstream of Derwent	
	and Soar confluences.	
	Trent River liable to flood.	
	Local mean discharge of Trent about 55 cumecs.	
	Plateau 30m above flood plain.	
	Site already earmarked by Rushcliffe Borough Council	
	for advanced manufacturing and green energy development.	
	Last UK Coal-fired power station.	

Table 13.7a
Characteristics of the Ratcliffe-upon-Soar Site

Name of Site:	East Halton	
Nearest Settlement:	Immingham	
Site Latitude:	53:37:23 N	
Site Longitude:	0:16:09 W	
Site Ordnance Survey NGR:	TA 14431 21745	
Elevation:	6	meters
Areal Extent of Site:	4	km^2
Distance to Navigation:	3600	meters
Distance to Railway:	3000	meters
Distance to Major Road:	6000	meters
Owner or Tenant (19 Aug 2024):	Able UK Limited, Billingham	
Historical User:	Messrs. John George Crowther and Charles Dishman	
Proposed Iron Smelting Method:	H_2DRF	
Site Condition:	Arable flat coastal reclaimed fen.	
Comments:	Center is site of demolished 19th Century farmstead.	
	Virgin agricultural site next to navigable water and	
	near industrial rail spurs and Immingham	
	Ore Terminal port (serves Scunthorpe).	
	Intending Able Energy Park offshore services and	
	port complex to complement other Able Port Company	
	investments in NE England:- Tenants sought.	

Table 13.7b
Characteristics of the East Halton Site

Name of Site:	Redcar	
Nearest Settlement:	Redcar	
Site Latitude:	54:37:23 N	
Site Longitude:	1:07:44 W	
Site Ordnance Survey NGR:	NZ 56347 25653	
Elevation:	9	meters
Areal Extent of Site:	6	km^2
Distance to Navigation:	On Site	meters
Distance to Railway:	On Site	meters
Distance to Major Road:	1400	meters
Owner or Tenant (19 Aug 2024):	Sahaviriya Steel Industries, Bangkok	
Historical User:	Dorman Long Steel Company	
Proposed Iron Smelting Method:	H$_2$DRF	
Site Condition:	Derelict Steel Mill Site	
Comments:	Mostly cleared severely polluted former smelting and steel mill site. I was unable to identify remaining blown-out furnace from aerial photography (site not visited). https://www.ssi-steel.com/en/main/ makes no mention of UK interests and on 19 August 2024 the intentions of Mr Nava Chantanasurakon and his colleague friends in regard to Redcar were not clear to me.	

Table 13.7c
Characteristics of the Redcar Site

Name of Site:	Neptune Bank	
Nearest Settlement:	Wallsend	
Site Latitude:	54:55:05 N	
Site Longitude:	1:32:13 W	
Site Ordnance Survey NGR:	NZ 29689 65707	
Elevation:	16	meters
Areal Extent of Site:	0.0144	km^2
Distance to Navigation:	150	meters
Distance to Railway:	350	meters
Distance to Major Road:	10	meters
Owner or Tenant (1) (19 Aug 2024):	Lhyfe, Paris	
Owner or Tenant (2) (19 Aug 2024):	Shepherd Offshore, Newcastle	
Historical User:	Newcastle upon Tyne Electric Supply Company	
Proposed Iron Smelting Method:	H$_2$DRF	
Site Condition:	Grassy shallow soil	
Comments:	Former power station site	
	20MW for 8 tonnes of H$_2$ per diem.	
	Flat grassy-scrubby plot 120×120 meters.	
	Power Station closed 1915AD.	
	Succeeded by Viriosil Glass Works	
	This site is of substantial archaeological interest.	

Table 13.7d
Characteristics of the Neptune Bank Site

The soon-to-be-vacated Ratcliffe power station site is especially suggestive for the development of a small 100-furnace MOE farm that could theoretically generate enough steel to build an Irena-weight container ship every 130 days, outputting a theoretical 197 thousand tonnes of hot iron per year. The existing Ratcliffe Power Station has an electric output in excess of 2GW.

East Halton and Redcar are good H2DRF sites with plenty of spare land for gas storage and ready sea access for ore and flux delivery and the removal of spent slag, a substance that can be used for fertiliser, engineering brickmaking or for civil engineering fill.

Table 13.8 summarises the potential yield characteristics of existing or potential BF, MOE and H2DRF estates.

By analogy with wind "farms" a furnace farm is a local group of current-splitting electrolytic cells or "pots" similar in concept to the sixty aluminum electrolytic cells of the Lochaber Works.

The furnace farm may be fed electricity from an SMR or an actual wind farm (with gas storage) or indeed any reliable and adequate source of electric power.

Calculating the Time Needed for a Required Yield of Iron

For the sake of simplicity we will calculate for an entire farm or installation of n identical furnaces or electrolytic cells.

The Total Power supplied is P which for example may be the 200MW supplied by a re-purposed submarine engine similar to the Rolls-Royce SMR unit, or the nominal 219MW supplied by the Humber Gateway wind farm of the Holderness Coast of Yorkshire.

K_m is the Energy required to Smelt One Tonne of Iron in the given cell or furnace type; and m_{Fe} is the Mass of Iron Required, which in the case of a ship like the MSC Irena would be 70000 tonnes. ξ is the Overall Efficiency.

h is the number of Hours Needed to generate enough iron M_{Fe}.

Then:-

$$h = \frac{k_m \cdot m_{Fe}}{\xi P} = \frac{[L^2 T^{-2}][M]}{[ML^2 T^{-3}]} = [T]$$
Equation D.24

The number of Days d needed is accordingly:-

$$d = \frac{h}{24}$$
Equation D.25

Method Code	Source	Furnace Technology	Available Power (MW)	Furnace Farm Population (number of furnaces)	Power per Furnace (MW)	Farm Area (km²)	Single Furnace Value K_m (MWh/tonne)	Assumed Efficiency ζ (fraction)	Tonnes Fe per Hour Per Furnace	Single Furnace Annual Iron Yield Y (tonnes Fe)	Single Furnace Annual Energy Demand E_{sf} (MWh)	Farm Annual Iron Yield Y_{farm} (tonnes)	Farm Annual Energy Demand E_{farm} (MWh)	Annual Iron Requirement Rq (tonnes)	Years to Smelt Rq tonnes	Days to Smelt Rq tonnes (elab.)	Days to Smelt Rq tonnes (form.)	Annual Yield per Area (tonnes×10^4/km²)	Notes and References
a	Boston Metal	MOE	200	100	2	0.00523	4.00	0.45	0.23	1972.35	7889.40	197235	788940	70000.000	0.3549	129.6296	129.6296	37.6906	Ch13: Ref 13.2
b	European BF	BF	11.25	3	3.75	17	3.75	0.12	0.12	1051.92	3944.70	3156	11834	70000.000	22.1816	8101.8208	8101.8208	0.0002	Ch11: Eqn 11.46
c	Vogl	H₂DRF	219	100	2.19	1	3.48	0.53	0.33	2923.76	10174.70	292376	1017470	70000.000	0.2394	87.4472	87.4472	0.2924	Ch11: Table 11.8 Ref 11.5 (Vogl Et Al)
d	Mean			67.67	2.65	6.00	3.74	0.37	0.23	1982.68	7336.27	164255.70	606081.24	70000.00	7.59	2772.97	2772.97	12.66	#DIV/0!
e	JRW	MOE	200	100	2	0.00523	6.23	0.20	0.06	563.10	3506.10	56310	350640	70000.000	1.2431	454.0478	454.0478	10.7606	Ch12: Route 1: φ = 1: Eqn 12.38
f	JRW	MOE	200	100	2	0.00523	5.94	0.20	0.07	590.15	3506.40	59015	350640	70000.000	1.1861	433.2354	433.2354	11.2775	Ch12: Route 2: ψ = 1: Eqn 12.52
g	Mean		200	100	2	0.00523	6.08423	0.2	0.06578	576.6270777	3506.4	57662.71	350640	70000	1.2146245	443.6416	443.64	11.01905	

PSD(a,b)   6.25 h
PSD(a,c)   13.00 |
PSD(a,g)   -52.11 |

One MWyear = 24*365.25 = 8766 MWhours
ζ H₂DRF: Figure for Water Electrolysis Only (Bhasker Et Al)

Table 13.8
Yield Potentials of Existing or Potential
BF, MOE and H₂DRF Sites

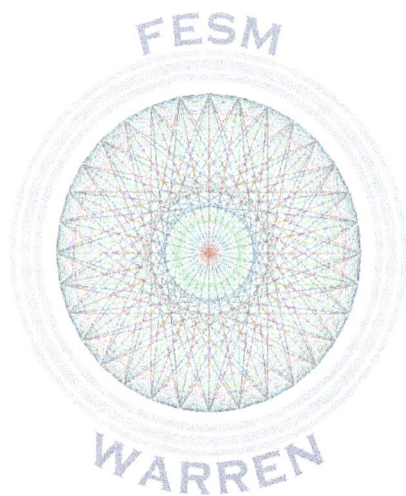

FESM

WARREN

EPILOG

The people of Argentina, a vast country, know, and have known since Versailles, that their splendid land is nothing without the resources of Antarctica. In 1982AD Argentina fought Britain for control of the Falkland Islands and the Falkland Islands Dependencies. The Argentines deployed excellent French weapons against the British fleet dispatched to contest occupation.

France itself withheld the further supply of arms to Argentina, as France is a NATO member, and collective security from Russia was more important to France than the victory, or even the gold, of its Latin American customer.

When the Argentines bombed British ships in the South Atlantic the Royal Navy ships promptly caught fire, exploded and sank. This was because the ships were made of aluminum, which burns readily in air whereas steel does not. The British owed a narrow victory to the gallantry and professionalism of their Army.

Without iron you can make no viable gun, tank or warship. Without iron you can build no bunker or dockyard, because, as the Periclean Greeks discovered, without iron you can reinforce neither ashlar nor concrete.

Without iron you can make no nuclear containment assembly, whether for peaceable or warlike use.

An island nation will not survive long without an adequate Navy. But by the same token, an island nation cannot import iron in wartime.

No ships: No say

Without iron there is no independence.

Both Russia and Ukraine struggle to maintain their respective sovereignties, even from one another. So they fight over their coal and iron resources, though I am insolent enough to write that they are one and the same race and people, as indeed are the Scottish and the English, or the British and Americans of Franklin's time.

To relinquish iron is to surrender the self-determination that befits an adult, and lends a temporal grace to the man or woman of honor, substantiating the Free Will vouchsafed him by God.

Without iron you inhabit a Bronze Age, or even a Stone Age, as savages do even today.

If you can have eco-friendly, Net Zero iron then that is excellent. But if you have to have dirty, carbon-intensive iron have that you must, for as long as mankind is a troupe of apes.

I have no financial interest and frankly little intellectual interest in any steel corporation. I have no investment in any metallurgical firm, except for Rolls Royce, whose efforts to restore civil nuclear power in Britain I applaud.

The Loves of God and Country are firm and strong. The love of mortal things brings suffering. The love of language is highly dangerous for language makes thought a railway and an airline a loft ladder, just as your pen does not make you a poet. A country is not a tract of land but a concept, whilst a doctrine is not an ideal but a folly, and a crown no kingdom but a bauble. Love lightly: Love rightly. Love the wind and the water, though they too shall perish, for a world without either is unthinkable..

I hope not to profit from the misfortune of others. I am motivated by my love for my country. Why should I love my country? Like all nations, it has the glories of its past in the endeavors of its people. I hope that it profits and persists in splendors that may be imagined but shall always be other than anticipated.

Love, the least rational of emotions, is as unlooked-for as it is ineluctable.

My love of Science is a selfish love, basking in the God-given glory of my own skills and fortitudes, as if reflected in a mirror. I did not look for that love as a nine-year-old at Cadgwith.

Like all mature love my love of Britain is given freely in the recognition of its object's many defects, and in gratitude for its nurture.

APPENDIX A
BASIC ELECTRICS

Bear with me one moment. I want to talk about something else.

A small aluminum smelter is fed by a reservoir high in the hills behind. You could look upon this lake as a store of energy, E, waiting to serve the factory far below. Or you could view it as a pent-up Charge, Q, waiting to throw itself down the brae.

Hydraulicians think in terms of the energy of this resource, Potential Energy, E_p, and calculate it using:-

$$E_P = m_{water} \times g \times \Delta h$$
Equation A.1

Where m_{water} is the Mass of Stored Water; g is the Acceleration Due to Gravity and Δh is the Change in Elevation between the Reservoir Surface and the Turbines.

When the water gets on the move it is helpful to think of its Kinetic Energy, E_k, as Power, P_{water}:-

$$P_{water} = \frac{m.g.\Delta h}{t} = \frac{\rho V.g.\Delta h}{t} = \rho Dg.\Delta h = (\rho.g.\Delta h) \times D = p.D$$
Equation A.2

where m is Unit Mass of Water (one kilogram); ρ is the Density of Water (1000 kg/m^3); and t is Unit Time (one second). V is the actual Volume of Water flowing through the Unit Second, a parameter called Discharge, D, measured in m^3/s. p is a Constant equivalent to the product of Water Density, Gravitational Acceleration and Elevation Height Change, all of which, *in context*, are constants.

p is Pressure.

$$p = \rho.g.\Delta h = [ML^{-3}.LT^{-2}.L] \equiv [ML^{-1}T^{-2}]$$
Equation A.3

So the Power at the business end of the penstock is the product of the pressure there and the discharge.

It is the same with electricity, which early electrical scientists viewed as a "fluid" by analogy with hydraulic fluids.

The Current I is the discharge measured in Amperes $[I^{+1}]$; the Potential Difference (Voltage) $[ML^2T^{-3}I^{-1}]$ is the pressure measured in Volts; and there is a third parameter, resistance, measured in Ohms $[M^{-1}L^{-2}T^3I^2]$. The resistance is whatever impedes the current sapping energy from it, in our aluminum smelter this is the actual melt bath with its heating and splitting of chemicals, together with any attendant paraphernalia such as rectifiers, etcetera.

So analogously:-

$$P = IV$$
Equation A.4

where P is Power in Watts; I is Current in Amperes; and V is Potential Difference in Volts.

As we know, Energy, E, is Power, P, multiplied by Time, t in seconds. So:-

$$E = Pt = VI.t = V.\frac{Q}{t}.t = VQ$$
Equation A.5

where Q is Charge measured in Coulombs $[I^{+1}T^{+1}]$.

The reciprocal of Resistance, Ω, is Conductivity measured in siemens.

Ohm's Law

Ohm's Law states that:-

$$V = I\Omega$$
Equation A.6

where V is Voltage; I is Current and Ω is Resistance. Electricians often use R for resistance as in V=IR, but I shall use Ω because I do not want to cause confusion with R, the Universal Gas Constant. Correspondingly, I shall use ω for quantities in Ohms.

Notwithstanding that, in the labelling of individual resistant devices I shall use R1, R2, R3, ... , Rn.

Winding Loss ("Copper Loss")

Winding Loss is any electrical energy loss which expresses through a resistance as heat. It may be deliberate, as in the case of radiant electric fires which are intended to use electricity to give out heat.

Winding Loss, P_{loss} is given by:-

$$P_{loss} = I^2\Omega = \frac{V^2}{\Omega}$$
Equation A.7

Winding Loss is often informally known as " i squared r heat loss ".

Equation A.7 applies directly to steady Direct Currents (DC) from for example batteries or dynamos. In the case of AC (alternating current) we need to use the Root Mean Squares of Current or Voltage levels.

Direct Current Power Drop

The Direct Current Power Drop, P_{drop}, across a resistance is given by:-

$$P_{drop} = I(V_A - V_B)$$
Equation A.8

where V_A is the Input Volage and V_B is the Output Voltage.

Serial and Parallel Resistors

The net resistance of two or more resistant devices depends upon their arrangement in the electrical circuit. In particular, the Current-Voltage-Resistance effects differ greatly as to whether the resistors are in Series or in Parallel.

The Series arrangement is nose-to-tail like a train or processionary caterpillars as illustrated in Figure A.1:-

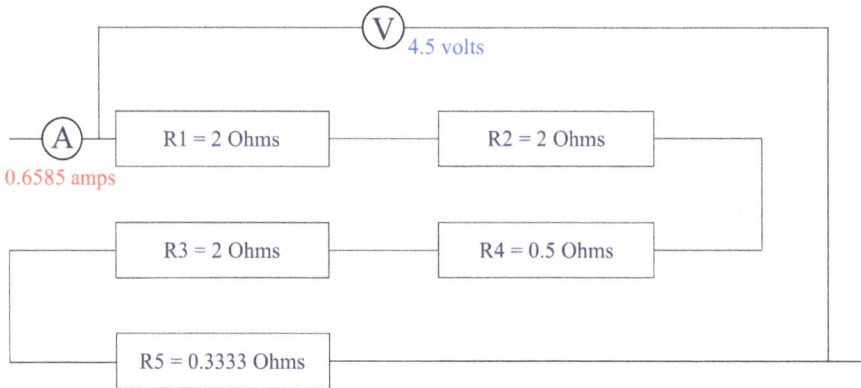

4.5 volts

A

0.6585 amps

R1 = 2 Ohms   R2 = 2 Ohms

R3 = 2 Ohms   R4 = 0.5 Ohms

R5 = 0.3333 Ohms

Figure A.1
Serial Resistors

Note that you measure Voltage across the array, and Current in-line with the array. Theoretically speaking, a Voltmeter should in itself have infinite resistance and an Ammeter zero resistance.

On the other hand, a Parallel conformation of resistors places them side-by-side splitting the current between them as shown in Figure A.2. Again in theory, the Voltage Drop across each of the resistors is the same in a parallel array, but it is wise to measure Voltage across the entire set as shown.

Allow that the Voltage across each array is 4.5 Volts. In each array there are five resistors. Resistor R1 has a value of 2 ohms, as does R2 and R3. Resistor R4 has a value of 0.5 ohms, and R5 has 0.3333 ohms.

We expect rated circuit board resistors made of coiled nichrome wire or some other compact poor conductor to dissipate energy as heat, though lights are resistors as are all other electrical gadgets.

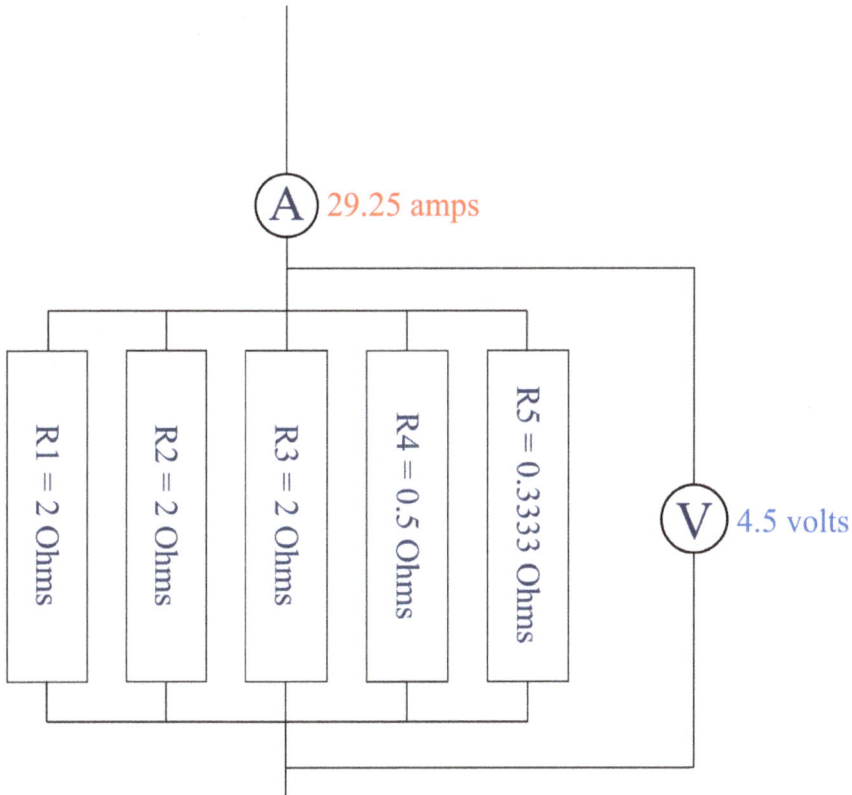

Figure A.2
Parallel Resistors

In the Series case the Total Resistance of n Resistors, Ω_{series}, is given by:-

$$\Omega_{series} = \Omega_1 + \Omega_2 + \Omega_2 + \cdots + \Omega_n = \sum_{i=1}^{n} \Omega_n = n.\,\mu(n, \Omega)$$

Equation A.9

where n is the Number of Resistors; Ω_i is the Resistance (in Ohms) of the ith. Resistor; Ω_{series} is the Net Resistance of the Entire Series of Resistors; and $\mu(n,\Omega)$ is the Arithmetic Mean of the Resistances.

For the Parallel case the Net Resistance, $\Omega_{parallel}$, of the entire array is given by:-

$$\Omega_{parallel} = \frac{1}{X} = \frac{H(n, \Omega)}{n}$$

Equation A.10

where X is the Reciprocal of $\Omega_{parallel}$, and $H(n,\Omega)$ is the Harmonic Mean of the Resistances.

X is given by:-

$$X = \frac{1}{\Omega_1} + \frac{1}{\Omega_2} + \frac{1}{\Omega_3} + \cdots + \frac{1}{\Omega_n} = \sum_{i=1}^{n} \frac{1}{\Omega_i}$$

Equation A.11

Resistance Ratio and the Descriptive Statistics of Resistance Arrays

The Resistance Ratio, RR, is defined by:-

$$RR = \frac{\Omega_{series}}{\Omega_{parallel}}$$

Equation A.12

In the illustrated examples of Figures A1 and A2, the Number of Resistances is 5; the Voltage is 4.5 volts; and the respective Resistances are $\Omega_1 = 2$, $\Omega_2 = 2$, $\Omega_3 = 2$, $\Omega4 = 0.5$ and $\Omega3 = 0.3333$ ohms.

Meanwhile, the Arithmetic Mean of these Resistances, $\mu(n,\Omega)$, is given by:-

$$\mu(n, \Omega) = \frac{\sum_{i=1}^{n} \Omega_i}{n} = 1.36666$$

Equation A.13

whilst the Harmonic Mean, $H(n,\Omega)$, is:-

$$H(n, \Omega) = \frac{n}{\sum_{i=1}^{n} \frac{1}{\Omega_i}} = 0.769195264360389$$

Equation A.14

The Parallel Array Resistance Reciprocal, X, is given by:-

$$X = \sum_{i=1}^{n} \frac{1}{\Omega_i} = \frac{n}{H(n,\Omega)} = 6.500300030003$$

Equation A.15

Accordingly, for the Series Case:-

$$\Omega_{series} = \sum_{i=1}^{n} \Omega_n = n.\mu(n,\Omega) = 6.8333$$

Equation A.16

whilst the Current, I_{series}, flowing through the series resistors is:-

$$I_{series} = \frac{V}{\Omega_{series}} = 0.658539797755111$$

Equation A.17

For the Parallel Case:-

$$\Omega_{parallel} = \frac{1}{X} = \frac{1}{\sum_{i=1}^{n} \frac{1}{\Omega_i}} = \frac{H(n,\Omega)}{n} = 0.153839052872078$$

Equation A.18

whilst the Current, $I_{parallel}$, flowing through the parallel resistors is:-

$$I_{parallel} = \frac{V}{\Omega_{parallel}} = 0.658539797755111$$

Equation A.19

From these results it follows that the Resistance Ratio, RR, is given by:-

$$RR = \frac{\Omega_{eries}}{\Omega_{parallel}} = X.\Omega_{series} = \sum_{i=1}^{n} \Omega_i \sum_{i=1}^{n} \frac{1}{\Omega_i} = n^2 . \frac{\mu(n,\Omega)}{H(n,\Omega)}$$

Equation A.20

In this example, RR has the value 44.4185001950195

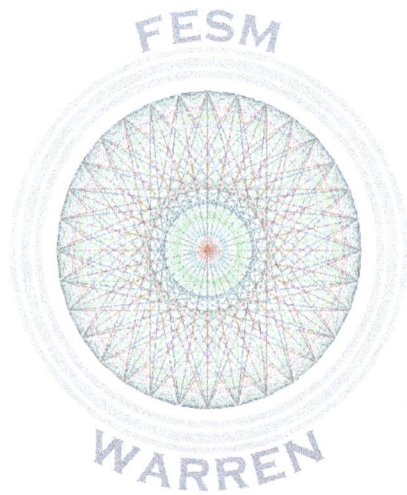

ALTERNATING CURRENT AND DIRECT CURRENT

As its name suggests, the flow of electrons in an Alternating Current goes back and forth along its conductor. It rises to a maximum positive voltage, V_m or V_{max}, in one direction, falling to stasis at zero volts, then surging in the opposite direction to -V_{max} volts.

On the other hand, Direct Current flows in a constant direction with positive electron flow (confusingly, this is conventionally a negative discharge). The actual level of DC voltage is not *necessarily* steady: Half-Wave DC is sinusoidal in profile, but only exists half the time. Steady, constant-voltage DC is usually what is desired and many electrical devices have been developed to achieve steady DC or something closely approximating it.

AC and DC each have their advantages and drawbacks.

Industrially, it is expensive and impracticable to transmit DC over distances in excess of a kilometer due to resistance and other impedance losses which dissipate power as lost heat. DC can be transmitted very locally, such as in an electric smelter, using heavy copper or silver Busbars, themselves expensive and inconvenient, and sometimes requiring active cooling. For extended transmission AC must be used, and such AC electricity can be transmitted for hundreds of kilometers underground or overhead using cheap steel-reinforced aluminum cables.

Tiny DC currents can be used in electronics and instrumentation. Batteries, clocks, radios, televisions and computers are examples of light equipment that must always work on DC. The AC public supply must always be adjusted with regard to voltage and current (Transformed) and then Rectified to DC before use in those appliances and also in much heavy industrial plant.

It is easy to Transform AC electro-magnetically using a Transformer which involves no moving parts (except possible cooling circuits) but incurs I^2R heat losses. Transforming DC is traditionally expensive and troublesome involving a Solenoid (magnetic coil) supported by a vulnerable make-and-break apparatus. For light electronic work, solid state Oscillators are employed to generate tunable alternating signals.

Converting AC to DC and *vice versa* is unsurprisingly called Conversion. Converting AC to DC is called Rectification and is done using a Rectifier of which there are many distinct designs. Converting DC to AC is called Inversion and again there are many types of Inverter on the market. Perhaps the crudest and costliest

Inverter is a DC motor mechanically yoked to an AC-generating Alternator. Static Inverters are favored today because they conserve power, involve no moving parts, and are consequently reliable.

All AC and DC electricity beyond the flashlight battery level is highly dangerous. It is the current that does the actual damage, but the voltage "pushes" that current forward. AC is five times more dangerous than DC. ***Do not test this assertion***. One of the reasons why is that AC tends to make the human hand grip the tormenting live cable, whereas DC tends to jolt the hand or body clear. Even half an amp of current can stop the heart.

In Britain, provincial high-speed trains work on 25000 volt AC drawn from overhead catenaries. Foolish boys who climb or jump onto these catenary cables are usually killed instantly. London or Liverpool tube (subway) and other suburban trains of the metropolitan hinterlands use 750 volt DC third rails. Intending suicides who bridge the rails with their bodies often survive, but with severe burns.

You can only use DC for electrolysis, so either you generate the DC on site with, for example, water power; or else you Rectify AC drawn from elsewhere.

In many contexts the Maximum Voltage of an AC current is of little interest. The AC voltage level which represents the steady power of a sinusoidally-varying AC current is the Root Mean Square (RMS) Voltage, V_{RMS}, and it is this V_{RMS} average which also references the steady power of a DC current Rectified from the AC.

The relation between V_{max} and V_{RMS} is:-

$$V_{RMS} = \frac{V_{max}}{\sqrt{2}}$$
Equation B.1

or transposed for $V_{max} \equiv V_m$:-

$$V_{max} = \sqrt{2}.V_{RMS}$$
Equation B.2

For example, in Britain the public domestic electricity supply is maintained at 240V AC. This is V_{RMS}.

Accordingly, V_{max} is 339.411254969543 volts. As per usual the 15-figure precision is for checking purposes only: In practical terms the maximum domestic supply voltage is 340 volts.

Voltage Profiles

Table B.1 displays the input and output parameters for the AC and related DC electric wave profiles illustrated in Figures B.1, B.2 and B.3

Figure B.1 is a composite picture in which the original AC signal is half overlain by the resulting full-wave rectified Direct Current, plus the smoothing Ripple Wave which, in concept, invaginates itself over the raw rectification humps.

AC voltage variation is shown as a solid blue sinusoidal line; the DC full-wave rendition as a solid red line; and the Ripple Voltage as a solid green line.

The Highest crests of the AC and DC represent V_{max} (V_m) whilst the horizontal pecked purple line traces the Root Mean Square Voltage V_{RMS}.

Table B.1

RMS AC Supply Voltage (V_{RMS})	240
Analytic Integral for V_{RMS}	240
AC Supply Current	13
Peak Voltage (V_{max})	339.411255
Supply Frequency	50
Supply Period	0.02
Supply Angular Frequency	314.1592654
Peak Voltage Check	339.411255
AC RMS Power	3120
Ludolphine Constant	3.141592654
Intervals per Cycle	256
Number of Cycles	2
Number of Intervals, n	1612
Intervals per Cycle	806
Interval Width	2.48139E-05
Mean of AC^2 (μAC^2)	57564.29014
Root Mean of AC^2 (AC_{RMS})	239.9255929
$(1/2^{0.5})*A$	240
$PSD(AC_{RMS},(1/2^{0.5})*A)$	-0.031012561
t_0	0
t_n	0.04
t_n-t_0	0.04
$AC_{RMS}*(t_n-t_0)$	9.597023717
$V_{RMS}=(\pi/2^{(3/2)})*V_{DC}$	240
R	100
V_{DC}	216.0759159
I_{DC}	2.160759159
f_{ripple}	100
V_{ripple}	0.029462783
$C_{implied}$	0.366692989
Simpson h	2.48139E-05
Simpson Sum Of Products	1044937.837
Simpson Integral (SI)	8.642992865
$SI/(t_n-t_0)$ (V_{DC})	216.0748216
$PSD(V_{RMS},SI/(t_n-t_0))$	9.96882432
$2^{(3/2)}/\pi$	0.900316316
$(2^{(3/2)}/\pi)*V_{RMS}$	216.0759159
$PSD((2^{3/2}/\pi*V_{RMS},SI/(t_n-t_0))$	0.000506418
Root Simpson Integral (RSI)	2.939896744
AC_{RMS}/RSI	81.61021077
$(\pi/2)-1$	0.570796327
$(\mu AC^2/SI)^{0.5}$	81.61021077

Table B.1

Table B.2

	Amplitude RMS	Amplitude A	Frequency f (cycles/s)	Period T	Angular Frequency ω	Phase ϕ	Shift $\phi.T.\pi$	Symmetry Constant κ	Lift Constant λ
AC	240	339.411255	50	0.02	314.1592654	0	0	0	0
Smoothing Function	7.07106781	10	100	0.01	628.3185307	-65	-2.04204	0.3	330

Table B.2

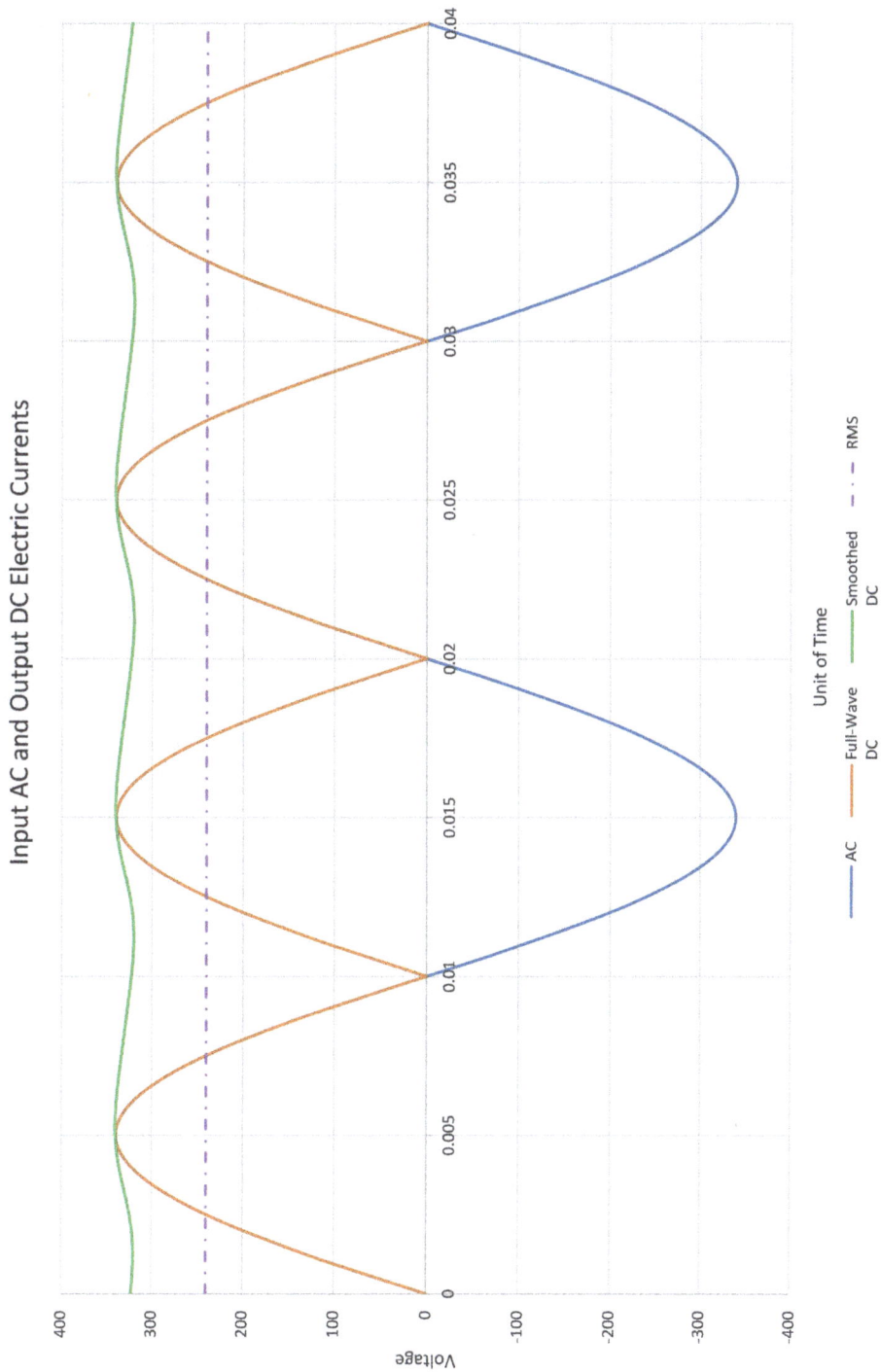

Figure B.1

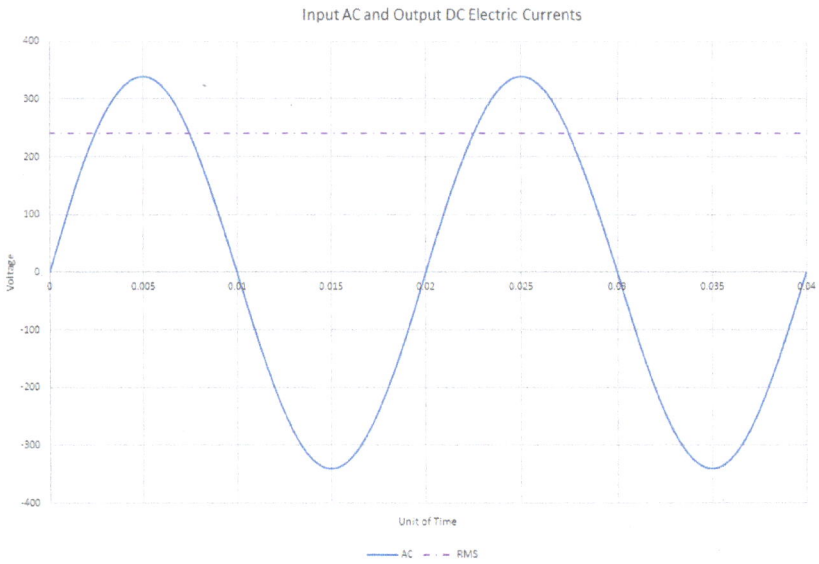

Figure B.2

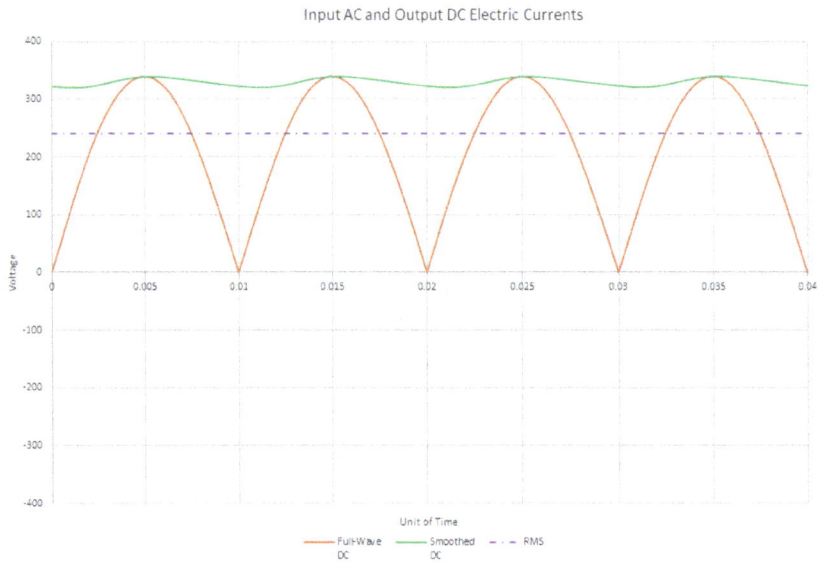

Figure B.3

Amplitude, Frequency and Power

The Amplitude of a sinusoidal wave is the height of the wave in meters or volts *above the zero line* whilst the Period is the regular peak-to-peak horizonal distance in meters or whatever.

Frequency is the reciprocal of Period (treated as time $[T^{+1}]$). Frequency is dimensioned $[T^{-1}]$ and measured in cycles per second, called Hertz in the SI system.

The Angular Frequency, ω, is given by:-

$$\omega = 2\pi f$$

Equation B.3

where f is the Frequency in Hertz.

The Total Potential and Kinetic Power of a wave-train cycle is given by:-

$$P_{ave} = \frac{E_\lambda}{T} = \frac{1}{2}.\mu v.A^2\omega^2$$

Equation B.4

where P_{ave} is the Averaged Power; E_λ is Total Energy; T is the Wave Period; μ is some Contextual Factor (linear density for a vibrating string); v is Wave Velocity; A is Wave Amplitude and ω is Wave Angular Frequency.

To be pedantic, waves as such do not possess speed or velocity: Only particles of matter possess velocity or speed.

Waves possess Celerity, also measured in meters per second, but usually much faster than the motion of any transmitting medium.

The Wave Speed (i.e. Celerity), v, is given by:-

$$v = f\lambda$$

Equation B.5

where λ is the Wavelength (meters L^{+1}).

A little calculation confirms that the Contextual Factor μ has the dimensions $[ML^{-1}]$.

In the context of DC the Power P_{DC} is given by:-

$$P_{DC} = IV$$
Equation B.6

Where P_{DC} is the Direct Current's Power; I is the Current flowing; and V is the Voltage (Electromotive "Force") across the device.

In the context of a single-phase AC supply the Power P_{AC} is given by:-

$$P_{AC} = I_{RMS}V_{RMS} \cdot \cos(\Phi) = \frac{R}{Z} = \frac{R}{\sqrt{R^2 + jX^2}}$$
Equation B.7

where Z^2 is the Impedance, a combined effect of the circuit Resistance and Reactance, which latter is itself a function of circuit storage (Capacitance). Φ is the Phase Angle and j is the Square Root of Minus One, which electricians use in preference to i, which is too easily confused with current, etcetera.

R and Z are in Ohms.

Derivation of Root Mean Square Voltage[B.1]

If you examine Table B.1 you will discover that:-

$$V_{DC} = \frac{\int (FullWave\ DC)}{t_n - t_0}$$
Equation B.8

and that the AC Root Mean Square Voltage, V_{RMS}, (i.e. 240 volts) is given by:-

$$V_{RMS} = \frac{\pi}{2^{\frac{3}{2}}} \cdot V_{DC}$$
Equation B.9

On my spreadsheet RECTCHOKd.xlsx I used a simple n = 1612-interval Simpson Integration with $t_0 = 0$ and $t_n = 0.04$ for the 4 available cycles of full-wave DC rectified from 2 cycles of AC.

A more systematic approach to the cycle integral is offered by the analytic integral:-

$$V_{RMS} = \sqrt{\frac{1}{T} \cdot \int_0^T V_{max}^2 \cdot [\cos(\omega t)]^2 \cdot dt} = 240$$

Equation B.10

which solves to:-

$$V_{RMS} = \sqrt{\frac{1}{T} \cdot \left\{ V_{max}^2 \cdot \left[\frac{2T\omega + \sin(2T\omega)}{4\omega} \right] \right\}} = 240$$

Equation B.11

and noting that:-

$$2T\omega = 4\pi = 12.5663706143592$$
Equation B.12

we may simplify Equation B.11 to:-

$$V_{RMS} = \sqrt{\frac{1}{T} \cdot \left\{ V_{max}^2 \cdot \left[\frac{4\pi + \sin(4\pi)}{4\omega} \right] \right\}} = 240$$

Equation B.13

Observing that $\sin(4\pi)$ is zero, and that $T\omega = 2\pi$, we may move forward with:-

$$V_{RMS} = \sqrt{\frac{1}{T} \cdot \left[V_{max}^2 \left(\frac{4\pi}{4\omega} \right) \right]} = \sqrt{\frac{1}{T} \cdot \left[V_{max}^2 \left(\frac{\pi}{\omega} \right) \right]} = \sqrt{\frac{V_{max}^2}{T} \cdot \frac{\pi}{\omega}}$$

Equation B.14

and therefore:-

$$V_{RMS} = \sqrt{\frac{V_{max}^2}{2}} = \frac{V_{max}}{2^{\frac{3}{2}}}$$

Equation B.15

The Ripple Voltage and the Implied Smoothing Capacitance[B.2]

　　　　The Ripple Voltage, V_{ripple}, is essentially the amplitude of the smoothed "wobble" that caps the resulting full-wave DC whose humps may be undesirable. The Ripple is represented by the green line in Figure B.3

　　　　A ripple voltage has twice the frequency of the AC signal whose DC full-wave output it smooths.

　　　　The Output DC Voltage is given by:-

$$V_{DC} = \frac{2.V_{max}}{\pi} = \frac{2^{\frac{3}{2}}}{\pi}.V_{RMS} = 216.075915877705$$

Equation B.16

　　　　where V_{max} is the Maximum AC Voltage and V_{RMS} is the AC Root Mean Square Voltage.

　　　　whilst the Output DC Current is:-

$$I_{DC} = \frac{V_{DC}}{R} = 2.16075915877705$$

Equation B.17

　　　　R is the Applied Resistance.

　　　　Allow that the Suppression Factor, F_{supp}, (essentially a function of ripple amplitude) is ten, then the Ripple Voltage, V_{ripple}, is:-

$$V_{ripple} = \frac{F_{supp}}{V_{max}} = 0.02946278254944$$

Equation B.18

　　　　The Implied Capacitance due to ripple suppression is:-

$$C_{implied} = \frac{I_{DC}}{2f_{ripple}V_{ripple}} = 0.366692988883727 \ farads$$

Equation B.19

The Speed of Waves and the Speed of Electrons

The Celerity of a Wave and the Velocity of its constituent particles (if any) are distinctly different.

If a child holds one end of a suspended rope and you wave the other end up and down a wave travels along the rope, changes phase at the other held end, and quickly returns to you. Meanwhile the parts of the rope go nowhere.

If you are old enough to remember the action of railway goods wagons loose-coupled in a train by chains you possibly recall that when the locomotive hit one end a rapid wave would repercuss along the train whilst individual cars sluggishly moved a few inches forward and returned a few inches to their starting point.

Another homely illustration for those who have seen a beach is the shoaling water wave which breaks upon the shore at about a meter a second whilst the particles of water within the wave describe a circular motion, with a net travel nowhere, displayed by the languid to-and-fro motion of flotsam.

The wave travels with celerity whilst its constituents move with Velocity, a Group Velocity which electricians call a Drift Velocity (of electrons in a conductor). Conventionally-speaking, electrons drift *against* the current. Drift Velocity is proportional to the current.

The Celerity of Light *in vacuo*, c, is thought to be the maximum attainable speed and has a value of 2.99792458×10^8 meters per second.

The Celerity of an electric signal in a conductor depends upon the nature of the conductor, its Form Factor (i.e. geometry, such as flat or cylindrical, woven ribbon or coaxial, etcetera), temperature, and upon the dielectric properties of its insulator including air or vacuum. The celerity of a conducted wave, such as AC or onsetted DC, varies from, 0.75c to 0.985c according to these complicated factors.

The Speed of Light in Copper at 75°C is 0.985c.

On the other hand, the Drift velocity of the carrying electrons is a matter of millimeters per second and is given by:-

$$u = \frac{m\sigma . \Delta V}{\rho e f \ell}$$

Equation B.20

where u is the Drift Velocity; m is the Molecular Mass; σ is the Conductivity; ΔV is the Voltage across the conductor; ρ is the Mass Density of the medium; e is the Charge on the Electron (the Elementary Charge) ($1.602176634 \times 10^{-19}$ Coulombs); f is the Number of Free Electrons per Atom (related to valency); and ℓ is the Length of the Conductor.

The drift speed of an electric signal in a copper wire suspended in air and having a cylindrical radius of one millimeter is about 0.000023 meters per second, which is a lot different to 0.985c

Rectification Circuitry

A Rectifier is a device that converts AC electricity to DC. Modern rectifiers depend upon the action of a Diode for small currents. A diode is any electrical device which will allow electrons to flow in one direction but block them in the other. A single Diode in series can yield Half-Wave Rectification. For Full-Wave Rectification you have to arrange four Diodes in a Rectification Bridge.

Full-Wave Rectification is still crude and "lumpy" but can be smoothed using a Choke Circuit which despite its name is not a full loop but essentially an Inductor coil arranged in series *along* one arm of the DC circuit, whilst being flanked by two Capacitors in parallel *between* the arms of the circuit.

An Inductor can pass Direct Current but blocks Alternating Current: A Capacitor passes AC but blocks DC.

Heavy duty rectifiers such as those used in industrial electrolysis usually involve Thyristors, thick solid-state components rather like Triodes. A Triode is a valve which can control the passage of current commensurately with a static charge upon its Gate. Two Thyristor gates can be linked in a cunning way so that the total component imitates a chunky, heavy-duty Diode.

Put your Choke Circuit across the DC output of a Rectification Bridge to obtain an acceptably-smooth DC current for an electrolyser.

Figure B.4 is a diagram of the symbol for a Diode showing the permitted electron flow direction:-

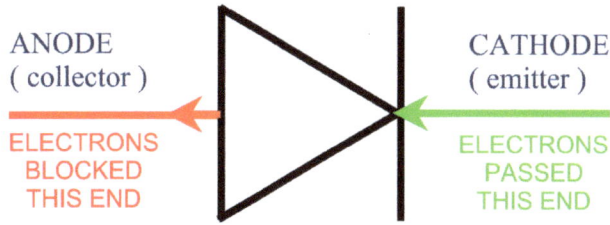

Figure B.4
Diode Action

ANODE (collector)

CATHODE (emitter)

ELECTRONS BLOCKED THIS END

ELECTRONS PASSED THIS END

Figure B.5
Conventional Symbol of a Capacitor

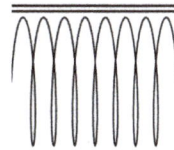

Figure B.6
Conventional Symbol of an Inductor

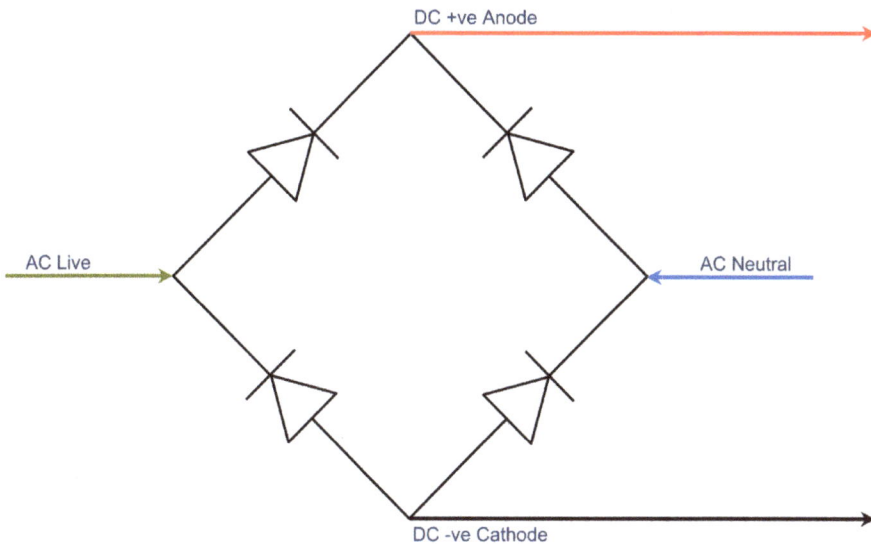

DC +ve Anode

AC Live

AC Neutral

DC -ve Cathode

Figure B.7
Scheme of a Rectification Bridge

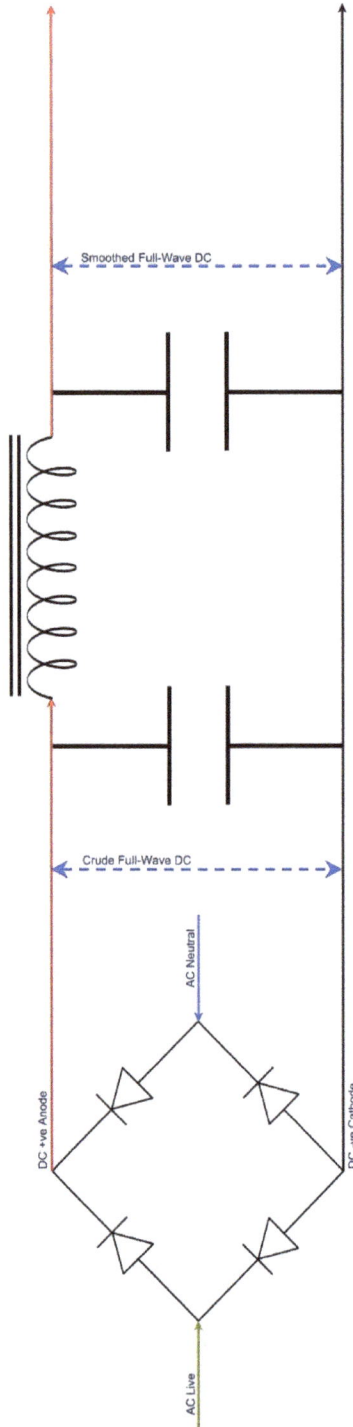

Figure B.8
A Rectification Bridge with a Smoothing Choke

APPENDIX C
RECTIFICATION AND POWER LOSS AT AN
ALUMINUM SMELTER

Earlier we discussed some of the characteristics of a hypothetical hydro-electric aluminum smelter whose power, P, at the turbine gate totalled approximately 82MW.[C.1,C.2, C.3]

The Voltage applied to each of the hypothetical n = 80 pots (electrolysis baths) was V = 4.5 Volts from which we deduce that the Root Mean Square (AC) Voltage is $V/2^{0.5}$ = 3.18198051533946. Now bear in mind that the root mean square may be largely irrelevant since there will almost certainly be a DC direct feed to the works and that the 4.5 volts is the thermodynamically-calculated overvoltage necessary for electrolysis.

So unless there is a switch to Grid power for some reason (such as drought) then we are advised to work assuming supply Voltage V at 4.5 volts.

Accordingly, the Total Feed Current is:-

$$I = \frac{P}{V} = 18222222.2222222 \; Amperes$$
Equation C.1

In any given hour the Energy consumed is:-

$$E_{hour} = P.t = 3600.P = 295200000000 \; Joules$$
Equation C.2

or 295.2 GJ

If the current is equally divided between eighty pots (cells) we can define the current per cell as:-

$$I_{cell} = \frac{I}{n} = \frac{18222222.2222222}{80}$$
$$= 227777.777777778 \; Amperes \; per \; Cell$$
Equation C.3

The Process Efficiency η = 0.95 and so the Effective Current per Cell, I_{eff}, is:-

$$I_{eff} = \eta.I_{cell} = 216388.888888889$$
Equation C.4

Therefore, the Current Loss, I_{loss} is:-

$$I_{loss} = I_{cell} - I_{eff} = I_{cell}(1 - \eta) = 11388.8888888889$$
Equation C.5

Given that the Anode Carbon Resistance, R_{ACR}, is 0.000038 Ω we may calculate the Cell Power Loss due to Anode Carbon Resistance (Watts), P_{loss}, using:-

$$P_{loss} = I_{loss}^{2}.R_{ACR} = 4928.85802469137 \ Watts$$
Equation C.6

This gives an Anode Carbon Power Loss for the Smelter, $P_{smelter}$, as:-

$$P_{smelter} = P_{loss}.n = 394308.64197531$$
Equation C.7

Meanwhile the Smelter Power Loss, P_{SPL} is:-

$$P_{SPL} = P(1 - \eta)$$
Equation C.8

so that the Proportion of Total Power Loss due to Anode Carbon Resistance is:-

$$\frac{P_{smelter}}{P_{SPL}} = 0.096172839506173$$
Equation C.9

Therefore, about 10% of the power losses in the smelter are due to Anode Carbon Resistance. Researchers have identified eight other sources of Hall-Héroult Cell Power Loss.

An Application of Faraday's Second Law

The half-reaction germane to this context may be written:-

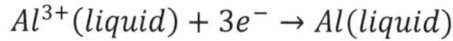

$$Al^{3+}(liquid) + 3e^- \rightarrow Al(liquid)$$
Pathway C.1

so that v, the Valency, is 3. Faraday's Constant, F, is the amount of Coulombs of Charge Q in a mole of electrons and has the value 96485.33212331 Coulombs/mol. Also, the Formula Weight of Aluminum, FW_{Al}, is 26.9815384 grams/mol

The question arises "How many tonnes of aluminum does this smelter produce in one year, assuming the continuous operation of all eighty pots?"

Firstly, the Duration, t, is the product t = 365.25×24×60×60 = 31557600 seconds representing the more than thirty-one and a half million seconds in an average year, corrected for the extra day once every four years.

Given the I = 18222222.2222222 amperes coursing through the smelter Faraday's Second Law allows us to compute the adjusted Mass Yield of metallic aluminum *in grams* as:-

$$m_{Al} = \eta . \frac{I.t.FW_{Al}}{v.F} = 50922892964.5372 \ grams$$
Equation C.10

This may be rendered as:-

$$m_{Al} = \eta . \frac{I.t.FW_{Al}}{1000.v.F} = 50922892.9645372 \ kilograms$$
Equation C.11

or:-

$$m_{Al} = \eta . \frac{I.t.FW_{Al}}{1000000.v.F} = 50922.8929645372 \ tonnes$$
Equation C.12

The estimate given by Fort William Ltd. for this smelter is 40000 tonnes per annum, very light of the computed figure with a PSD of -27.3072324113429 percent. The rather more reliable 48000 advised by Alvance British Aluminium plc, the owners of the Lochaber Works, has a PSD of -6.08936034278578 percent and the margin is probably accounted for by pot closures for maintenance and for market response.

<u>The Consumption of Alumina and Power (KWh)</u>
<u>per Kilogram of Aluminum</u>

The Amount of Alumina

The fecund part of the electrolyte in aluminum smelting is Alumina (Al_2O_3) refined from Bauxite, the crude aluminum ore. The useful pseudoreaction may be specified as:-

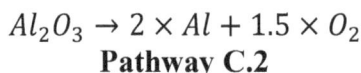

$$Al_2O_3 \rightarrow 2 \times Al + 1.5 \times O_2$$
Pathway C.2

We need the Formula Weights FW_x for Aluminum, Oxygen and Al_2O_3. The former may conveniently be taken from tables as $FW_{Al} = 26.9815384$; $FW_O = 15.999$; and we will compute the Formula Weight of Alumina, FW_{Al2O3}:-

$$2 \times FW_{Al} + 3 \times FW_O = 101.9600768 \ grams/mol$$
Pathway C.3

The Weight Ratio, W_{rAl2O3} is then:-

$$W_{rAl2O3} = \frac{2 \times FW_{Al} + 3 \times FW_O}{2 \times FW_{Al}} = 1.88944150048909$$
Equation C.13

Therefore, 1.88 kilograms of alumina are consumed for every kilogram of aluminum produced. The Kvande and Drabløs[C.4] figure is 1.93 kilos of alumina, due possibly to process losses (it is not impossible for aluminum to volatilise at high temperatures as does, for example, zinc or arsenic). The US Department of Energy figure is 1.89. The PSD(1.89,W_{rAl2O3}) is 0.029550238672488

The Consumption of Power per Kilogram of Aluminum Produced

Although modern pots have an operating voltage as low as 4.2V, we shall persist in our assumption that V_{cell} is 4.5V

We are seeking to define the Mass Yield of Aluminum m_{AlKg} as 1 kilogram whilst we are going to use FW_{AlKg} the Formula Weight of Aluminum in Kilograms by scaling the gram FW_{Al} by 0.001

One Ampere-Hour is t = 60×60 Coulombs

Therefore, by transposition of Faraday's Second Law the Ampere-Hours expended per Kilogram of Aluminum reduced is:-

$$AhperKg = \frac{1000.\,F.\,v}{t.\,FW_{Al}} = 2979.97994940465$$

Equation C.14

The U.S. Energy[C.5] figure is 2980 Ah/Kg giving a PSD(2980,AhperKg) of 0.000672838770084 a discrepancy of one part in 1486.24015806217

Moving forward, the number of Kilowatt-Hours expended per Kilogram of Aluminum smelted, KWhperKgAl, is:-

$$KWhperKgAl = \frac{AhperKg \times V_{cell}}{1000} = 13.4099097723209$$

Equation C.15

This figure of 13.41 Kilowatt-hours for each kilo of Aluminum neglects thermal demands (e.g. sensible heat) and also Heat Losses due to Anode Resistance, etcetera. The U.S. Energy figure in their report at Section 5 (Page 24) is 15.58 KWh/kg. The relevant PSD(15.58,KWhperKgAl) is 13.9286920903663 percent.

Consolidation

P is 82000000 Watts and there are n = 80 electrolysis pots. Because we have defined Duration, t, as 60×60 = 3600 seconds and have discovered v = 3 we know that Power per Pot, P_{cell}, is:-

$$P_{cell} = \frac{P}{n} = \frac{82000000}{80} = 1025000$$

Equation C.16

and:-

$$I_{cell} = \frac{P_{cell}}{V_{cell}} = 227777.777777778$$

Equation C.17

In real life, I believe that the Lochaber Works has sixty electrolysis pots, but I am unsure, never having visited the Works, though I have frequently passed through the town of Fort William in my youth.

By Faraday's Second Law:-

$$m_{Al} = \frac{I_{cell}.t.FW_{Al}}{v.F} = \frac{P_{cell}.t.FW_{Al}}{V_{cell}.v.F} = 76436.010189694$$

Equation C.18

from which:-

$$m_{Al} = \frac{KWhperKgAl.t.FW_{Al}}{V_{cell}.v.F} = 1$$

Equation C.19

A useful check by substitution is:-

$$m_{Al} = \frac{1000.F.v}{t.FW_{Al}} \times \frac{V_{cell}}{1000} \times \frac{t.FW_{Al}}{V_{cell}.v.F} = 1$$

Equation C.20

APPENDIX D
THE IRON CONTENT OF
MODERN CONTAINER SHIPS

The biggest container ship (boxship) plying the seas today (at this moment it is somewhere East of Natal) is the motor vessel MSC IRENA which was constructed in 2023AD by the Jiangsu Yangzi Xinfu Shipbuilding Company, Taizing City, China.

It has a Gross Registered Tonnage of 233,328, an Overall Length of 400 meters and a Midship Beam of 61.3 meters. Its capacity is 24,346 TEU (Twenty-foot Equivalent Unit) meaning that it can carry a maximum of 24,346 standard twenty-foot shipping containers of mixed goods, about half stowed in holds and the other half stored as a pile on the weather deck.

As I attempt to explain in Chapter Thirteen the Gross Registered Tonnage (GT or GRT) is not the mass of the vessel in tonnes: It is basically a measure of capacity to carry, and does not relate directly to the mass of the ship, which though mostly steel is to our intents and purposes the mass of the iron of which the ship is mostly made. This actual ship mass is termed by mariners and naval architects the Lightweight Tonnage (LWT).

Unfortunately, no-one and in particular the builders, have sought fit publicly to divulge this LWT figure for the MV Irena.

Accordingly, we must attempt by indirect means roughly to estimate the amount of iron needed to construct a vessel like the Irena.

Ms Wafaa Souadji[13.7] has commendably contributed her design for a smaller Polish container ship B-178 the particulars of which I present in Table D.1

Crucially, it is necessary to itemise each elementary component of the ship build in order to calculate the total weight of steel to be erected. After all, one is not a leviathan and could hardly pick up the product and plonk it onto a weighing machine!

x is a given species of Ship's Component and n_x is the number of such things installed in the build. (x is weather deck, cell guide, transverse bulkhead or other convenient component).

L_x, B_x, H_x, τ_x are respectively the cuboidal Length, Breadth, Height and Thickness of Component x such that the Volume V_x of a Transverse Component is $B_x.H_x.\tau_x$, etcetera, and of a Longitudinal Component L_x, H_x, t_x and so forth.

These parameters are listed as Height, Width, Length and Thickness on the right-hand side of Table D.1

The Overall Length of the whole Ship is L_{ov}, the Midship Beam is B, and the Loaded Height, H. The (Structural) Draught is T. The loaded Draught T_{load} is clearly greater than T, but it is T that pertains to LWT and hence to the mass of iron that makes up the completed but empty ship.

Therefore, the Total Volume of Metal used in the fabrication is:-

$$V_{total} = \sum_x n_x \tau_x \| L_x B_x H_x \|$$

Equation D.1

where the expression in double vertical bars has one term absent for each x, that term being assumed unity.

Assuming that the steel has a uniform density (this is not quite realistic: Ms Souadji uses different grades of steel for different stress regimes within her design) we can compute the mass of steel assembled using:-

$$M_{total} = \rho_{Fe} \cdot \sum_x n_x \tau_x \| L_x B_x H_x \|$$

Equation D.2

where M_{total} is the Total Empty Ship Mass due to Structural Steel and ρ_{Fe} is the Density of Iron (near to the density of structural steels).

We identify M_{total} and LWT.

A second key parameter is the Displacement Tonnage, Δ. It is important to understand that this is *not* identical to LWT, as myself and others have often erroneously thought. Actually, it is easy to appreciate this fact when you think about floating an old-fashioned tin tray in a pond or bath. It becomes clear that the mass of water displaced differs from the mass of metal in the tray.

$$\Delta = C_B . L . B . T . \rho_{sea} (1 + R_{plat})$$

Equation D.3

CONTAINER SHIP B-178
Concept Ship
The New Stettin Shipyard

SCANTLINGS AND CAPACITIES

	Meters	Tons
Ship OverAll Length, L_{oa}	220.5	
Length between Perpendiculars, L_{pp}	210.5	
Midship Beam, B	32.24	
Ship Height, H	20.5	
Draught Scantling, T	12.15	
Block Coefficient, C_B	0.67	
Deadweight Capacity, t		41850
Total Gross Tonnage, GT		35881
Total Net Tonnage, NT		14444
Number of Containers in Holds		1408
Number of Containers on Deck		1683
Total Containers		3091
Density of Iron (tonnes per m³), ρ_{Fe}	7.85	
Density of Seawater (tonnes per m³), ρ_{sea}	1.025	
Moulded Displacement, Δ_{mould}	57870.16326	
Plating Displacement, Δ_{plat}	1858.6451	
Plating Ratio, R_{plat}	0.032117503	
Displacement Tonnes Δ	61222.0286	
Lightweight Tonnage, LWT	14590.3643	
Indicated Displacement Tonnes, t+LWT	56440.3643	
PSD(Δ,t+LWT)	7.810365674	

WALL ANALYSIS (meters)

	Number	Height	Width	Length	Thickness	Volume (m³)	Mass (tonnes)	
Larboard Hull Side: Wet	1	20.5		220.5	0.03	135.6075	1064.5189	
Starboard Hull Side: Wet	1	20.5		220.5	0.03	135.6075	1064.5189	
Bottom: Wet	1		32.24	220.5	0.03	213.2676	1674.1507	
Double-Bottom Void Width			1.73					
Larboard Hull Side: Dry	1	18.77		220.5	0.015	62.0818	487.3419	
Starboard Hull Side: Dry	1	18.77		220.5	0.015	62.0818	487.3419	
Bottom: Dry	1		28.78	220.5	0.015	95.1899	747.2403	
Weatherdeck	1		32.24	220.5	0.06	426.5352	3348.3013	
Transverse Bulkheads	11	20.5	32.24		0.0125	90.8765	713.3805	
Longitudinal BulkHeads	3	20.5		220.5	0.0125	169.5094	1330.6486	
Anti-shift Cell Guides	520	1.5	1.5	20.5	0.015	239.8500	1882.8225	
Longitudinal Girders	23	1.73		220.5	0.014	122.8317	964.2291	
SubTotal						1753.4388	13764.4946	
Superstructure at 6%						105.2063	825.8697	
Total						1858.6451	14590.3643	a
Dewanto-Nazirudin Equ.(2) Vol						97641.02		b
Dewanto-Nazirudin Equ.(2) GT						29272.06		c
Dewanto-Nazirudin Equ.(3) LWT						25291.97		d

Table D.1
The Scantlings and Parameters of Container Ship Design B-178

where Δ is the Displacement Tonnage of the Ship; C_B is the Block Coefficient, a function of the hull geometry, always less than unity (in our cases 0.67); T is the unloaded Structural Draught (the depth to which the bottom of the unburdened ship sinks into the water); ρ_{sea} is the Density of Sea Water (1025 tonnes per cubic meter) and R_{plat} is the Plating Ratio.

When Δ_{plat} is the Plating Displacement (i.e. V_{total}) and Δ_{mould} is the geometrical Volume of the Hull we can say that R_{plat} is given by:-

$$R_{plat} = \frac{\Delta_{plat}}{\Delta_{mould}} = \frac{V_{total}}{C_B.L_{oa}.B.T}$$
Equation D.4

R_{plat} should lie between 0.03 and 0.05: In the case of B-178 it is 0.032

The Indicated Displacement Tonnage, Δ_{indic}, is given by:-

$$\Delta_{indic} = t + LWT$$
Equation D.5

where t is the Deadweight Capacity (the Archimedean limit of the cargo mass) and LWT is the Lightweight Tonnage (the mass of the hull and its fittings).

For B-178, the PSD(Δ,Δ_{indic}) is theoretically zero, but in practice 7.81%

The Motor Vessel MSC Irena

The MSC Irena is the largest container ship in the World at 236,184 GRT. It was commissioned and is currently owned by the Italo-Swiss shipping company Mediterranean Shipping Company S.A.

It has a maximum capacity of 24,346 TEU. I estimated from a recent photograph that 4,536 containers were loaded above the weather deck, though the ship was docked and some containers may have either been unloaded or remained to be added.

I assume for computations that the B-178 and the Irena are geometrically similar, though this is not exactly true as the ships differ in detail. The containers have a standard and invariant size so accordingly the Irena has many more bulkheads and especially many more cell guides than the smaller ship.

For estimating the weight of steel in the MSC Irena we may specify similarity in these terms:-

$$\frac{\Delta_{mould}(B178)}{\Delta_{mould}(MSC\ Irena)} = \frac{LWT(B178)}{LWT(MSC\ Irena)}$$

Equation D.6

from which it follows that:-

$$LWT(MSC\ Irena) = LWT(B178).\frac{\Delta_{mould}(MSC\ Irena)}{\Delta_{mould}(B178)}$$

Equation D.7

The computed LWT of the MSC Irena is 38194.6965 tonnes of iron. Because of our neglect of the extra cell guides, etcetera, we should round this figure up to 50000 tonnes of iron for planning purposes.

These MSC Irena particulars are summarised in Table D.2

For MSC Irena, the PSD(Δ,Δ_{indic}) is 5.4749%

Despite Ms Wafaa Souadji's meticulous care and our own best efforts LWT(MSC Irena) is still a gross underestimate. We have forgotten to take into account the (factory-fresh) engines and plenty of other iron things that pertain to the MSC Irena.

MSC IRENA
Container Ship in Plying (2024)
Jiangsu Yangzi Xinfu Shipbuilding Company

SCANTLINGS
AND
CAPACITIES

	Meters	Tons
Ship OverAll Length, L_{oa}	399.5	
Length between Perpendiculars, L_{bp}		
Midship Beam, B	61.3	
Ship Height, H	33.5	
Draught Scantling, T (structural)	17	
Block Coefficient, C_B	0.67	
Deadweight Capacity, t		240739
Total Gross Tonnage, GT		236184
Total Net Tonnage, NT		
Number of Containers in Holds		
Number of Containers on Deck		
Total Containers	24346	
Year	2023	
Density of Iron (tonnes per m^3), ρ_{Fe}		7.85
Density of Seawater (tonnes per m^3), ρ_{sea}		1.025
Moulded Displacement, Δ_{mould}		278933.6965
Plating Displacement, Δ_{plat} (pro rata)		8958.6538
Plating Ratio, R_{plat}		0.032117503
Displacement Tonnes Δ		295089.659
Lightweight Tonnage, LWT		38194.6965
Indicated Dispacement Tonnes, t+LWT		278933.6965
PSD(Δ,t+LWT)		5.474933483
Service Speed , v_{knots} (knots)		22.5
Deck Area (stated) (m^2)	24000	
Deck Area ($L_{oa}{\times}B$) (m^2)	24489.35	

Table D.2
The Scantlings and Parameters of
Container Ship MSC Irena

Empirical Equations by Dewanto and Nazirudin

Literature abounds with empirical equations formulated to relate Δ, GT, LWT and other design parameters for specific kinds of ships.

As a general statement these should be treated with extreme caution.

For Indonesian Ro-Ro ferries Dewanto and Nazirudin[13.6] offered:-

$$\Delta = C_B. L. B. H$$
Equation D.8

as an estimate of geometrical Displacement tonnage. Unfortunately the use of ship height, H, rather than Draught, T, essentially doubles the estimate and is especially invidious for boxships with cargo stacked on deck.

Further to this an equation for GT of a form often seen is:-

$$GT = \Delta[0.2 + 0.02 \times \log_{10}(\Delta)]$$
Equation D.9

which grossly underestimates GT and a linear regression for LWT:-

$$LWT = 118 + 0.86. GT$$
Equation D.10

which grossly overstates LWT to the potential embarrassment of both shipbuilders and shipbreakers.

But remember this work is special to Indonesian Ro-Ro Ferries.

Regression Equations by Abramowski, Cepowski and Zvolenský[13.12]

Abramowski Et Al (ACZ) developed their empirical regression equations *specifically for container ships* which carry considerable cargo outwith their holds.

ACZ Equation 28 has an excellent R^2 Coefficient of Determination of 0.979 (unity would be perfection of fit to the feed data) and is:-

$$LWT = 228.81 + 0.979DWT^{0.9}$$
Equation D.11

where LWT is Lightweight Tonnes (tonnes of steel) and DWT is Deadweight tonnes.

Transposing this for DWT gives:-

$$DWT = \left(\frac{LWT - 228.81}{0.979}\right)^{\frac{1}{0.9}}$$
Equation D.12

ACZ Equation 14 has an $R^2 = 0.992$ and expresses the relation of Gross Tonnage GT to DWT:-

$$GT = 2826.289 + 0.0881DWT^{1.2}$$
Equation D.13

Taken together, Equations D.11 and D.13 give rise to these local constants:-

$$a = \frac{7.57158}{1.02386} = 7.39513214697322$$
Equation D.14a

$$b = \frac{5}{6} = 0.833333$$
Equation D.14b

$$c = 2826.29$$
Equation D.14c

$$d = \frac{10}{9} = 1.111111$$
Equation D.14d

$$f = 228.81$$
Equation D.14e

For the case of MV MSC Irena of 240739 Deadweight Tonnes (DWT) ACZ Eqn. 14 ≡ Equation D.13 we may achieve the transpositions:-

$$DWT = \left(\frac{LWT - 228.81}{0.979}\right)^{\frac{1}{0.9}} = \left(\frac{GT - 2826.289}{0.0881}\right)^{\frac{1}{1.2}}$$

$$\approx 1.02386(LWT - 228.81)^{1.11111}$$

Equation D.15

from which:-

$$1.02386(LWT - 228.81)^{1.11111} \approx 7.57158(GT - 2826.29)^{0.833333}$$

Equation D.16

or substituting symbolic constants and assuming equality:-

$$(LWT - f)^d = a(GT - c)^b$$

Equation D.17

Taking Logarithms:-

$$d.\ln(LWT - f) = \ln(a) + b\ln(GT - c)$$

Equation D.18

where:-
$$\ln(a) = 2.00082196577636 \approx 2$$

Equation D.19

Hence:-

$$d.\ln(LWT - f) = 2 + \frac{5}{6}\ln(GT - c)$$

Equation D.20

Moving d to the RHS:-

$$\ln(LWT - f) = \frac{1}{d}\left(2 + \frac{5}{6}\ln(GT - c)\right)$$

Equation D.21

and antilogaritmatising both sides:-

$$LWT - f = exp\left[\frac{1}{d}\left(2 + \frac{5}{6}\ln(GT - c)\right)\right]$$

Equation D.22

So that:-

$$LWT = exp\left[\frac{1}{d}\left(2 + \frac{5}{6}\ln(GT - c)\right)\right] + f$$

Equation D.23

At this juncture you may find it convenient to define a computational function of the form:-

$$LWT(GT) = exp\left[\frac{1}{d}\left(2 + \frac{5}{6}\ln(GT - c)\right)\right] + f$$

Function D.1

so that several lightweight tonne values may be addressed.

Table D.3 shows the Iron Tonnage Values (in metric tonnes) of the British Merchant Navy at several dates together with the value of the Empire merchant marines at December 1939 and Lord Teynham's target of an additional UK tonnage of 2¾ million GRT per annum. The LWT steel mass of MV MSC Irena is also given:-

LOCAL CONSTANTS

Symbol	Value
a	7.395132147
b	0.833333333
c	2826.29
d	1.111111111
e	(omitted for clarity)
f	228.81

Entity	Date	Gross Tonnage	Light Weight Tonnage (Tonnes Fe)	Notes
UK Merchant Fleet	December 1939	17891134	1664242.94	
British Empire Merchant Fleets	December 1939	3110791	448031.94	
UK Merchant Fleet	December 2023	10400000	1107914.49	
Lord Teynham	9 December 1953	2750000	408449.17	Lord Teynham: 2¾ million tonnes per annum
MV MSC Irena	October 2022	236184	64460.11	

Table D.3
Iron Tonnages for Shipping

Additional Iron Production

Table D.3 indicates the additional iron production required to bring current UK merchant tonnage up to December 1939 levels in terms of:-

$$LWT(UK\ GT\ Dec\ 1939) - LWT(UK\ GT\ Dec\ 2023)$$
$$= 556328.4482252$$
Equation D.24

So that we see that 556329 additional tonnes of iron supply is needful.

This represents an increase of some 50.21% as shown by:-

$$\frac{LWT(UK\ GT\ Dec\ 1939)}{LWT(UK\ GT\ Dec\ 2023)} = 1.50214023895079$$
Equation D.25

If we adopt Lord Teynham's modest proposal of 1953AD that we should build an extra 2¾ million GT per annum then the Number of Years y to produce the necessary steel will be:-

$$y = \frac{LWT(UK\ GT\ Dec\ 1939) - LWT(UK\ GT\ Dec\ 2023)}{LWT(Lord\ Teynham)}$$
$$= 1.36205063565144$$
Equation D.26

Hence is will take 1 year 132 days and 6 hours to make enough iron to restore the British Merchant Fleet to its tonnage before the depredations of the Axis enemy.

BIBLIOGRAPHY AND REFERENCES

CHAPTER ONE

    1.1

Howard Carter, Daniela Comelli1, Massimo D'orazio, Luigi Folco, et al. - Comelli, Daniela; d'Orazio, Massimo; Folco, Luigi; et al. (2016). "The meteoritic origin of Tutankhamun's iron dagger blade". Meteoritics & Planetary Science. Wiley Online. doi:10.1111/maps.12664.

Photograph featured in:-

Wikipedia contributors. (2024, March 19). Tutankhamun's meteoric iron dagger. In *Wikipedia, The Free Encyclopedia*. Retrieved 09:53, March 31, 2024, from https://en.wikipedia.org/w/index.php?title=Tutankhamun%27s_meteoric_iron_dagger&oldid=1214473188 Image rotated by the Author

    1.2

"Economic History of the British Iron and Steel Industry"
Alan Birch
First Edition   1967
Routledge of London
ISBN 9781315020396
(30 September 2013 Edition)
pp 432
https://doi.org/10.4324/9781315020396

    1.3

Port Talbot Steelworks
https://www.geograph.org.uk/photo/41552
Seen at:-
https://commons.wikimedia.org/wiki/File:Port_Talbot_Steelworks_-_geograph.org.uk_-_41552.jpg

Grid Square
  SS7788, 115 images   (more nearby 🔍)
Photographer
  Chris Shaw   (more nearby)
Date Taken
  August 2005   (more nearby)
Submitted
  Monday, 22 August, 2005
Subject Location
  OSGB36: geotagged! SS 77 88 [1000m precision]
  WGS84: 51:34.9203N 3:46.1632W
Camera Location
  OSGB36: geotagged! SS 765 913
View Direction
  South-southeast (about 157 degrees)

1.4

Neath Port Talbot : Steel Works
https://www.geograph.org.uk/photo/5502264
Seen at:-
https://commons.wikimedia.org/wiki/
File:Neath_Port_Talbot_,_Steel_Works_-
geograph.org.uk-_5502264.jpg
Grid Square
  SS7787, 18 images   (more nearby 🔍)
Photographer
  Lewis Clarke   (more nearby)
Date Taken
  Sunday, 13 August, 2017   (more nearby)
Submitted
  Friday, 18 August, 2017
Subject Location
  OSGB36: geotagged! SS 7701 8795
[10m precision]
  WGS84: 51:34.6173N 3:46.5758W
Camera Location
  OSGB36: geotagged! SS 78017 89922
View Direction
South-southwest (about 202 degrees)

1.5

World Steel Association

https://worldsteel.org/data/
world-steel-in-figures-2023/

CHAPTER TWO

2.1

Wikipedia contributors. (2024, April 9).
Mass.
In *Wikipedia, The Free Encyclopedia*.
Retrieved 12:52, August 24, 2024, from
https://en.wikipedia.org/w/index.php?
title=Mass&oldid=1218095693

2.2

Wikipedia contributors. (2024, August 13).
Energy.
In *Wikipedia, The Free Encyclopedia*.
Retrieved 12:54, August 24, 2024, from
https://en.wikipedia.org/w/
index.php?title=Energy&oldid=1240158954

2.3

Wikipedia contributors. (2024, July 17).
Voltage.
In Wikipedia, The Free Encyclopedia.
Retrieved 12:54, August 24, 2024, from
https://en.wikipedia.org/w/index.php?
title=Voltage&oldid=1235101223

2.4

Wikipedia contributors. (2024, August 1).
Electric potential.
In *Wikipedia, The Free Encyclopedia*.
Retrieved 12:56, August 24, 2024, from
https://en.wikipedia.org/w/index.php?
title=Electric_potential&oldid=1237990834

2.5

Wikipedia contributors. (2024, August 4).
Electronvolt.
In *Wikipedia, The Free Encyclopedia*.
Retrieved 12:58, August 24, 2024, from

https://en.wikipedia.org/w/
index.php?title=Electronvolt&oldid=1238638080

CHAPTER THREE

3.1   Adapted from Figure 2 page 59 of:-

Ayush Bhattacharya, Sadhasivam Muthusamy,
Static Heat Energy Mathematical Model for an Iron
Blast Furnace.
International Journal of
Mineral Processing and Extractive Metallurgy
Vol.2, No.5, 2017, pp.57-67.
 doi: 10.11648/j.ijmpem.20170205.11

3.2

This document is downloaded from
DR-NTU (https://dr.ntu.edu.sg)
Nanyang Technological University, Singapore.
"Simulations of melting in fluid-filled packed
media due to forced convection
with higher temperature"
Soon, Genevieve; Zhang, Hui; Yang, Chun;
Law, Adrian Wing-Keung
Heat and Mass Transfer, 175, 121358-.   2021
https://dx.doi.org/10.1016/
j.ijheatmasstransfer.2021.121358
https://hdl.handle.net/10356/155834
https://doi.org/10.1016/
j.ijheatmasstransfer.2021.121358
© 2021 Elsevier Ltd. All rights reserved.
This paper was published in
International Journal of Heat and Mass Transfer
and is made available with permission of Elsevier Ltd.
Downloaded on 04 Apr 2024 18:29:36 SGT

3.3

Transactions of the
 Korean Nuclear Society Autumn Meeting
Gyeongju, Korea, October 27-28, 2016
"Forced Convection Heat Transfer of a sphere
 in Packed Bed Arrangement"

Dong Young Lee and Bum Jin Chung
Department of Nuclear Engineering,
Kyung Hee University
#1732 Deokyoung
daero, Giheung gu, Yongin si, Gyeonggi do,
17104, Korea *
Corresponding author:bjchung@khu.ac.kr

3.4

Y. M Ferng, K. Y. Lin,
"Investigating Effects of BCC and FC Arrangements
on Flow and Heat Transfer Characteristics in
Pebbles through CFD Methodology"
Nuclear Engineering Design,
Vol. 258, pp. 66-75, 2013.

CHAPTER FOUR

4.1

LABTERMO:
METHODOLOGIES FOR THE CALCULATION
OF THE
CORRECTED TEMPERATURE RISE
IN ISOPERIBOL CALORIMETRY
LMNBF Santos , MT Silva, B. Schröder and L Gomes
Journal of Thermal Analysis and Calorimetry
Vol. 89 (2007) 1, 175–180

4.2

SOLVING BOMB CALORIMETER PROBLEMS
https://mrskubacki.weebly.com/uploads/
8/7/3/0/87301008/bomb-calorimetry.pdf

CHAPTER FIVE

5.1

National Institute of Standards and Technology
(NIST)
U.S. Department of Commerce
NIST Chemistry WebBook, SRD 69
https://webbook.nist.gov/

5.2

"The Blast Furnace Mathematical Model"
Gurprit Singh
First Edition: May 2021
Second Edition: September 2022
Published by Gurprit Singh and
Printed in Great Britain by Amazon
ISBN 979-872574-96-70   paperback
pp 281

CHAPTER SIX

6.1

Wikipedia contributors. (2023, October 27).
Activation energy.
In *Wikipedia, The Free Encyclopedia*.
Retrieved 09:48, April 24, 2024,
from https://en.wikipedia.org/w/index.php?
title=Activation_energy&oldid=1182119303

6.2

Glucose Metabolism Diagram by
Jerry Crimson Mann, Tutmosis and Fvasconcellos
Wikipedia contributors. (2023, October 27).
Activation energy.

6.3

Thermodynamic Symbolism attaching to
Activation and Reaction Energies.
Anonymous
Wikipedia contributor. (2023, October 27).
Activation energy.

CHAPTER SEVEN

(no references)

CHAPTER EIGHT

8.1

"blast furnace"
Encyclopaedia Britannica

Encyclopaedia Britannica, Inc.   23 April 2024
https://www.britannica.com
https://www.britannica.com/technology/blast-furnace
https://www.britannica.com/print/article/69019

8.2

Wikipedia contributors. (2024, April 27).
Blast furnace.
In Wikipedia, The Free Encyclopedia.
Retrieved 08:00, May 16, 2024, from
https://en.wikipedia.org/w/index.php?
        title=Blast_furnace&oldid=1220991679

8.3

Ayush Bhattacharya, Sadhasivam Muthusamy.
Static Heat Energy Balance Mathematical Model for
an Iron Blast Furnace.
*International Journal of Mineral Processing and
Extractive Metallurgy.*
Vol. 2, No. 5, 2017, pp. 57-67.
doi: 10.11648/j.ijmpem.20170205

8.4

"The Blast Furnace Mathematical Modelling"
Gurprit Singh
First Edition: May 2021
Second Edition: September 2022
ISBN 9798725749670
pp 281

8.5

Tosaka
13 June 2008
Blast_furnace_NT.PNG

Wikipedia contributors. (2024, April 27).
Blast furnace.
In *Wikipedia, The Free Encyclopedia.*
Retrieved 10:56, May 18, 2024, from
https://en.wikipedia.org/w/index.php?
        title=Blast_furnace&oldid=1220991679

8.7

energiesArticle
Investigation of the Effects of Coke Reactivity and
Iron Ore Reducibility on the Gas Utilization
Efficiency of Blast Furnace
Fanchao Meng, Lei Shao, and Zongshu Zou
27 September 2020
https://www.mdpi.com/journal/energies
Energies 2020, 13, 5062; doi:10.3390/en13195062
pp 14

8.8

12.5 Iron And Steel Production
https://www3.epa.gov
/ttnchie1/ap42/ch12/final/c12s05.pdf
Nov-86
For 1 ton of Iron

8.9

"Study of Combustion Properties for
Cokes with Various Grain Size Composition"

V. I. Matukhin, S. Ya. Zhuravlev, A. V. Khandoshka,
and A. V. Matukhina

TIM'2018
VII All-Russian Scientific and
Practical Conference of Students,
Graduate Students and Young Scientists on
"Heat Engineering and Computer Science in
 Education, Science and Production"
Volume 2018
Conference Paper

This article is distributed under the
terms of the Creative Commons
Attribution License, which permits unrestricted use and
redistribution provided that the
original author and source are credited
17 July 2018

KnE Engineering, pages 259–265.

DOI 10.18502/keg.v3i5.2678

8.10

"Mathematical Model of Blast Furnace"
Iwao Muchi
Transactions ISIJ, Vol. 7, 1967
Pages 223-237
* Originally published in Journal of
The Japan Institute of Metals
(REFERENCES 11)- 14)) in Japanese.
** Department of Iron and Steel Engineering, Faculty
of Engineering, Nagoya University, Chigusa-ku,
Nagoya.
https://doi.org/10.2355/isijinternational1966.7.223
https://www.jstage.jst.go.jp/article/
        isijinternational1966/7/5/7_223/_article

8.11

"Atomic-Scale Structure of the Hematite
α-Fe2O3(1$\bar{1}$02) "R-Cut" Surface"
Florian Kraushofer, Zdenek Jakub, Magdalena Bichler,
Jan Hulva, Peter Drmota, Michael Weinold, Michael
Schmid, Martin Setvin, Ulrike Diebold, Peter Blaha,
and Gareth S. Parkinson
The Journal of Physical Chemistry C
2018 122 (3), 1657-1669
DOI: 10.1021/acs.jpcc.7b10515

8.12

Modern Instrumentation, 2013, 2, 68-73
http://dx.doi.org/10.4236/mi.2013.24010
Published Online October 2013
(http://www.scirp.org/journal/mi)
Copyright © 2013 SciRes. MI
"Hot Blast Flow Measurement in
  Blast Furnace in Straight Pipe"
Ricardo S. N. Motta , Edson C. Bortoni , Luiz E. Souza
1CSN/USS/UNIFEI, Volta Redonda, Brazil
2ISEE/UNIFEI, Itajubá, Brazil
3IESTI/UNIFEI, Itajubá, Brazil
Email:nadur@CSN.com.br, bortoni@unifei.edu.br,
edival@unifei.edu.br

August 22, 2013
Copyright © 2013 Ricardo S. N. Motta et al.
This is an open access article distributed under the
Creative Commons Attribution License, which permits
unrestricted use, distribution, and reproduction in any
medium, provided the original work is properly cited.

8.13

"The Theory of the Hot Blast"
I Lowthian Bell
From "Journal of the Iron and Steel Institute"

Quoted in:-
Van Nostrand's Eclectic Engineering Magazine
(1869-1879)
New York Vol. 6, Iss. 41, (May 1, 1872): page 545.

CHAPTER NINE

9.1

"Evolution of Blast Furnace Process toward
 Reductant Flexibility
 and Carbon Dioxide Mitigation in Steel Works"
Tatsuro ARIYAMA, Michitaka SATO,
Taihei NOUCHI and Koichi TAKAHASHI
25 August 2016
ISIJ International, Vol. 56 (2016), No. 10,
pp. 1681–1696

9.2

National Institute of Standards and Technology
United States Department of Commerce
NIST Chemistry WebBook, SRD 69
https://webbook.nist.gov/cgi/cbook.cgi?
ID=C7727379&Units=SI&Mask=1#Thermo-Gas

9.3

https://www.engineeringtoolbox.com/
gas-density-d_158.html
https://www.engineeringtoolbox.com/
standard-heat-of-combustion-energy-content-
d_1987.html

The Engineering ToolBox
https://www.engineeringtoolbox.com/

Nitrogen Gas - Specific Heat vs. Temperature
https://www.engineeringtoolbox.com/
nitrogen-d_977.html

Oxygen Gas - Specific Heat vs. Temperature
https://www.engineeringtoolbox.com/
oxygen-d_978.html

Gases - Specific Heats and Individual Gas Constants
https://www.engineeringtoolbox.com/
        specific-heat-capacity-gases-d_159.html

9.4

Citation: Fang, J.; Pang, Z.; Xing, X.; Xu, R.
"Thermodynamic Properties, Viscosity,
 and Structure of
 $CaO–SiO_2–MgO–Al_2O_3–TiO_2$–Based Slag."
Materials 2021, 14, 124.
https://doi.org/10.3390/ma14010124
Published: 30 December 2020
Copyright: © 2020 by the authors
https://doi.org/10.3390/ma14010124
https://www.mdpi.ccom/journal/materials

CHAPTER TEN

10.1

"Gross Calorific Value in Metallurgical Coke"
LECO Corporation, Saint Joseph, Michigan, USA
Instrument: AC600
pp 2

https://eu.leco.com/images/Analytical-Application-Library/AC600_GROSS_HEAT_METALLURGICAL_COKE_203-821-473.pdf

10.2

"Theoretical Minimum Energies to Produce Steel"
March 2000
RJ Fruhan, O Fortini, HW Paxton and R Brindle
Carnegie Mellon University, Pittsburgh PA
Prepared By: Energetics, Inc., Columbia MD
U.S. Department of Energy
Office For Industrial Technologies
Washington DC
pp 43
Table E-1. Comparison of Theoretical Minimum Energy and Actual Energy Requirements for Selected Processes
Table 11 Practical Theoretical Minimum and Energy for Selected Processes (Page 20)
Table 12 Comparison of Theoretical Minimum Energy and Actual Energy Requirements for Selected Processes

10.3

European Steel Association
Map-20191113_Eurofer_SteelIndustry_Rev3
-has-stainless
https://www.eurofer.eu/assets/Uploads/Map-20191113_Eurofer_SteelIndustry_Rev3-has-stainless.pdf
Year 2022AD?

10.4

Ayush Bhattacharya, Sadhasivam Muthusamy.
"Static Heat Energy Balance Mathematical Model for an Iron Blast Furnace"
International Journal of Mineral Processing and Extractive Metallurgy. Vol. 2, No. 5, 2017, pp. 57-67.
doi: 10.11648/j.ijmpem.20170205.11
pp 11
Table 20: Page 67

10.5

Energy Reports 3 (2017) 29–36
Contents lists available at ScienceDirect
Energy Reports
journal homepage: www.elsevier.com/locate/egyr
"Assessment on the energy flow and carbon emissions
 of integrated steelmaking plants"
Huachun He, Hongjun Guan, Xiang Zhu, Haiyu Lee
9 January 2017
Table 3: Page 32
pp 8
https://www.researchgate.net/publication/
312658134_Assessment_on_the_energy_flow_and_ca
rbon_emissions_of_integrated_steelmaking_plants/
download?_tp=eyJjb250ZXh0Ijp7ImZpcnN0UGFnZS
I6Il9kaXJlY3QiLCJwYWdlIjoiX2RpcmVjdCJ9fQ

10.6

"Theoretical Minimum Energies to Produce Steel"
March 2000
RJ Fruhan, O Fortini, HW Paxton and R Brindle
Carnegie Mellon University, Pittsburgh PA
Prepared By: Energetics, Inc., Columbia MD
U.S. Department of Energy
Office For Industrial Technologies
Washington DC
pp 43
Table E-1. Comparison of Theoretical Minimum
Energy and Actual Energy Requirements for Selected
Processes
Table 11 Practical Theoretical Minimum and Energy
for Selected Processes (Page 20)
Table 12 Comparison of Theoretical Minimum Energy
and Actual Energy Requirements for Selected
Processes

CHAPTER ELEVEN

11.1

Citation: He, J.; Li, K.; Zhang, J.; Conejo, A.N.
"Reduction Kinetics of Compact Hematite with
 Hydrogen from 600 to 1050 °C"

Metals 2023, 13, 464,
https://doi.org/10.3390/met13030464
Academic Editor: Man Seung Lee
Published: 23 February 2023
Copyright: © 2023 by the authors.
Licensee MDPI, Basel, Switzerland.
This article is an open access article
distributed under the terms and
conditions of the Creative Commons
Attribution (CC BY) license (https://
creativecommons.org/licenses/by/4.0/).

11.2

"Reduction of Iron Oxides with Hydrogen—A Review"
Daniel Spreitzer and Johannes Schenk
Steel Research International
www.steel-research.de
steel research int. 2019, 90, 1900108 1900108 (1 of 17)
pp17
Published by WILEY-VCH
Verlag GmbH & Co. KGaA, Weinheim

11.3

"Reduction behavior of iron oxides in hydrogen and
carbon monoxide atmospheres"
W.K. Jozwiak, E. Kaczmarek, T.P. Maniecki,
W. Ignaczak, W. Maniukiewicz
Applied Catalysis A: General 326 (2007) 17–27
20 March 2007

11.4

"Green steel: design and cost analysis of hydrogen-based
direct iron reduction"
Fabian Rosner, Dionissios Papadias, Kriston Brooks,
Kelvin Yoro, Rajesh Ahluwalia,
Tom Autrey, Hanna Breunig

11.5

"Assessment of hydrogen direct reduction for fossil-
free steelmaking"
Valentin Vogl , Max Åhman, Lars J. Nilsson
Journal of Cleaner Production

Volume 203: 1 December 2018: Pages 736-745
https://doi.org/10.1016/j.jclepro.2018.08.279

11.6

"Energy Efficiency in Iron and Steel Making"
ABB
pp11   2022AD

11.7

"The Disruptive Potential of Green Steel"
Thomas Koch Blank
Insight Brief:- The Rocky Mountain Institute
pp7   September 2019

11.8

"How much clean energy does it take to make green steel"
Maria Gallucci
22 May 2024
Canary Media

11.9

"Materials and Energy Balance:-
Heat Balance in Pyrometallurgical Processes II"
(Carbon Blast Only)
Common Futures, Utrecht, Nederland   2024
Ijmuiden Works

11.10

"Results of Hydrogen Reduction of Iron Ore Pellets at
  Different Temperatures"
Oleksandr Kovtun, Mykyta Levchenko,
Mariia O. Ilatovskaia,
Christos G. Aneziris, and Olena Volkova
Dedicated to the 150th anniversary of the
Institute for Iron and Steel Technology
steel research international; 2024, 2300707
found in Advanced Science News
Equations (3), (4) and (5) on Page 2 of 11

11.11

"Toward a Fossil Free Future with
  HYBRIT: Development of Iron and
  Steelmaking Technology in
  Sweden and Finland"

Martin Pei, Markus Petäjäniemi,
Andreas Regnell and Olle Wijk
Metals 2020, 10, 972   18 July 2020
MDPI
pp11

© 2020 by the authors.
Licensee MDPI, Basel, Switzerland.
This article is an open access
article distributed under the terms and conditions of the
Creative Commons Attribution
(CC BY) license
(http://creativecommons.org/licenses/by/4.0/).

11.12

"ArcelorMittal Eisenhüttenstadt steel plant"
https://www.gem.wiki/w/index.php?
title=ArcelorMittal_Eisenhüttenstadt_steel_plant&
oldid=603751
11 April 2024
pp5

11.13

"HYBRIT produces hydrogen-reduced
 sponge iron on pilot scale"
https://www.greencarcongress.com/
2021/06/20210622-hybrit.html

CHAPTER TWELVE

12.1

"Comparative study of electroreduction of
iron oxide using acidic and alkaline electrolytes for
sustainable iron production"
Akmal Irfan Majid, Niels van Graefschepe,
Giulia Finotelloc, John van der Schaaf,
Niels G. Deen, Yali Tang
Electrochimica Acta 467 (2023) 142942
pp10

12.2

"Deoxidation Electrolysis of Hematite in
Alkaline Solution:
Impact of Cell Configuration and Process Parameter
Reduction Efficiency"
Reza Fayaz, Ingmar Bosing, Fabio La Mantia,
Michael Baune, Md lzzuddin Jundullah Hanafi,
Thorsten M. Gesing, and Jorg Thoming
ChemElectroChem 2023, 10, e202300451 (1 of 8)
© 2023 The Authors. ChemElectroChem
published by Wiley-VCH GmbH

12.3

"MOLTEN OXIDE ELECTROLYSIS
FOR IRON PRODUCTION:
IDENTIFICATION OF
KEY PROCESS PARAMETERS
FOR LARGE SCALE DEVELOPMENT"
Antoine Allanore, Luis A Ortiz and Donald R Sadoway
Energy Technology 2011
Edited by: Neale R. Neelameggham, Cynthia K. Belt,
Mark Jolly, Ramana G. Reddy, and James A. Yurko
TMS (The Minerals, Metals & Materials Society), 2011
pp121-129

12.4

"Electrolysis of iron in a molten oxide electrolyte"
Jan Wiencke, Hervé Lavelaine, Pierre-Jean Panteix,
Carine Petitjean, Christophe Rapin
Received: 9 July 2017 / Accepted: 26 December 2017

/ Published online: 5 January 2018
© The Author(s) 2018.
This article is an open access publication
Journal of Applied Electrochemistry
(2018) 48:115–126
https://doi.org/10.1007/s10800-017-1143-5

12.5

"Towards Carbon-free Metals Production by
Molten Oxide Electrolysis"
Deepak Khetpal, Andrew Ducret and
Donald R. Sadoway
Department of Materials Science & Engineering
Massachusetts Institute of Technology
Cambridge, Massachusetts
TMS Meeting, Charlotte, NC
March 17, 2004
pp29
(PowerPoint Lecture)

12.6

"Electrical Conductivity and Transference Number
 Measurements of
 FeO - CaO - MgO - SiO$_2$ Melts"
Andrew Ducret, Deepak Khetpal, Donald R Sadoway
Massachusetts Institute of Technology
ECS Proceedings Volumes 2002-19(1)
January 2002
DOI:10.1149/200219.0347PV
pp8

12.7

"Candidate Anode Materials for Iron Production
by Molten Oxide Electrolysis"
James D Paramore
Submitted to the
Department of Materials Science and Engineering
in Partial Fulfillment of the Requirements for the
Degree of Master of Science in
Materials Science and Engineering
at the
Massachusetts Institute of Technology

September 2010
pp63

12.8

"The Impact of Iron Oxide Concentration on the
  Performance of Molten Oxide Electrolytes
  for the Production of Liquid Iron Metal."
Jan Wiencke, Hervé Lavelaine, Pierre-Jean Panteix,
Carine Petitjean, Christophe Rapin.
Metallurgical and Materials Transactions
B, 2020, 51 (1), pp.365-376.
10.1007/s11663-019-01737-3. hal-04010934

12.9

"Electrical Conductivity and Transference Number
 Measurements of FeO - CaO - MgO - SiO 2 Melts"
Andrew Ducret, Deepak Khetpal, Donald R Sadoway
Massachusetts Institute of Technology
ECS Proceedings Volumes 2002-19(1)
January 2002
DOI:10.1149/200219.0347PV
pp8

12.10

"HIGH-TEMPERATURE PROPERTIES OF
 NUCLEAR GRAPHITE"
Tracy L. Albers, Lionel Batty, and David M. Kaschak
GrafTech International Holdings Inc.
Parma, OH, USA
Proceedings of the 4th
International Topical Meeting on
High Temperature Reactor Technology HTR2008
September 28-October 1, 2008, Washington, DC USA
HTR2008-58284
2pp

12.11

"Features and Challenges of
 Molten Oxide Electrolysis"
Antoine Allanore
25 November 2014
Journal of The Electrochemical Society,

162 (1) E13-E22 (2015)

12.12

NIST Chemistry WebBook, SRD 69
National Institute of Standards and Technology
U.S. Department of Commerce
https://webbook.nist.gov/chemistry/faq/

CHAPTER THIRTEEN

13.1

"A Boolean Calculus of Moiety for Two Arguments"
Pages 427-471
in

"Mathematical Explorations"
James R Warren
Midland Tutorial Productions
Bloxwich   2022
ISBN 978-1-7396296-6-3
pp471

13.2

"Steel Production through electrolysis:
  impacts for electricity consumption"
Adam Rauwerdink
Boston Metal
https://iea.blob.core.windows.net/assets/
imports/events/
288/S5.4_20191010BostonMetalIEA
Decarbonization2019.pdf
October 18 2019
pp14

13.3

Re-Equipment Of The Merchant Fleet
debated on Wednesday 9 December 1953
Hansard: Volume 184
https://hansard.parliament.uk/lords/
953-12-09/debates/
9bd67fd4-10bb-45c6-9424-ada8f2798fdd/
Re-EquipmentOfTheMerchantFleet

13.4

"The top 10 largest container ships in the world"
Peter Nilson
April 4, 2024
Ship Technology
https://www.ship-technology.com/features/
the-top-10-largest-container-ships-in-the-world/

13.5

MSC Irena
Vessel Finder
https://www.vesselfinder.com/vessels/details/9929429
Tangier to Singapore: 1000GMT 16 August 2024

13.6

SHIP LIGHTWEIGHT ESTIMATION AT
CONCEPT DESIGN STAGE:
CASE OF INDONESIA SINGLE ENDED
RO-RO FERRIES
Christoforus Chandra Dewanto and Ahmad Nasirudin
Journal of Marine-Earth Science Technology,
Volume 3 Issue 3
ISSN: 2774-5449 Page I 86
pp86 to 89   June 7 2023

13.7

"Structural design of a container ship approximately
3100 TEU according to the concept of general ship
design B-178"
Wafaa Souadji
Master Thesis
West Pomeranian University of Technology, Szczecin
For the Double Degree of the
University of Liege with Ecole Centrale de Nantes
Ref. 159652-1-2009-1-BE-ERA MUNDUS-EMMC
https://matheo.uliege.be/bitstream/2268.2/6105/1/
MAster%20thesis-
%20Wafaa%20Souadji%20%28ZUT%29.pdf
pp151

13.8

    "PRELIMINARY SHIP DESIGN
     PARAMETER ESTIMATION"
    https://www.slideshare.net/slideshow/
    preliminary-shipdesign/74998098
    pp44

13.9

    Wikipedia contributors. (2024, August 12).
    List of largest container ships.
    in Wikipedia, The Free Encyclopedia.
    Retrieved 09:45, August 18, 2024,
    from https://en.wikipedia.org/w/index.php?
    title=List_of_largest_container_ships&
    oldid=1239865557

13.10

    Shipnext Fleet Monitor
    MV MSC IRENA
    https://shipnext.com/vessel/9929429-msc-irina
    1200 BST 18 August 2024 (1100 GMT)

13.11

    "Latest MSC container giant makes her
     European debut"
    Ships Monthly"
    https://shipsmonthly.com/news/16018/

13.12

    "Determination of Regression Formulas for
     Key Design Characteristics of
     Container Ships at Preliminary Design Stage"
    Article in New Trends in Production Engineering
    October 2018
    DOI: 10.2478/ntpe-2018-0031
    Tomasz Abramowski, Tomasz Cepowski
    and Peter Zvolenský
    Maritime University of Szczecin
    https://www.researchgate.net/publication/
    329418173_Determination_of_Regression_Formulas
    _for_Key_Design_Characteristics_of_Container_Ship
    s_at_Preliminary_Design_Stage

13.13

UK OFFSHORE WINDFARM MAP 2023
Supplied by
Wind Energy Network Magazine and
4C Offshore, a TGS Company
https://www.windenergynetwork.co.uk/
wp-content/uploads/2023/03/
UK-Offshore-Windfarm-Map-2023-V2.pdf

13.14

Wikipedia contributors. (2023, May 9).
Humber Gateway Wind Farm.
In *Wikipedia, The Free Encyclopedia.*
Retrieved 16:19, August 22, 2024, from
https://en.wikipedia.org/w/index.php?
title=Humber_Gateway_Wind_Farm&oldid=1153965
340

APPENDIX A

(No References)

APPENDIX B

B.1

Electronics Tutorials
RMS Voltage Tutorial
RMS Voltage of a Sinusoidal Waveform
https://www.electronics-tutorials.ws/
accircuits/rms-voltage.html

B.2

Electronics Tutorials
Full Wave Rectifier
https://www.electronics-
tutorials.ws/diode/diode_6.html

APPENDIX C

C1

"Analysis of anode current distribution in
aluminum electrolysis cell
based on equivalent circuit numerical simulation"
Yifan Wang, Peixin Zhou, Jun Tie
E3S Web of Conferences 385, 03022 (2023)
ISESCE 2023
https://doi.org/10.1051/e3sconf/202338503022
https://www.e3s-
conferences.org/articles/e3sconf/pdf/2023/22/
e3sconf_isesce2023_03022.pdf
5 pp

C2

Alvance British Aluminium
Who We Are
https://alvancebritishaluminium.com/about-us/

C3

"Liberty British Aluminium smelter in Fort William"
Fort William Ltd.
https://visitfortwilliam.co.uk/pages/rio-tinto-alcan-
aluminium-smelter-in-fort-william-at-rio-tinto-alcan-
works-33bd8985

C4

"The Aluminum Smelting Process and Innovative
 Alternative Technologies"
Halvor Kvande, PhD and Per Arne Drabløs, MD
Journal of Occupational and Environmental Medicine
Volume 56, Number 5S, May 2014
Pages S23-S32
pp10

C5

"U.S. Energy Requirements for
 Aluminum Production"
U.S. Department of Energy
February 2007
pp150

APPENDIX D

Please consult references for Chapter Thirteen

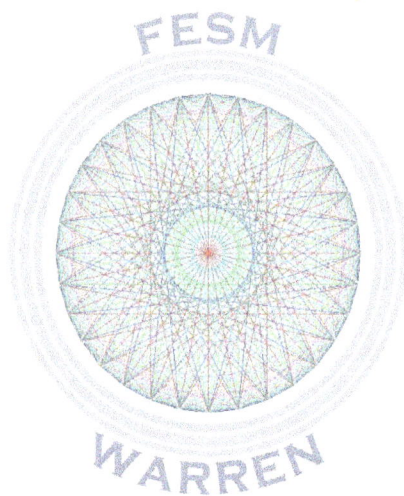

INDEX

3

3D, 81, 149

A

ABB, 225, 226, 228, 355
Abbey, 24, 25, 26
Aberdeen, 171
Abraham, 18, 21
Abramowski, 290, 335, 362
Abramowski Et Al, 335
Absolute, 114, 115, 116, 215, 229, 259
AC, 119, 303, 309, 310, 311, 316, 318, 319, 320, 323
AC600, 180, 351, 352
Acceleration, 37, 39, 42, 301
Acetylene, 60
Act, 2, 272, 274
Activation, 111, 112, 113, 114, 115, 117, 203, 346
Active, 258
African, 274
Age, 13, 14, 15, 35, 287, 299
Agency, 13
Ages, 272
AGR, 288
Air, 82, 100, 147, 148, 158
Akhenaten, 13
Al2O3, 133, 174, 242, 326, 351
Alan, 19, 341
Algebraic, 201
Alkali, 29, 64
Alkali Earth, 29, 64
Alkaline, 31, 50, 357
Allanore, 241, 357, 359
Alloying, 23
Alternating, 119, 309, 320
Alternator, 310
Alumina, 133, 134, 326
Aluminium, 136, 137, 364
Aluminum, 30, 133, 134, 135, 263, 325, 326, 327, 364
Alvance British Aluminium plc, 326
American, 19, 36, 47, 133, 223, 272, 299
Amount, 36, 45, 68, 131, 260, 262, 268, 326
Amperes, 36, 122, 134, 261, 262, 302
Amplitude, 315

Analysis, 36, 345, 364
Ancients, 15, 242
Anglesey, 288
Angular Frequency, 315
Anode, 119, 125, 255, 324, 327, 358
Antarctica, 299
Anteros, 133
Anthracite, 150, 175
Antiquity, 35
Antrim, 34
Appendix, 289, 290
Aqueous, 241
ArcelorMittal, 234, 356
Archaeological, 20
Archimedean, 289, 332
Arctic, 234
Area, 28, 35, 44, 72, 243, 248
Argentina, 27, 299
Argon, 64, 82, 98, 170
Arguments, 274, 360
Ariyama, 166, 170, 175, 176
Ariyama and Sato, 175, 176
Army, 299
Arrays, 254, 306
Arrhenius, 114, 115, 161, 201, 203, 205, 211
Ashby, 20
Association, 28, 342
ASTM D5865, 180
Astute, 288
Asymptote, 208
Atmosphere, 81, 165
Atom, 46, 48, 320
Atomic, 47, 51, 55, 56, 60, 61, 67, 120, 135, 136, 259, 264, 349
Attractant, 38
Attraction, 38
Australia, 27, 28, 282
Autothermal, 258
Avogadro, 46, 55, 61, 114, 122, 123, 124, 135
Avro, 44
Ayrshire, 287

B

B-178, 329, 331, 332, 361
Backward, 113

Balance, 66, 71, 95, 103, 104, 105, 106, 108, 109, 123, 153, 156, 163, 176, 187, 347, 352, 355
Bank, 287
Barium, 50, 120
Baryons, 46, 47
Basic, 23, 223, 226, 244, 246, 247, 281
Battle, 20, 272
Bauxite, 326
BBP, 8
BCC, 79, 345
Belavezh Accords, 272
Benzoic Acid, 180
Bessemer, 23
Best Iron Bar, 16
BF, 16, 29, 30, 34, 85, 157, 163, 167, 168, 170, 176, 183, 225, 233, 234, 236, 239, 243, 259, 267, 269, 282, 291, 296, 297
BFG, 100, 142, 144, 152, 157, 164, 166, 167, 168, 176, 183
Bhattacharya, 153, 186, 187, 344, 347, 352
Bio, 236
Birch, 341
Blast, 15, 16, 17, 18, 20, 24, 28, 29, 33, 45, 65, 77, 78, 85, 95, 96, 98, 100, 103, 104, 105, 106, 108, 109, 141, 142, 143, 144, 145, 146, 149, 151, 152, 154, 155, 157, 158, 159, 160, 161, 163, 164, 166, 169, 170, 175, 176, 183, 186, 189, 194, 223, 225, 227, 235, 236, 238, 262, 267, 275, 281, 282, 344, 346, 347, 348, 349, 350, 352, 355
Blast Furnace Gas, 106, 141, 142, 143, 152, 157, 164, 176
Blessed, 26
Block Coefficient, 332
Bloomery, 14, 16
Bloxwich, 2, 3, 12, 360
Body Centered Cubic, 79
BOF, 288
Bohr Model, 47, 49
Bolton, 27
Boltzmann, 46, 114, 115
Bomb Calorimeter, 87
Boolean, 274, 360
Booth Aluminium Company, 139
Boris, 272
Boston, 86, 261, 264, 267, 360
Bosworth, 272
Botfield, 159

Boulby, 48
Braggins, 5, 12
Brazil, 239, 282, 349
Breadth, 329
Bremen, 234
Bridge, 22
Bristol, 18
Britain, 4, 11, 19, 20, 24, 27, 32, 34, 133, 138, 150, 151, 159, 171, 237, 239, 272, 274, 282, 286, 287, 288, 291, 299, 300, 310
Britannica, 141, 346, 347
British, 13, 19, 26, 32, 34, 36, 42, 67, 123, 138, 223, 237, 274, 286, 287, 288, 289, 290, 291, 299, 338, 340, 341, 364
Brittany, 286
Bronze, 13, 14, 299
Broseley, 21
Busbars, 309
Butterley, 179, 180, 287

C

Cadgwith, 300
Calcium, 31, 34, 50, 61, 62, 64, 120
Calculated, 96, 107
Calculus, 37, 274, 360
Calder Hall, 288
Calorimeter, 87, 88
Cambridge, 12, 123, 358
Canada, 239
Canal, 20, 138
Canary Media, 228, 355
CaO, 199, 242, 249, 351, 358, 359
Capacitance, 316, 318
Capacitors, 320
Capacity, 88, 98, 141, 186, 332
Carbon, 15, 30, 51, 64, 65, 67, 69, 82, 87, 89, 90, 91, 96, 103, 104, 105, 106, 107, 142, 143, 161, 164, 165, 170, 200, 223, 235, 236, 324, 350, 355, 358
Carbon Dioxide, 15, 65, 82, 89, 91, 103, 142, 170, 235, 350
Carbon Monoxide, 15, 82, 142, 143, 161, 164, 165, 200
Carnarvon, 13
Carolina, 242
Carter, 13, 341
Cartesian, 149
Casio Z1200, 24
Cast, 16, 20, 21, 31

Cathode, 119, 123, 125, 255

Cathode Rays, 123

Causeway, 34

Caves, 287

CD-MELT, 75

Celerity, 49, 55, 315, 319

Celestial, 40

Cell, 119, 125, 126, 127, 128, 129, 131, 136, 241, 254, 265, 323, 324, 357

Cell Potential, 125, 126

Cement, 34

Centers, 38

Cepowski, 290, 335, 362

Chapter, 287, 329, 365

Characteristic, 76, 248

Charcoal, 15, 17, 18

Charge, 36, 40, 47, 48, 51, 52, 119, 120, 123, 125, 130, 131, 135, 147, 148, 149, 216, 217, 261, 301, 302, 320, 325

Charlcotte, 17

Charles, 133, 272

Charles Martin Hall, 133

Chemical, 36, 47, 49, 65, 74, 83, 93, 103, 232, 259

Chemistry, 36, 254, 345, 349, 350, 360

Chicago, 123

China, 27, 275, 279, 329

Chinese, 16, 180

Chloride, 133

Choke, 320

Choke Circuit, 320

Christ, 16

Church, 26

Circuit, 119

Circuitry, 320

Cistercian, 26

Clapeyron, 45

Clare, 5

Classical, 48, 149, 150, 199

Clearwell, 287

Clio, 274

Clyde, 137, 138, 139, 160, 161

Clyde Iron Works, 160, 161

Co, 14, 274, 354

CO2, 15, 65, 67, 91, 98, 142, 235

Coal, 16, 28, 149, 152, 175, 179, 236, 267, 275, 278, 279, 280, 281

Coal Tar Liqors, 152

Coalasnacon, 137

Coalbrookdale, 18, 179, 180, 182, 274

Coalfield, 20

Cobalt, 14

Code, 2

Coefficient, 61, 66, 77, 78, 90, 115, 172, 208, 217, 254, 275, 335

Coke, 16, 18, 20, 30, 147, 148, 149, 152, 175, 179, 180, 181, 183, 191, 348, 351

Coke Rate, 179, 183, 191

Coking, 16, 28, 149

Cold, 15, 16, 17, 18, 20, 160

College, 5

Colliery, 26

Combustion, 88, 93, 142, 149, 164, 165, 171, 175, 179, 180, 348

Comminution, 72

Common Futures, 229, 355

Company, 26, 179, 287, 329, 363

Component, 66, 169, 329

Compressed Air, 151

Computer, 195, 274, 348

Concentration, 72, 73, 87, 208, 359

Concept, 41, 60, 64

Conduction, 77, 96

Conductive, 76

Conductivity, 76, 247, 249, 251, 252, 302, 320, 358, 359

Conductor, 248, 320

Conformation, 144, 195, 197, 198, 244, 247

Conservation, 15, 35, 67, 68, 95, 201, 267

Constant, 38, 45, 46, 48, 74, 96, 104, 114, 115, 122, 124, 125, 128, 129, 134, 135, 163, 165, 172, 203, 217, 254, 261, 268, 301, 302, 325

Constants, 53, 245, 351

Container, 331, 334, 362

Contents, 7, 353

Continental, 19, 138, 272

Contributory, 62

Control, 195

Convection, 96, 344

Convective, 76

Conversion, 309

Copeland, 32, 287

Copper, 303, 319

Copper Loss, 303

Copyright, 2, 349, 350, 351, 354

Corporation, 26

Correction, 88

Corus, 26

Cost, 269

Coulombs, 47, 51, 120, 123, 124, 135, 216, 302, 320, 325, 327

Country, 300
Cowper, 15, 142, 143, 144, 147, 152, 157
Crisis, 286
Critical, 28, 231
Cross, 2
Crude, 275, 278, 279, 280, 282, 284, 285
Cryolite, 133
Crystals, 81
Cube, 56, 60, 149
Cubic, 35, 79, 137, 159, 164
Cultures, 271
Cumberland, 139, 287
Cumecs, 43
Current, 36, 40, 83, 119, 122, 125, 130, 131, 136, 255, 258, 261, 262, 302, 303, 304, 307, 309, 311, 316, 318, 320, 323, 324

D

Dalton, 51, 52, 53, 54
Darby, 18, 19, 21, 274
DC, 119, 303, 309, 310, 311, 315, 316, 318, 319, 320, 323, 352, 353, 359
De Loutherbourg, 179
Deadweight, 290, 332, 336, 337
Dean, 18, 287
Defect, 54, 55, 128, 163, 229
Degrees, 36, 45, 74, 96, 100, 144, 164, 168, 171
Densimetric, 72
Density, 43, 48, 56, 59, 72, 122, 137, 149, 165, 169, 258, 301, 320, 330, 332
Deposited, 135
Depression, 286
Derbyshire, 287
Design, 2, 331, 345, 362
Determination, 256, 275, 335, 362
Deuterium, 29
Devon, 24
Dewanto and Nazirudin, 335
Diatomic, 91, 127, 130, 158
Dielectric, 119
Difference, 42, 97, 255
Diffusivity, 76
DIH Barr, 60
Dimensionless, 75
Dimensions, 36, 201, 202, 247
Dimers, 61
Diode, 320, 321
Dioxide, 31

Direct, 77, 85, 86, 119, 195, 200, 201, 208, 223, 229, 239, 267, 269, 303, 309, 311, 316, 320
Discharge, 43, 301
Displacement, 289, 330, 332, 335
Dissociation, 254
Distance, 38, 39, 243
Divalent, 64
Doctrine, 271
Dollar, 267, 286
Don, 5, 12
Draught, 330, 335
Drax, 287
DRI, 239
Drift Velocity, 319, 320
Ducret, 242, 358, 359
Dufrenoy, 159, 160
Duralumin, 139
Duration, 36, 125, 131, 209, 261, 325, 327
DWT, 290, 336, 337
Dynamic, 76

E

EAF, 237, 240
Earl, 13, 19
Earth, 30, 31, 37, 38, 39, 40, 41, 42, 50, 133, 361
East Halton, 294, 296
Eastern, 15, 234
Economic, 19, 32, 341
EDF, 288
Edition, 2, 3
Efficacy, 267
Efficiencies, 81, 137, 268
Efficiency, 43, 127, 128, 129, 176, 257, 259, 261, 264, 265, 267, 268, 296, 323, 348, 355, 357
Egypt, 13
Eighteenth-Century, 16
Eighth, 272
Einstein, 49, 52
Eisenhüttenstadt, 234, 356
Electric, 26, 36, 40, 47, 52, 119, 122, 125, 197, 216, 240, 281, 343
Electric Arc Furnace, 27, 240
Électricité de France, 288
Electricity, 127, 269, 275, 280, 281, 282, 284, 285
Electrochemical, 124, 261, 359
Electro-chemical, 120
Electrode, 126

Electrodes, 128, 246

Electrolysis, 85, 86, 119, 121, 122, 126, 195, 216, 241, 243, 244, 247, 259, 267, 269, 357, 358, 359

Electrolyte, 119, 249, 251, 252, 255, 258

Electrolytic, 119, 124, 130, 254

Electron, 48, 51, 52, 120, 123, 135, 320

Electron Volt, 51, 52

Electrons, 47, 119, 120, 125, 128, 129, 130, 131, 135, 261, 319, 320

Element, 47, 61, 135

Elemental, 61, 281, 282

Elevation, 301

Eleven, 287

EMF, 51

Empire, 19, 338

Empirical, 97, 164, 172, 335

Empty, 330

Endothermic, 90, 92, 143

Energetics, 112, 352, 353

Energy, 15, 28, 35, 36, 39, 40, 42, 43, 49, 51, 52, 54, 55, 57, 61, 67, 87, 95, 100, 101, 109, 111, 112, 113, 114, 115, 127, 128, 129, 131, 153, 156, 158, 163, 175, 183, 186, 189, 191, 203, 224, 225, 228, 258, 259, 262, 263, 265, 267, 268, 296, 301, 302, 315, 323, 327, 343, 344, 347, 352, 353, 355, 357, 363, 364

Engine, 274

England, 16, 17, 18, 22, 24, 26, 33, 48, 179, 272, 287

English, 16, 19, 26, 67, 119, 239, 272, 299

Enlightenment, 35

Enquiry, 271

Enthalpy, 89, 113, 116, 117, 126, 253

Entropy, 87, 88, 116, 117, 128, 129, 253, 254

Environmental, 267, 364

Epigraph, 7

Equation, 37, 38, 39, 40, 41, 42, 43, 44, 45, 46, 48, 49, 50, 51, 52, 55, 56, 59, 60, 61, 62, 66, 68, 69, 70, 72, 73, 74, 75, 76, 77, 78, 79, 83, 88, 89, 90, 91, 92, 95, 96, 97, 98, 100, 103, 104, 105, 106, 107, 114, 115, 116, 117, 120, 122, 123, 124, 126, 127, 128, 129, 130, 131, 132, 135, 136, 137, 141, 148, 149, 152, 153, 158, 159, 161, 163, 164, 165, 166, 168, 169, 171, 172, 174, 175, 176, 201, 203, 204, 208, 209, 213, 214, 215, 216, 217,

218, 220, 221, 222, 223, 224, 225, 226, 228, 229, 248, 252, 253, 254, 255, 257, 258, 259, 260, 261, 262, 263, 264, 265, 268, 275, 280, 281, 290, 296, 301, 302, 303, 305, 306, 307, 310, 315, 316, 317, 318, 319, 323, 324, 325, 326, 327, 328, 330, 332, 333, 335, 336, 337, 338, 340

Equilibrium, 74, 81, 113

Equivalent, 98, 100, 105, 120, 122, 125, 129, 131, 261

Era, 23, 287

Étaples, 272

Euclidean, 47, 48

Europe, 19, 26, 27, 33, 152, 183, 225, 237, 272

European, 13, 14, 15, 137, 183, 185, 186, 188, 189, 190, 191, 193, 225, 227, 234, 262, 263, 269, 274, 352, 362

European Steel Association, 183, 352

European Union, 234, 274

Exothermic, 92, 104, 142

Explorations, 4, 274, 360

Exponent, 172

Exports, 27

Extent, 201, 208

Eyring, 114, 115, 161

Eyring-Polanyi, 114, 161

F

Factors, 245, 246

Falkland, 299

Fang, 163, 173, 351

Faradaic, 257, 259, 261, 262, 263, 264, 265, 267

Faraday, 119, 120, 122, 124, 125, 128, 129, 130, 135, 136, 216, 217, 254, 259, 260, 261, 262, 265, 325, 327, 328

Farm, 363

Farnolls, 21

Father, 48, 287

Fayaz Et Al, 241

FCC, 79

Fe, 14, 30, 59, 64, 67, 68, 83, 105, 171, 215, 242, 243, 330

Fe2O3, 32, 67, 69, 82, 143, 150, 161, 241, 243, 349

Fe3O4, 32, 241, 242

Feedstocks, 149, 267

FeO, 15, 32, 33, 143, 150, 161, 200, 242, 249, 251, 358, 359

Ferng and Lin, 79
Ferric, 241
Ferrous, 224
Ferruginous, 17
Field, 272
File, 2, 341, 342
Finnish, 233, 235, 236
First, 2, 3, 41, 103, 120, 136, 137, 201,
    259, 261, 272, 281, 341, 346, 347
Fission, 53, 54, 55, 56
Flame Calorimeter, 88
Fleet Restoration, 290
Florence, 287
Flourite, 34, 151
Flow, 43, 76, 77, 345, 349
Fluid, 76
Flux, 96, 147, 148, 152
Force, 35, 37, 39, 40, 44, 316
Forest, 18, 287
Form Factors, 245
Formation, 87, 89, 90, 91, 103, 171, 218,
    221, 232
Formula, 59, 60, 61, 66, 67, 73, 83, 90,
    91, 97, 101, 102, 103, 107, 122, 172,
    209, 212, 217, 258, 325, 326, 327
Fort William Ltd., 325, 364
Forth, 138, 159
Forth and Clyde, 138
Forward, 113
Fossil, 281, 355
Foulness, 286
Four, 4
Fraction, 69, 75, 98, 100, 158, 164, 166,
    174, 209
Fractional, 72, 152
France, 159, 171, 269, 274, 281, 286,
    299
Free Will, 299
French, 19, 20, 133, 288, 299
Frequency, 114, 115, 315
Fruehan, 183, 184, 191, 193, 259
Fruehan Et Al, 191, 259
Fruit Fly, 50
Fuel, 85, 142, 157, 175
Fukuyama, 274
Full-Wave, 320
Fundamental, 36, 245
Furnace, 15, 16, 17, 18, 19, 20, 23, 33,
    45, 75, 77, 85, 86, 95, 100, 104, 105,
    108, 109, 141, 142, 143, 144, 145,
    146, 149, 151, 152, 159, 163, 166,
    175, 176, 183, 186, 189, 191, 194,
    195, 223, 225, 226, 229, 235, 236,

238, 243, 244, 245, 246, 247, 261,
    262, 267, 275, 281, 344, 346, 347,
    348, 349, 350, 352
Fusion, 81, 82, 83, 101, 170
Future, 271, 355

G

G20, 274, 277
G20+, 277
G20PLUS, 276
G22, 282
Gaelic, 138
Gap, 257
Garage, 12
Gas, 45, 46, 50, 82, 95, 96, 98, 99, 100,
    101, 114, 115, 122, 134, 161, 164,
    165, 166, 168, 169, 176, 203, 208,
    281, 288, 302, 348, 350, 351
Gasoline, 82
Gate, 320
GDP, 285, 286
Geometrical, 245, 246, 247
George, 13
Geothermal, 281, 282
Germain, 15, 66
Germany, 237, 274
Ghost Prism, 45
Gibbs, 115, 116, 126, 128, 129, 254
Gibbs Free Energy, 116, 126, 128, 129,
    254
GigaJoule, 44, 225
Gilchrist, 33, 237
Gilchrist-Thomas, 33, 237
GJ, 44, 164, 166, 175, 183, 186, 225, 323
Glasgow, 138
Glass, 81
Global, 28
Glucose, 112, 346
God, 35, 299, 300
Gold, 14, 133, 242, 286
Golden, 13
Government, 17, 138, 191, 286, 288
Gram, 105
Grams, 165
Graphite, 30, 252
Gravitational, 37, 38, 42, 301
Gravity, 38, 39, 42, 301
Great Britain, 21, 286, 346
Grid, 17, 18, 20, 22, 269, 287, 288, 323,
    342
Gross, 143, 180, 225, 272, 282, 289, 290,
    329, 336, 351

Group Velocity, 49, 319
GRT, 290, 329, 332, 338
GT, 290, 329, 335, 336, 340
Gulf of Bothnia, 234
Gurprit Singh, 107, 346, 347
GW, 288

H

H2DRF, 29, 30, 85, 195, 196, 197, 198,
  223, 229, 231, 233, 234, 236, 239,
  243, 267, 268, 282, 288, 291, 296,
  297
Half-Wave, 309, 320
Hall-Héroult, 133, 134, 135, 324
Hallstatt, 271
Harmonic Mean, 306
Hartlepool, 49
Harvey Fletcher, 123
HBI, 239
He Et Al, 189, 190, 203, 205, 209, 211
Heat, 35, 76, 77, 78, 79, 81, 82, 83, 84,
  85, 87, 88, 90, 91, 93, 96, 97, 98, 99,
  100, 101, 103, 105, 106, 113, 115,
  127, 128, 142, 149, 156, 164, 165,
  166, 169, 170, 171, 172, 173, 174,
  175, 176, 179, 180, 181, 183, 186,
  187, 195, 197, 198, 220, 221, 222,
  223, 224, 234, 235, 236, 238, 257,
  258, 327, 344, 345, 347, 348, 352,
  355
Heat Exchange, 195, 198
Heat of Formation, 103, 220, 222, 235
Heat of Reaction, 82, 90, 91, 93, 103,
  115, 142, 220, 221, 223, 224, 234
Heat Transfer, 76, 77, 78, 79, 344, 345
Heating, 164
Height, 42, 301, 329, 330
Helium, 64
Hematite, 32, 67, 69, 73, 82, 89, 90, 96,
  103, 104, 105, 106, 142, 143, 150,
  161, 199, 200, 201, 208, 209, 214,
  221, 232, 233, 241, 243, 349, 353,
  357
Henry, 26, 114, 272
Herbert, 13
Hertz, 186, 315
Hess, 15, 66, 124, 201, 267
Highland, 138
Hill, 34
History, 19, 159, 341
HMG, 286
Hodgebower, 21

Holderness, 296
Holy, 272
Hot, 15, 16, 28, 29, 65, 77, 78, 85, 95, 96,
  98, 100, 103, 104, 105, 108, 142, 144,
  151, 157, 158, 159, 160, 161, 170,
  183, 185, 223, 239, 349, 350
Howard, 13, 341
Hull, 332
Humanities, 11
Humber Gateway, 296, 363
Humphry Davy, 120
Hundredweight, 160
HYBRIT, 233, 234, 235, 236, 239, 355,
  356
Hydraulic, 75
Hydrocarbon, 152, 267
Hydrogen, 29, 47, 48, 60, 77, 83, 85, 86,
  119, 126, 127, 129, 130, 131, 132,
  142, 143, 164, 195, 196, 199, 212,
  213, 214, 216, 218, 219, 223, 229,
  232, 233, 239, 267, 269, 353, 354,
  355
Hydropower, 4, 282, 288

I

I Lowthian Bell, 159, 350
I2R, 134, 257, 309
Ice, 81
Ideal, 45, 96, 98, 119, 215, 216
Igneous, 34
Iguazu, 282
Ijmuiden, 229, 355
Impedance, 316
Incas, 239
India, 27, 179
Inductor, 320, 321
Industrial, 16, 19, 27, 31, 33, 179, 195,
  352, 353
Industry, 19, 341
Inert, 50, 64, 82
Information, 7
Infrastructure, 27
Input, 127, 175, 303
Integrated, 24
Intensity, 37
Intercentric, 38
Internal Energy, 116
International, 36, 40, 42, 344, 347, 350,
  352, 354, 359
Invergordon, 286
Inversion, 309
Inverter, 309

Iodine, 81
Ion Bridge, 195
Ionised, 122, 125
Ireland, 286, 291
Iridium, 242
Irish, 19, 34, 138
Iron, 14, 15, 16, 19, 20, 21, 22, 23, 26,
    27, 28, 30, 31, 59, 60, 64, 68, 69, 71,
    72, 73, 74, 77, 82, 86, 89, 93, 95, 98,
    99, 103, 104, 105, 106, 107, 108, 141,
    142, 145, 150, 159, 163, 170, 171,
    172, 173, 175, 195, 196, 199, 200,
    212, 219, 225, 228, 229, 232,233,
    239, 241, 243, 258, 259, 260, 261,
    262, 264, 265, 268, 270, 287, 290,
    291, 292, 293, 296, 330, 338, 339,
    340, 341, 344, 347, 348, 349, 350,
    352, 354, 355, 358, 359
Iron Mad, 21
Iron Ore, 72, 73, 86, 103, 142, 150, 195,
    196, 232, 243, 348, 355
Ironbridge, 21
Ironworks, 25, 144, 159
Isle of Wight, 286
Isotope, 47
Itaipu, 282
Italy, 269, 274
Iwao Muchi, 152, 349

J

James, 1, 2, 3, 4, 17, 18, 20, 22, 25, 35,
    122, 159, 357, 358, 360
Japan, 28, 349
Japanese, 23, 349
Jerry Crimson Mann, 112, 346
Jiangsu Yangzi Xinfu, 329
John Turner, 234
Joswiak, 199
Joule, 35
Joules, 39, 42, 43, 44, 49, 52, 55, 88, 105,
    116, 131, 166, 174, 175, 179, 223,
    224, 228, 229, 258, 259, 263, 265
Jurassic, 33

K

Kelvin, 36, 45, 74, 96, 97, 100, 144, 164,
    168, 171, 172, 209, 354
Kenfig, 25
Khetpal, 242, 358, 359

Kilograms, 36, 45, 106, 120, 122, 125,
    131, 160, 169, 171, 175, 183, 215,
    216, 218, 265, 327
Kinematic, 76
Kinetic, 40, 45, 82, 98, 122, 161, 301,
    315
King, 12, 14, 26, 123
Kingdom, 2
Kinlochleven, 138
Kirkintilloch, 138
KMS, 212
Kovtun Et Al, 232
Krypton, 50
KTA, 78, 79
Kumar and Himabindu, 126
Kumar and Lim, 125
Kvande and Drabløs, 326
KW, 39

L

Laboratory, 48
Lambda, 208
Laminar, 77
Lancashire, 27
Lancaster, 44
Laplace Equations, 255
Larne, 34
Larnite, 34, 60, 61, 62
Lat, 17, 18, 20, 22
Latent, 81, 83, 101, 166, 170, 171, 174
Latin, 299
Law, 15, 38, 45, 67, 68, 75, 77, 78, 95,
    120, 122, 124, 130, 135, 136, 201,
    216, 217, 248, 255, 257, 259, 260,
    261, 262, 265, 267, 302, 325, 327,
    328, 344
Lead, 64
LECO Corporation, 180, 351
Lee and Chung, 78
Leicestershire, 20
Length, 35, 36, 59, 60, 76, 137, 248, 320,
    329, 330
Leptons, 46, 47
Leven, 137
LHS, 46, 53, 67
Liberty, 289, 364
Libyan, 28
Light, 49, 55, 319
Limestone, 33, 151, 152
Limitations, 96
Limited, 5, 12
Limonite, 32, 33, 150

Linear, 202
Liquid, 16, 81, 106, 183, 252, 255, 359
Liverpool, 310
LKAB, 234
Loaded Height, 330
Lochaber, 263, 296, 326, 328
Log, 279, 284, 285
Log-Log, 279, 284, 285
London, 87, 119, 133, 182, 310, 341
Long, 17, 18, 19, 20, 22, 160, 223
Longitudinal, 329
Lorraine, 237
Loschmidt, 122, 123
Loss, 75, 257, 324
Losses, 96, 327
Love, 300
Loves, 300
Lowry and Cavell, 12, 134, 139
Luleå, 234, 235
Luminous, 37
Luxemburg, 33
LWT, 289, 290, 329, 330, 332, 333, 335,
    336, 338

M

Magnesium, 120
Magnetite, 32, 150, 199, 200, 233, 241
MAGNOX, 288
Majid Et Al, 241
Makeup, 173, 177
Making, 163, 355
Manchester, 123
Mann Et Al, 112
Mansel, 26
Margam, 26
Maria Gallucci, 228, 355
Maritime, 362
Mary, 26
Maryhill, 138
Maryport, 139
Mass, 23, 35, 36, 38, 42, 43, 45, 48, 51,
    52, 54, 56, 59, 60, 61, 62, 68, 72, 75,
    77, 83, 84, 88, 97, 98, 100, 101, 120,
    122, 123, 124, 125, 126, 130, 132,
    135, 136, 149, 163, 164, 165, 169,
    171, 172, 174, 175, 215, 216, 217,
    218, 224, 259, 265, 267, 296,
    301,320, 325, 327, 330, 343, 344
Mass Transfer, 344
Mass Yield, 215, 216, 325, 327
Materials, 351, 355, 357, 358, 359

Mathematical, 4, 274, 344, 346, 347,
    349, 352, 360
Matter, 36, 45, 48, 49, 81, 201
Maximum, 310, 318
Maxwell, 122, 123
Mayfair, 119, 133
Mean, 98, 122, 158, 164, 168, 171, 172,
    180, 183, 186, 225, 305, 306
Mean Free Path, 122
Mechanical, 39
Mechanics, 40
Mechano-Electric, 195
Medieval, 26
Mediterranean Shipping Company, 332
MegaWatt, 225, 264, 265, 267, 268
Megawatts, 43, 44, 179, 186, 192, 230
Melt, 75, 253
Melting, 171
Merchant, 289, 290, 338, 340, 360
Mercian, 60
Mercury, 82
Meredith Gwynne Evans, 114
Merlin, 44, 179
Mesozoic, 237
Metabolism, 112, 346
Metal, 56, 86, 135, 152, 183, 185, 188,
    190, 193, 243, 244, 247, 255, 259,
    261, 267, 269, 282, 330, 359, 360
Metallic, 89
Metallurgical, 28, 149, 175, 180, 181,
    183, 351, 359
Metals, 5, 12, 64, 261, 349, 354, 356,
    357, 358
Meter, 36, 212, 247, 252
Meters, 35, 36, 159, 164, 247
Methane, 82, 142, 151
Metric, 223
MgO, 171, 242, 249, 351, 358, 359
Michael Faraday, 119
Middle, 272
Midland, 2, 3, 360
Midship Beam, 329, 330
Mild, 23, 281
Mine, 48, 287
Mines, 159
Minimum, 127, 128, 183, 254, 259, 352,
    353
Mixture, 98, 100, 164, 165, 172, 174,
    242
Model, 48, 103, 108, 109, 163, 175, 176,
    186, 187, 188, 191, 194, 344, 346,
    347, 349, 352
Modelling, 96, 105, 108, 141, 347

Modern, 16, 31, 45, 183, 191, 320, 349
Modular, 78, 282, 288
MOE, 29, 30, 241, 242, 244, 245, 247,
    257, 261, 265, 267, 268, 282, 288,
    291, 295, 296, 297
MOEF, 86
Moiety, 274, 360
Moira, 19, 20
Mol, 46, 217
Molar, 59, 62, 98, 101, 105, 114, 115,
    122, 125, 126, 158, 163, 165, 169,
    172, 208, 212, 232, 259
Molarity, 36, 40
Mole, 45, 46, 97, 100, 106, 114, 122, 165
Molecular, 46, 88, 98, 100, 114, 115,
    120, 127, 130, 132, 158, 164, 165,
    168, 320
Molecules, 46, 60, 82
Molten, 82, 86, 241, 358, 359
Momentum, 76
Monatomic, 132, 196
Monoxide, 142
Monuments, 287
Morfa, 26
Mother, 12, 13, 139
Motor, 332
Ms Wafaa Souadji, 329, 333
MSC Irena, 290, 296, 332, 333, 337, 338,
    361
Museum, 182
Muthusamy, 186, 187, 344, 347, 352
Mutiny, 286
MV Irena, 329
MW, 43, 44, 98, 100, 179, 187, 191, 192,
    288

N

NaOH, 241
Napoleon III, 133
Napoleonic, 19
Narvik, 234
Natal, 329
National, 267, 269, 272, 282, 287, 345,
    350, 360
Native, 19
NATO, 299
Nature, 195
Naval, 286
Navier-Stokes, 75, 78
Navigation, 274
Navy, 20, 41, 299, 338
Nazi, 138

Near, 15
Nederland, 355
Neilson, 159
Nelson, 20
Neolithic, 15
Neon, 50, 64
Neptune Bank, 295
Net, 125, 299, 305
Net Zero, 299
Netherlands, 26, 274
Neutron, 46
Neutrons, 46, 47
New, 15, 239, 282, 350, 362
New Zealand, 239, 282
Newton, 38, 75, 78
Newtons, 38
Nickel, 14
Nineteenth, 33, 35, 47, 133, 286
NIST, 165, 166, 171, 172, 254, 345, 350,
    360
Nitrogen, 64, 142, 169, 170, 351
Nobel Prize, 123
Noble, 64
Nominal, 272, 282, 285, 286
Non-destructive, 14
North, 26, 138, 234, 286
North America, 138
Northern, 25, 26, 48, 234, 286, 291
Northumberland, 287
Norway, 234, 269
Notation, 65, 202
Nubians, 13
Nuclear, 49, 50, 52, 53, 78, 252, 281,
    282, 288, 344, 345
Nucleus, 47
Number, 46, 47, 55, 61, 68, 69, 75, 76,
    78, 79, 96, 114, 122, 123, 124, 125,
    128, 129, 131, 132, 135, 159, 164,
    172, 218, 261, 263, 268, 305, 306,
    320, 340, 358, 359, 364
Nusselt, 75, 76, 77, 78, 79

O

Occidental, 15
Ochre, 32
Octane, 60
Ohm, 247, 302
Ohms, 302, 305, 316
Oil, 236, 281
Olivine, 34
One, 4, 54, 129, 131, 132, 137, 159, 160,
    215, 218, 220, 222, 226, 228, 262,

265, 268, 272, 274, 286, 296, 310, 327
Operant, 126, 128
Operating, 152, 154, 155
Optimise, 195
Order, 123, 201
Ore, 27, 147, 148, 150, 152, 264

Ø

Ørsted, 133

O

Ortiz, 241, 258, 357
Oscillators, 309
Output, 88, 127, 175, 183, 187, 229, 258, 275, 282, 303, 318
Ovens, 16
Overall, 268, 296, 329, 330
Overall Length, 329, 330
Overpotential, 126, 265
Overvoltages, 254
Oxelösund, 234
Oxidation, 91, 96, 103, 106
Oxide, 86, 133, 174, 241, 243, 244, 247, 267, 269, 282, 358, 359
Oxygen, 23, 29, 31, 60, 61, 62, 82, 87, 91, 96, 103, 105, 107, 119, 125, 126, 131, 132, 134, 151, 158, 159, 161, 163, 170, 195, 199, 223, 226, 241, 243, 281, 326, 351

P

Page, 7, 8, 9, 327, 352, 353, 355, 361
Pakistan, 28
Palmer, 48
Paraguay, 282
Parallel, 303, 304, 305, 306, 307
Parameters, 201, 331, 334
Paris, 159
Parliament, 21
Partial, 358
Particle, 53, 72
Pascal, 158
Past, 271
Pathway, 50, 51, 53, 54, 65, 66, 67, 68, 69, 74, 83, 89, 91, 103, 105, 113, 134, 142, 143, 199, 200, 201, 208, 209, 212, 214, 218, 220, 221, 223, 224, 234, 253, 259, 261, 264, 325, 326

PBMR, 78
PCI, 150, 179, 236
Peace, 272
Pebble Bed, 78
Pei Et Al, 234, 235, 236, 237, 239
Pelletised, 151
Pelletised Ore Mix, 151
Pellets, 355
PEM, 125
Pentrich, 179
Percentage, 128, 163, 174, 229
Periclean Greeks, 299
Period, 315
Permian, 48
Permittivity, 48
Permo-Triassic, 17
Petroleum, 152
pH, 241
Pharoah, 13
Phase, 75, 316
Phase Angle, 316
Phi, 4
Phosphorous, 18
Phosphorus, 64
Physical, 36, 53, 84, 349
Physics, 36
Pi, 4
Pi and Phi, 4
Piccadilly Circus, 133
Pig, 16, 20, 225
Planck, 48, 115
Plating Ratio, 332
Plimsoll, 87
Plutonium, 288
Point, 36, 171, 239
Polish, 329
Polyhalite, 48
Polynomial, 201
POM, 151
Population, 28, 282, 284
Porosity, 72, 73, 78, 147, 149
Port, 24, 26, 287, 341, 342
Portland, 34, 286
Potassium, 120, 133
Potent, 2
Potential, 40, 42, 52, 126, 128, 255, 297, 301, 302, 315, 355
Potential Difference, 52, 255, 302
Potentials, 124, 297
Pound, 286
Power, 39, 40, 43, 49, 55, 64, 129, 130, 131, 191, 194, 223, 258, 262, 265,

277, 281, 296, 301, 302, 303, 315, 316, 324, 326, 327
Power Drop, 303
Prandtl, 75, 76
Predicament, 271, 272, 286
Prefix, 2
Prescott, 35
Present, 231, 271
Pressure, 35, 36, 44, 45, 81, 82, 87, 89, 116, 119, 122, 158, 163, 165, 214, 301
Pressurised, 288
Pritchard, 21
Problem, 75
Process, 90, 91, 95, 133, 134, 135, 137, 163, 183, 184, 237, 323, 350, 357, 364
Product, 62, 68, 74, 124, 125, 152, 224, 268, 272, 282
Production, 27, 139, 183, 195, 216, 227, 275, 278, 279, 280, 281, 284, 285, 287, 340, 348, 354, 358, 359, 360, 362, 364
Productions, 2, 3, 360
Proof, 41
Protectionism, 27
Proton, 46, 47
Protons, 46, 47
Province, 34
PSD, 98, 163, 165, 169, 172, 174, 217, 223, 229, 230, 258, 259, 267, 325, 326, 327, 332, 333
Publication, 7
Publishers, 2
Pulverised Coal Injectant, 150
PWRs, 288
Pyrolysis, 149

Q

Quaker, 18
Qualitative, 112
Quantification, 96

R

Raahe, 235
Radiation, 96
Radius, 38, 48, 141, 149
Range, 168, 171, 172
Ratcliffe, 294, 295
Ratcliffe-upon-Soar, 294

Rate, 43, 114, 115, 191, 201, 204, 207, 208, 271, 272, 274, 281, 282, 286, 287
Rate of Reaction, 201, 204, 208
Rateable, 191
Rated, 273
Ratio, 123, 208, 229, 264
Ratios, 148, 149, 215
Rautaruukki, 235
Reactance, 316
Reactants, 89
Reaction, 65, 74, 81, 88, 89, 96, 103, 104, 105, 106, 113, 114, 115, 124, 125, 126, 128, 129, 135, 142, 143, 161, 195, 201, 208, 209, 218, 221, 346
Reactivity, 149, 161, 348
Reactor, 78, 288, 359
Reactors, 78, 282, 288
Reagent, 70, 74, 88, 105, 122, 208
Reagents, 89, 103, 105, 208
Reciprocal, 206, 306
Recovery, 195, 197
Rectification, 309, 320, 321, 322
Rectification Bridge, 320, 321, 322
Rectified, 309, 310
Rectifier, 119, 309, 320, 363
Rectify, 310
Redcar, 26, 295, 296
Reduce, 265, 268
Reduced, 48, 239, 265
Reduction, 77, 85, 86, 89, 90, 96, 103, 104, 195, 199, 200, 201, 208, 214, 223, 229, 232, 267, 269, 282, 353, 354, 355, 357
Reference, 17, 18, 20, 22, 235
References, 363
Region, 258
Registered, 2, 289, 329
Regnault-Pfaundler, 88
Regression, 202, 335, 362
Relative, 40, 209, 215
Renewable, 281
Requirement, 68, 69, 183, 186, 225, 268, 290
Requirements, 68, 70, 183, 215, 352, 353, 358, 364
Research, 5, 12, 354
Researches, 4, 229
Reservoir, 301
Resistance, 248, 255, 302, 303, 305, 306, 307, 316, 318, 324, 327
Resistance Ratio, 306, 307

Resistivity, 246, 247, 251, 252, 255
Resistor, 304, 305
Resistors, 303, 304, 305
Rest, 48, 51
Restitution, 289
Retorts, 16
Review, 354
Revolution, 16, 19, 31, 33, 161, 179
Reynolds, 75, 76, 78
RHS, 46, 50, 54, 66, 67, 337
Ripley, 287
Ripple Voltage, 311, 318
Ripple Wave, 311
River, 18, 21
RMS, 23, 310, 363
Robert Millican, 123
Rolls-Royce, 44, 179, 287, 288, 296
Roman, 81, 287
Rome, 26, 272
Root Mean Square, 303, 310, 311, 316,
   318, 323
Ro-Ro, 335
Rosner Et Al, 223
Rotation, 41
Royal, 21, 119, 139, 299
Royal Institution, 119
Royal Navy, 21, 139, 299
Russia, 28, 183, 272, 274, 299
Russian, 28, 348

S

S1, 242, 249
Sadoway, 242, 357, 358, 359
Safety, 78, 195
Sahara, 14
Salt, 64, 241
Sand, 31, 81
Sandstone, 17
Scale, 268, 349
Scantlings, 331, 334
Scawt, 34
Scheduled, 287
Schenk, 200, 354
Science, 11, 182, 274, 300, 341, 348,
   355, 358, 361
Scotland, 26, 34, 137, 159, 160, 263
Scottish, 138, 299
Scunthorpe, 26
Second, 36, 75, 78, 103, 122, 130, 135,
   136, 138, 139, 183, 212, 216, 259,
   260, 261, 262, 265, 274, 281, 301,
   325, 327, 328, 346, 347

Second World War, 138, 139
Security, 267
Sensible, 83, 106, 166, 169, 170, 171,
   175, 195
Separation, 243
Serial, 303, 304
Seven, 186
Seventh, 272
Severn, 18, 21
Shaft, 43
Sheet, 156
Ship, 329, 330, 331, 332, 334, 361
Shipbuilding, 329
Ships, 289, 362
Shomate, 83, 97, 163, 168, 172, 174,
   253, 254, 256
Shropshire, 17, 18, 22
SI, 36, 38, 40, 41, 122, 152, 160, 186,
   203, 205, 209, 211, 212, 247, 261,
   315, 350
Side, 56, 59, 60, 137
Siderite, 33, 150
Siemens, 247, 252
Silica, 152
Silicon, 31, 61, 62, 64
Simpson Integration, 316
Sinks, 85
Sinter, 150, 152
SiO2, 31, 174, 242, 249, 351, 358
Six, 142
Sizewell, 288
Skomer, 286
Slag, 82, 96, 144, 173, 174, 175, 177, 351
Small, 282, 288
Smelter, 77, 103, 263, 324
Smelting, 28, 55, 93, 98, 133, 141, 145,
   199, 212, 219, 270, 277, 291, 292,
   293, 364
Smoothing, 318, 322
SMRs, 282
Sodium, 64, 120
Solenoid, 309
Solid, 72, 81, 83, 144, 161, 209
Solution, 4, 241, 357
Soon Analysis, 77
Sources, 27, 85, 277
South, 24, 26, 27, 274, 299, 342
South Africa, 27, 274
South Atlantic, 299
South Wales, 24
Soviet, 272
Space, 48, 149
Spain, 274, 282

Species, 122, 164, 215, 232
Specific, 76, 83, 88, 97, 98, 100, 101,
    128, 163, 166, 168, 169, 170, 171,
    172, 173, 174, 177, 229, 258, 351
Specific Heat, 76, 83, 88, 97, 98, 100,
    101, 163, 166, 168, 169, 170, 171,
    172, 173, 174, 177, 351
Speed, 49, 319
Sphere, 149
Spitfire, 44
Spreitzer, 199, 200, 354
Square Root of Minus One, 316
SSAB, 234, 235
Stack, 75
Staffa, 34
Stainless, 23
Standard, 36, 81, 88, 89, 119, 126, 152,
    165, 171, 214, 253, 260, 264, 286
State, 13, 83, 147, 291
States, 81
Status, 271
Steam, 30, 164, 199, 274
Steel, 19, 23, 24, 26, 27, 28, 183, 184,
    225, 229, 275, 278, 279, 280, 281,
    282, 284, 285, 330, 341, 342, 348,
    349, 350, 352, 353, 354, 355, 360
Steelworks, 341
Sterling, 286
Stoichiometric, 33, 61, 63, 66, 71, 74, 89,
    90, 91, 125, 157, 208, 213, 214, 217,
    218, 229
Stoichiometric Coefficients, 74, 89, 90,
    91, 208, 213, 214, 218
Stoichiometry, 59, 212, 214
Stone, 15, 299
Stove, 144, 152, 166, 168
Stoves, 15, 142, 143, 144, 147, 157
STP, 29, 30, 64, 81, 82, 87, 89, 157, 164,
    165, 166, 169, 208, 213, 214
Street, 12, 287
Strontium, 50, 64, 120
Structural, 330, 332, 361
Structural Draught, 332
Submarine, 288
Subscript, 47, 61, 66
Substance, 36, 46, 59, 83, 84, 97, 114,
    122, 171, 172, 174
Suffolk, 288
Sulfur, 18
Summary, 176, 178, 194
Sun, 50
Suppression Factor, 318
Supremacy, 272, 274

Surface, 39, 40, 72, 301, 349
Sussex, 18
Swansea, 287
Sweden, 32, 196, 234, 237, 239, 355
Swedish, 233, 234, 236
Symbol, 47, 321
System, 43, 45, 96, 116, 176, 197, 198,
    268
Systéme, 36, 40, 42

T

Table, 7
Taizing City, 329
Talbot, 24, 26, 287, 341, 342
Tata, 26
Temperature, 36, 40, 45, 46, 81, 83, 87,
    88, 89, 96, 97, 100, 114, 115, 116,
    119, 122, 129, 161, 164, 165, 168,
    171, 172, 201, 206, 208, 214, 253,
    254, 351, 359
TeraWatt, 226
Terrestrial, 39, 42
Tertiary, 34
TEU, 329, 332, 361
Teynham, 289, 290, 338, 340
The Royal Navy, 17
Theory, 45, 82, 98, 122, 161, 350
Thermal, 45, 76, 82, 85, 86, 101, 102,
    105, 176, 178, 192, 195, 345
Thermochemical, 63
Thermodynamic, 36, 81, 113, 219, 346,
    351
Thermodynamics, 126, 218, 221
Thermo-Neutral, 127, 128
Thickness, 329, 330
Thirteen, 329, 365
Thomas, 21, 27, 159, 228, 237, 355
Thomas Koch Blank, 228, 355
Thomson, 123
Threadneedle, 287
Three, 4, 270, 272, 286, 287
Thyristors, 320
Time, 36, 39, 43, 122, 130, 210, 264,
    296, 301, 302
Tin, 64
TIP, 34
Titanic, 23
Title, 7
TK Blank, 228
Ton, 160, 223
Tonnage, 289, 290, 329, 330, 332, 336,
    338

Tonnes, 183, 225, 262, 336, 337
Tons, 19, 223, 289
Toolbox, 168, 169
Tosaka, 147, 347
Town Gas, 16
Trade, 289
Trademark, 2
Trafalgar, 20
Transfer, 76, 78, 79
Transformed, 309
Transformer, 309
Transition, 64, 82
Transmission, 115
Transverse, 329
Treaty, 272
Trent, 45
Triodes, 320
Tritium, 29
Tudor, 272
Tungsten, 242, 255
Turbines, 301
Turbulent, 77
Turner, 179
Tutankhamun, 13, 14, 341
Tutorial, 2, 3, 360, 363
Twentieth, 34, 47, 138
Twenty-foot Equivalent Unit, 329
Two, 4, 154, 155, 264, 265, 272, 274,
    276, 287, 289, 320, 360

U

UK, 26, 27, 28, 183, 269, 286, 287, 288,
    289, 290, 338, 340, 363
Ukraine, 28, 183, 299
Unified, 51
Uniformity, 271
Union, 272, 274
Unit, 51, 60, 72, 101, 301
United, 2, 191, 272, 286, 350
United Kingdom, 2, 286
United States, 191, 272, 350
Universal, 29, 38, 45, 46, 82, 96, 114,
    115, 122, 134, 158, 159, 165, 203,
    302
Universal Gas Law, 29, 45, 82, 114, 122,
    158, 159, 165
University, 123, 171, 344, 345, 349, 352,
    353, 361, 362
Uranium, 50
US Department of Energy, 326
US Government, 183
USA, 28, 351, 359

USS Nimitz, 45
Utrecht, 229, 272, 274, 355

V

Valency, 64, 122, 126, 130, 135, 254, 261
Valois, 272
Values, 338
Vapor, 82, 170
Vaporisation, 82, 83
Vattenfall AB, 234
Velocity, 37, 40, 76, 315, 319
Versailles, 299
Vessel, 290, 332, 361
Viability, 27
Victorian, 171, 287
Virgin, 26
Viscosity, 76, 351
Vogl, Åhman and Nilsson, 223, 225
Voltage, 40, 52, 127, 128, 129, 131, 254,
    255, 302, 303, 304, 306, 310, 311,
    316, 318, 320, 323, 343, 363
Volts, 52, 134, 242, 258, 302, 304, 323
Volume, 4, 35, 44, 45, 56, 59, 72, 75,
    116, 137, 147, 149, 164, 165, 166,
    169, 213, 214, 301, 329, 330, 332,
    348, 355, 360, 361, 364
Volumetric, 59, 72, 75, 78, 141, 164
Volvo, 45

W

Wales, 26, 138, 287
Walmsley, 27
War, 28, 289
Warren, 1, 2, 3, 4, 17, 18, 20, 22, 25, 360
Wars, 19, 34
Water, 29, 30, 36, 43, 82, 85, 88, 125,
    126, 132, 170, 195, 213, 216, 220,
    288, 301, 332
Watt, 40
Watts, 39, 44, 76, 131, 136, 226, 230,
    262, 302, 324, 327
Wave Speed, 315
Waves, 315, 319
Weald, 18
WebBook, 345, 350, 360
Weight, 35, 36, 46, 55, 56, 59, 60, 61, 66,
    69, 73, 88, 97, 100, 120, 122, 125,
    127, 130, 132, 135, 136, 158, 164,
    165, 168, 217, 258, 259, 264, 325,
    326, 327
Weight Ratio, 326

Weighted, 172
Weincke Et Al, 241
Welsh, 26
Western Isles, 138
Westinghouse, 288
Wikipedia, 11, 52, 54, 112, 113, 124,
    341, 343, 346, 347, 362, 363
Wilkinson, 21
William Jennings Bryan, 272
Wilsonstown, 159
Wind, 281, 363
Winding Loss, 303
Woodrow Wilson, 272
Works, 24, 25, 26, 179, 229, 296, 326,
    328, 342, 350, 355
World, 21, 27, 28, 34, 288, 289, 332, 342

Wrought, 15, 16, 21, 31
Wüstite, 15, 33, 143, 161, 200, 233, 243
Wylfa, 288

X

Xenon, 50

Y

Yeltsin, 272
Yield, 52, 135, 152, 175, 201, 229, 255,
    261, 264, 268, 270, 296, 297
Yorkshire, 48, 296

www.ingramcontent.com/pod-product-compliance
Lightning Source LLC
Chambersburg PA
CBHW040138200326
41458CB00025B/6307